Two Manuals Combined into One Book

NBC Decontamination

And

Marine Scouting and Patrolling

Department of Defense

Pentagon
Publishing

Field Manual
No. 3-5

Marine Corps Warfighting Publication (MCWP)
3-37.3

Headquarters,
Department of the Army

Commandant,
US Marine Corps

Washington, DC, 28 July 2000

NBC DECONTAMINATION

Contents

* This publication supersedes Field Manual (FM) 305/Fleet Marine Force Manual (FMFM) 11-10, 17 November 1993.

Preface

FM 3-5 integrates the nuclear, biological, and chemical (NBC) fundamentals published in FM 3-100 and implements Standardized North Atlantic Treaty Organization Agreement (STANAG) 2426.

The extent and timing of decon depends on the tactical situation, the mission, the area of contamination, and the decon resources available. Survivability and reducing the effect of any chemical threat are the ultimate goals of decon. This manual provides detailed guidance on conducting decon operations performed by chemical and nonchemical personnel. Individual soldiers and leaders must be familiar with the basic procedures and concepts in this manual.

Appendix A contains an English to metric measurement conversion chart.

The proponent of this publication is Headquarters TRADOC. To submit changes for improving this publication use Department of the Army (DA) Form 2028 (Recommended Changes to Publications and Blank Forms) and forward to Commandant, US Army Chemical School, ATTN: ATSN-CMZ, Fort Leonard Wood, Missouri 65473-8926.

Unless this publication states otherwise, nouns and pronouns do not refer exclusively to men.

Chapter 1

Introduction

The use of NBC weapons creates unique residual hazards that may require decon. In addition to the deliberate use of these weapons, collateral damage, natural disasters, and industrial emitters may require decon. The presence of contamination generally reduces the effectiveness of our combat power. Contamination forces us into protective equipment that degrades our ability to perform individual and collective tasks.

CONTAMINATION FORMS

1-1. The following are the different forms of contamination:

- Solids – radioactive particles, biological spores, or dusty agents that could appear as a fine dust.
- Liquids – liquid droplets that fall like rain. Droplets can range from thick and sticky to the consistency of water.
- Vapors or gases – created by bursting munitions or generators. These clouds are affected by the weather and can cover large areas.
- Aerosols – fine liquids or solid particles suspended in the air. They behave much like vapors.

CONTAMINATION-HAZARDS TRANSMISSION

1-2. Contamination hazards can be transmitted in the following manner:

- **Transfer**. Anything that touches a surface covered with liquid or solid contamination will tend to pick up that contamination and move it from one surface to another.
- **Spread**. Touching a surface covered with liquid chemical agent can spread contamination on the same surface, thereby, increasing the size of the contaminated area.
- **Vapor**. Vapors can be carried through the air in the form of a dust, atomized liquids (aerosols), or true gases. Vapors in an open/outdoor area disperse rapidly, so there is no need to decon.
- **Desorption**. Liquid-chemical contamination absorbs into porous surfaces. Once absorbed, it begins to desorb or give off gas; that is, low levels of vapor pass out of the contaminated surface into the air and can be transferred to any surface that contacts it, including bare skin.
- **Radiation**. Radiation is given off by radioactive dust or dirt, most of which appears as fallout. For decon purposes, radiation can be thought of as a solid.

AGENT CLASSIFICATION

1-3. Depending on the length of time agents will be a hazard, they are classified as—

- Nonpersistent – an immediate threat that lasts a few minutes. They rarely require decon.
- Persistent – takes a longer time to act and may last for days. In a protected environment, these agents can last for long periods of time. All agents are affected to some extent by the weather. Even radiological particles can "rain out" of the air and form hot spots on the ground. Biological organisms are greatly affected by sunlight. Chemical agents can be decontaminated by the weather. The duration of a hazard is a complex estimation that is based on numerous factors which include the—
  - Type of contamination.
  - Contamination density and droplet size.
  - Temperature.
  - Wind speed.
  - Sunlight.
  - Humidity and rain.
  - Composition of the contaminated surface.
  - Type of soil and terrain.

DECONTAMINANTS

1-4. Decontaminants are described as follows:

- Natural – weather (rain, wind, and humidity).
- Standard – supertropical bleach (STB) and decontaminating solution number 2 (DS2).
- Nonstandard – soaps and detergents.

1-5. For more information on the types of decontaminants, see Appendix B.

DECON CONCEPTS

1-6. Whenever soldiers are unable to avoid contamination and have to use protective measures, decon is necessary to allow them to remove their protective gear and resume normal operations. Weathering is the most desirable means of decon; however, time and operational needs may not permit that option.

1-7. This chapter describes when, where, and how much to decon. Protective clothing (mission-oriented protective posture [MOPP] gear), protective equipment, and collective protective shelters (CPSs) offer only a temporary solution. Decon is the removal, destruction, or naturalization of contamination. If you become contaminated, some decon must occur as soon as possible.

DECON DECISIONS

1-8. The decision to decon is a risk assessment and is made within the context of the mission, enemy, terrain, troops, time available, and civilian consideration (METT-TC) and the resources available. Decon must be considered if the contamination levels exceed the negligible risk levels as follows:

- Chemical and biological contamination causes mild incapacitation in 5 percent or less of unprotected soldiers operating for 12 continuous hours within 1 meter of contamination. For the chemical-agent monitor (CAM), this equates to a one-bar reading at a distance of 1 inch from the surface.
- Radiological contamination can cause mild incapacitation in 2.5 percent or less of unprotected, nonpreviously exposed soldiers. This equates to a reading of no more than 0.33 centigray per hour (cGy/hr).

1-9. MOPP-gear exchange provides excellent protection against field concentrations of agents; however, wearing the gear causes performance degradation. Decon is performed to restore the normal operating tempo, but the logistical support that is required to keep soldiers in MOPP gear impacts operations. Therefore, decon should be conducted as soon as practical. Table 1-1 provides comparison data for decon levels/techniques.

Table 1-1. Comparison Data for Decon Levels/Techniques

Levels	Techniques[1]	Best Start Time	Performed by	Advantages
Immediate	Skin decon Personal wipe down Operator's spray down	Before 1 minute Within 15 minutes	Individual Individual or crew	Stops agent from penetrating.
Operational	MOPP-gear exchange[2] Vehicle washdown[3]	Within 6 hours	Unit Battalion crew or decon platoon (-)	Provides possible temporary relief from MOPP4. Limits liquid agent spread.
Thorough	DED and DAD DTD	When mission allows reconstitution	Decon platoon Unit	Provides probable long-term MOPP reduction with minimum risk.

[1]The techniques become less effective the longer they are delayed.

[2]Performance degradation and risk assessment must be considered when exceeding 6 hours. See FM 3-4 for battle-dress overgarment (BDO) risk assessment.

[3]Vehicle washdown is most effective if started within 1 hour.

DECON PRINCIPLES

1-10. Decon immediately for a chemical agent on the skin. Perform higher levels of decon as a result of the risk assessment. Personnel should—

- Decon as soon as possible. The sooner the contamination is removed, the sooner MOPP levels can be reduced and combat power can be restored.
- Decon only what is necessary. Weathering is the least costly method of decon. Expend resources where they count.
- Decon as far forward as possible. While METT-TC dependent, performing decon as close to the point as possible reduces the spread of contamination and minimizes any transfer hazard.
- Decon by priority. The commander identifies which items are most critical to the mission. Restoring combat power and reestablishing operating tempo are key.

DECON LEVELS

1-11. The three levels of decon operations are immediate, operational, and thorough (see Table 1-1, page 1-3).

IMMEDIATE

1-12. Skin decon is a soldier's basic survival skill and should be performed within 1 minute of being contaminated using the skin decontaminating kit (SDK).

1-13. Personal wipe down should be performed within 15 minutes. This is done to remove contamination from individual equipment by using an individual equipment decon kit (IEDK). Use detector paper or a CAM to locate the agent. For radiological contamination, use a radiac set to locate it and then brush, wipe, or shake it off.

1-14. Operators' spray down should be done within 15 minutes. Operators use the on-board decon apparatus to decon surfaces that they must touch or contact to operate the equipment. Radiological contamination in the form of dust particles may be wiped, scraped, or brushed off.

OPERATIONAL

1-15. A MOPP-gear exchange should be performed within 6 hours of being contaminated when thorough decon cannot be done. Soldiers will continue to wear MOPP gear and the operating tempo will be reduced.

1-16. Vehicle washdown should be performed within 6 hour of being contaminated when the mission does not permit a thorough decon. This process removes gross contamination and limits the spread of it.

THOROUGH

1-17. Detailed equipment decon (DED) and detailed aircraft decon (DAD) restore items so that they can be used without MOPP gear. Normally, the DED and the DAD are conducted as part of a reconstitution or during breaks in combat operations. These operations require support from a chemical decon unit.

1-18. Detailed troop decon (DTD) normally takes place in conjunction with DED/DAD. The contaminated unit conducts this process and supports the DED/DAD operations.

CHEMICAL-AGENT-RESISTANT COATING (CARC)

1-19. Army equipment is painted with CARC, which precludes the absorption of chemical agents and, thus, facilitates decon. By preventing penetration of an agent, contamination is easier to remove. Thickened agents are still a problem, but detergent and water should remove them.

1-20. The proper CARC paint colors should be used to paint bumper numbers, spot painting, and so forth. Do not use other paints in lieu of CARC.

WEATHER EFFECTS

1-21. If METT-TC allows, leaving the contamination alone is the easiest method of decon. However, marking, reporting, and periodically rechecking the contamination are required.

TEMPERATURE

1-22. The higher the temperature, the faster the rate of evaporation. Contamination persistency decreases as the temperature rises. It has no effect on radiological contamination.

WIND

1-23. Aeration aids in decon. Agents are dispersed by the wind, thus reducing the concentration.

HUMIDITY AND RAIN

1-24. Moisture tends to break down chemical agents but does so slowly. Heavy rain physically removes contamination; however, contaminated runoff may build up in drainage areas. Biological agents dehydrate (dry out) in low humidity, thus reducing their persistency. Rain may prevent the desorption of aerosols and leach contamination into the soil.

SUNLIGHT

1-25. Sunlight hastens the evaporation and decomposition of agents. Ultraviolet and infrared radiation in sunlight rapidly kills most biological agents.

TIME

1-26. Only time will neutralize and destroy radiological contamination. Other techniques merely shield or remove the problem.

DECON IN COMBAT

1-27. Combat effectiveness is degraded by MOPP. Decon can remove the contamination and restore combat power. The detection and warning network is vital to ensure that soldiers and equipment are protected when contamination is encountered. Considerations of decon in combat are—

- Immediate decon. It saves lives and permits the use of individual equipment and key systems.
- Operational decon. It reduces the spread and the level of contamination. In some cases, when combined with weathering, MOPP levels may be reduced without further decon.
- Thorough decon. It removes the unit from the fight but allows it to return with restored effectiveness.

1-28. Figure 1-1 depicts the drop in effectiveness as the affected unit and soldiers react.

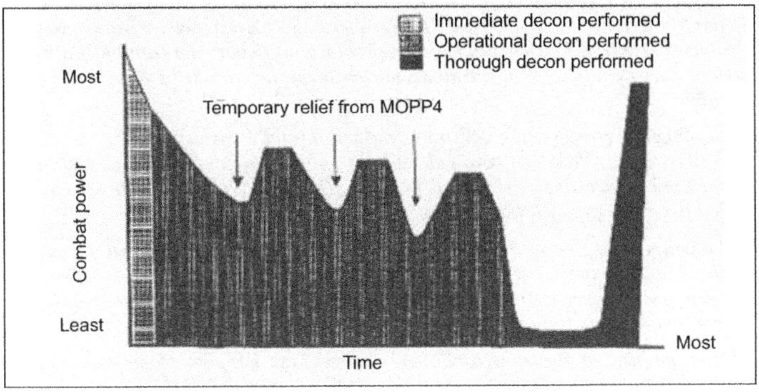

Figure 1-1. How Decon Affects Combat Effectiveness

Decon-Operation Scenario

You are the commander of the 155-millimeter Self-Propelled Howitzer Battery. Your unit is hit by a persistent nerve agent. Soldiers react by taking cover and going to MOPP4. The warning system provided sufficient alert and only a few have to conduct skin decon; most contamination is on the outside of your howitzers and vehicles.

The ongoing fire mission continues with minimal disruption, and the enemy is being detected. Your well-trained soldiers complete a personal wipe down and an operator's spray down as soon as possible. During this period, you are capable of providing fire support and moving, as necessary.

The temperature is 55°F, and humidity is low. While the work rate is strenuous, your troops drink water and suffer no heat casualties. However, after several hours, your response times to the fire mission are getting longer and soldiers have not eaten in more than 12 hours.

Upon making a decision to decon, you request support from your battalion. Coordination is made for an operational decon en route to an alternate position. You have 2 ½ hours to complete the movement and be in a firing position. The alternate position is 20 minutes away, with the decon site about halfway.

You send the advance party to meet the supporting power-driven decon-equipment (PDDE) crew and organize the site. The crew sets up and runs the vehicle washdown while the advance party, with battery supply personnel, prepares the MOPP-gear exchange.

The battalion chemical noncommissioned officer (NCO) directs the vehicle washdown as you rotate one platoon at a time through the site. The firing platoons go quickly as it is critical to have the artillery firing tables available. You complete the operation and are in position to fire at the prescribed time. Your soldiers are in MOPP4 because a vapor hazard still exists from the residual contamination. The advance party found the new area clear of contamination, and ammo resupply commences.

As sections begin improving their positions, gun crews one and two leave one soldier on each gun and move the rest of the soldiers 65 feet upwind. After a check on contamination proves negative, the battery NBC NCO directs unmasking procedures using two crew members. When no symptoms show up, the crews eat and get relief from the masks. The wind direction is monitored to ensure that the soldiers stay upwind of any vapors desorbing from their equipment.

Soldiers on the guns rotate into the clean area for rest and relief. Those returning to the guns assume MOPP4.

Chapter 2

Immediate Decon

Once a soldier is aware of chemical or biological contamination on his bare skin, he initiates immediate decon techniques, without command, by using his personal SDK. He decontaminates his hood, mask, gloves, and weapon using the IEDK or an additional SDK. To remove radiological contamination from equipment and personnel, brush it off and/or use soap and water.

NOTE: Throughout this manual, SDK refers to the M291 kit and IEDK refers to the M295 kit.

SKIN DECON

2-1. Start the skin-decon techniques within 1 minute of becoming contaminated. Some toxic chemical agents, especially nerve agents, kill in minutes.

CHEMICAL

2-2. Use the SDKs within 1 minute of contamination (see Figure 2-1). Instructions for use are listed on the outside of the kit itself, on the individual packet within the kit, and in STP 21-1-SMCT.

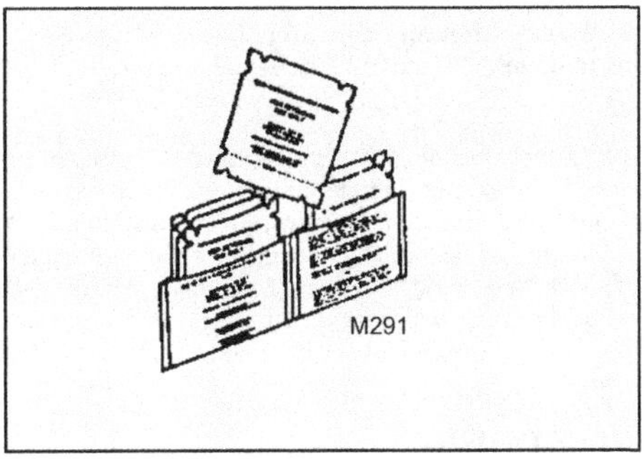

Figure 2-1. M291 SDK

2-3. If an SDK is not available, chemical contamination may be pinch-blotted from the skin with a cloth and flushed with water from a canteen. Soap, if available, can also be used to wash the agent from the skin. Washing with soap and water (or hot water) is the next best method for toxic-agent decon if SDKs are not available, but this method is not as effective as using the decon kits.

BIOLOGICAL

2-4. Washing with soap and water removes nearly all biological agents from the skin. A 0.5 percent chlorine (calcium hypochlorite [HTH] or household bleach) solution is an effective biological decontaminant. See Appendix B for details on how to make this solution.

RADIOLOGICAL

2-5. To remove radiological dust particles, brush, wash, or wipe them off.

PERSONAL WIPE DOWN

2-6. The personal wipe-down techniques are most effective when done within 15 minutes of being contaminated. The CAM and/or M8/M9 detector paper is used to detect and monitor equipment.

NOTE: Very high concentrations of DS2 can cause a CAM response in the G mode.

2-7. Wipe down your mask, hood, gloves, and other essential gear. Do not attempt to remove chemical contamination from your protective overgarment unless there are obvious clumps of the agent. In this case, scrape off the material. Brush off radiological or frozen chemical-agent contamination. See FM 3-4 for detailed information on your overgarment's protective qualities.

CHEMICAL

2-8. Decon individual equipment using the IEDKs. Wearing your Kevlar® helmet protective cover will prevent or reduce the adsorption of any liquid chemical agent. Washing with soap and water and bleach solutions is partially effective (see Appendix B).

BIOLOGICAL

2-9. Wash with soap and water. If water is not available, use SDKs in the same manner as described for chemical-agent decon.

RADIOLOGICAL

2-10. Locate radiological contamination with monitoring equipment and remove by brushing and shaking it off. Wash exposed areas of your skin and pay particular attention to your hair and fingernails. Avoid breathing the dust particles you shake off by wearing your protective mask or a piece of cloth over your nose and mouth. If wet, conduct a MOPP-gear exchange as soon as possible because brushing or shaking will not remove the contamination. Wipe off your equipment with warm, soapy water using rags or damp paper towels.

OPERATOR'S SPRAY DOWN

2-11. Decon other mission-essential portions of your equipment before continuing your mission. The CAM and/or M8/M9 detector paper is used to determine what surfaces require decon.

2-12. Operator's spray down is most effective when done within 15 minutes of contamination.

CHEMICAL

2-13. Decon those surfaces that you must touch on the exterior of the vehicle or equipment you must use to do your job with the on-board portable decon apparatus (such as the M11 or M13) (see Figure 2-2). The IEDK may be used on equipment that DS2 may cause damage to by corrosive action (see Figure 2-3).

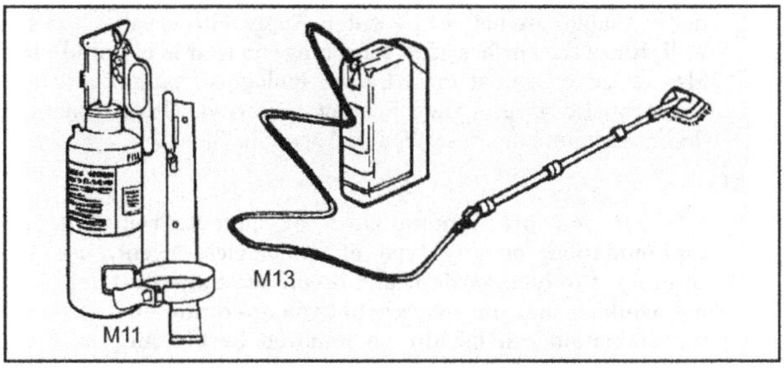

Figure 2-2. Portable Decon Apparatuses

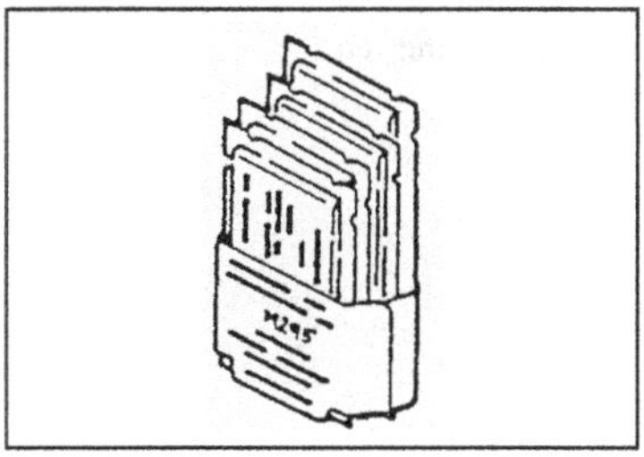

Figure 2-3. M295 IEDK

WARNING

Use extreme caution at all times when handling DS2. Do not mix DS2 and STB because it will cause a fire. Do not allow DS2 to be sprayed on personnel or protective clothing. DS2 is a combustible solution. Do not spray DS2 on hot engine blocks. Severe chemical burns can result if personnel fail to observe all safety precautions. DS2 can severely injure eyes and skin and, if inhaled, can cause illness. DS2 can damage the NBC protective overgarment. Long-term contact with DS2 can damage the NBC protective gloves, hood, and overboots.

2-14. Scrub the DS2 into the exterior surface with brushes. Wait 30 minutes, then wash off. If a decon apparatus is not available, use the field-expedient resources that are available to apply DS2 or STB from bulk containers. If necessary, use the nonstandard decontaminants that are discussed in Appendix B.

BIOLOGICAL

2-15. Use a bleach solution for biological contamination; however, if bleach is not available, use hot, soapy water. Apply with brushes and scrub the surface well. Rinse the surface after scrubbing (no wait is required). DS2 and STB are also effective against most known biological contamination, but because of their caustic nature, they are not preferred. Other nonstandard biological decontaminants are described in Appendix B.

RADIOLOGICAL

2-16. If you are contaminated by fallout, rain out, neutron-induced contamination, or any type of radiological agent, use your monitoring equipment to help locate it and decon as required. If detection equipment is not available and you suspect that you are contaminated, decon. Radiological contamination can usually be removed by brushing or scraping. Water is effective for flushing away radiological contamination; however, the runoff should be controlled by using drainage ditches that flow into a sump. Remember, you have not destroyed the contamination, it has just been moved. The runoff will still be hazardous. If you have time, brush or scoop away the top inch of soil from your fighting position to lower the amount of radiological contamination affecting you.

Chapter 3

Operational Decon

An operational decon limits the spread of contamination, allows temporary relief from MOPP4, and facilitates additional decon requirements. By speeding up the weathering process, the need for a thorough decon may be eliminated. This process requires about a 120-square-yard area and, depending on the number of contaminated vehicles, may not require a nearby water source. An operational decon consists of a vehicle washdown and a MOPP-gear exchange. The techniques that may be used to conduct an operational decon are decentralized company control, centralized battalion control, or centralized brigade control.

TECHNIQUES

3-1. Decentralized company control is provided by the unit commander. The commander requests PDDE support, selects the decon site, links up with the PDDE crew, and executes the operational decon.

3-2. Centralized battalion control is similar to the above except that the battalion chemical NCO controls the decon. The site is selected by the battalion commander, and the battalion chemical NCO directs site setup and security and provides command and control (C^2). The battalion chemical NCO travels with the PDDE and communicates via the radio.

3-3. Centralized brigade control is done when decon assets within the brigade have been consolidated. In this technique, the brigade chemical NCO performs those functions described for the battalion chemical NCO. The advantages and disadvantages of operational-decon techniques are listed in Table 3-1, page 3-2.

PHASES

3-4. The three phases for an operational decon are preparation, execution, and site clearance.

PREPARATION

3-5. The preparation phase starts with the decision to conduct an operational decon and ends with a site that is set up and ready for operation. (The decision to decon was described in Chapter 1.) The CAM is used to identify which vehicles need decontaminating, and PDDE support is requested. Coordination with higher headquarters (HQ) includes identifying the linkup point and the unit that will provide the PDDE support. Signal operating instructions (SOI) information is exchanged and the technique (decentralized company control, centralized battalion control, or centralized brigade control) is selected. The number and type of vehicles and the number of personnel for a MOPP-gear exchange are provided.

Table 3-1. Advantages and Disadvantages of Operational-Decon Techniques

	Decentralized Company Control	Centralized Battalion Control	Centralized Brigade Control
Advantages	Flexibility Dispersion	C^2 Flexibility Dispersion Planning Sustained operations	C^2 Synchronized support assets Rehearsal Time Planning Sustained operations
Disadvantages	C^2 Synchronized support assets Rehearsal Time Planning Sustained operations	Synchronized support assets Rehearsal Time	Flexibility Dispersion

3-6. Site selection is done by the controlling HQ. The site may have been designated in the operation order (OPORD) or selected based on the current METT-TC. For site characteristics, see Table 3-2.

3-7. Linkup at the designated point includes establishing security, positioning any necessary marshalling areas, camouflaging removal points, designating the direction of flow into and out of the site, and ensuring that unit supply has sufficient replacement MOPP gear.

3-8. Site setup requires the PDDE crew to position itself upwind from the entrance. Set up the MOPP-gear exchange about 164 feet upwind from the vehicle washdown at a 45° angle. Notify the contaminated unit when the site is ready.

EXECUTION

3-9. The execution phase starts with establishing positive control over the site and ends with processing the last contaminated vehicle and soldier through the site. Vehicles move from the predecon area to the site either on order or by watching the vehicle ahead. Contaminated soldiers leave the vehicle and proceed to the MOPP-gear-exchange point. The contaminated unit provides security.

3-10. Each vehicle receives a 2- to 3-minute washdown with hot, soapy water and moves to a point near the MOPP-gear exchange. Any remaining soldiers requiring MOPP-gear exchange will do so then. When the squad or crew has finished, it remounts its vehicles and moves into an after-decon assembly area (AA) to await further instructions or moves to its next battle position.

SITE CLEARANCE

3-11. The site-clearance phase includes cleanup, marking, and reporting. METT-TC will dictate the cleanup requirements. The PDDE crew buries or double bags and removes contaminated MOPP gear and other expendables. Standard NBC contamination markers are posted, and the PDDE crew submits an NBC 5 report.

Table 3-2. Operational-Decon Phases

Preparation	
Decon assessment	Identify soldiers and equipment to be decontaminated.
Coordination	Request PDDE support. The chemical section conducts coordination with the contaminated unit on the linkup point. Decon operations should be done between 1 and 6 hours after becoming contaminated.
Site selection	Is selected by the controlling HQ. Is off the main route but has easy access. Is a large enough area (120 square yards per site for a squad-size element). Has good overhead concealment. Has a good water source (plan for 100 gallons per vehicle). Has good drainage.
Linkup	Ensure that the noncommissioned officer in charge (NCOIC) knows where to link up with the contaminated unit and the location for site setup. Radio communication is essential for operations.
Site setup	Ensure that the PDDE is positioned properly and ready to dispense hot, soapy water. Ensure that the contaminated unit sets up and operates the MOPP-gear exchange at the same time as the vehicle washdown. Consider contamination runoff when positioning the PDDE.
Execution	
Site control and security	Ensure that the drivers of the contaminated vehicles know when to move into position at the washdown location. Ensure that the contaminated unit has provided site security.
Processing	Ensure that the decon site NCOIC is processing vehicles at a rate of 2 to 3 minutes per vehicle. Also, ensure that soldiers are going through the MOPP-gear exchange at the rate of 30 minutes per squad/crew.
Site Clearance	
Cleanup	Decon site NCOIC ensures that the MOPP-gear-exchange area is cleaned up.
Marking and reporting	Decon site NCOIC ensures that his team properly marks the decon site and sends the NBC 5 report forward.

VEHICLE WASHDOWN

3-12. A vehicle washdown may be conducted with or without PDDE and in either a one- or two-lane configuration. An unsupported washdown requires the contaminated unit to have washing equipment that can produce 60 to 120 pounds per square inch (psi) of water pressure. The capacity to heat water and inject soap increases effectiveness. A supported washdown requires PDDE organic to the unit or from a supporting chemical decon or dual-purpose company. A two-lane washdown is simply a one-lane washdown parallel with another (see Figure 3-1, page 3-4).

3-13. Other configurations are limited only by the constraints of the METT-TC. Figure 3-2, page 3-5, illustrates a dispersed operational-decon setup. Table 3-3, page 3-5, describes the vehicle-washdown process.

MOPP-GEAR EXCHANGE

3-14. A MOPP-gear exchange can be done with a buddy team, a triple buddy team, or an individual (emergency). For planning purposes, estimate about 60 minutes for a squad or a platoon to complete the exchange. Since individual performance is severely degraded after 6 hours at MOPP4, the MOPP-gear exchange should be scheduled before that time.

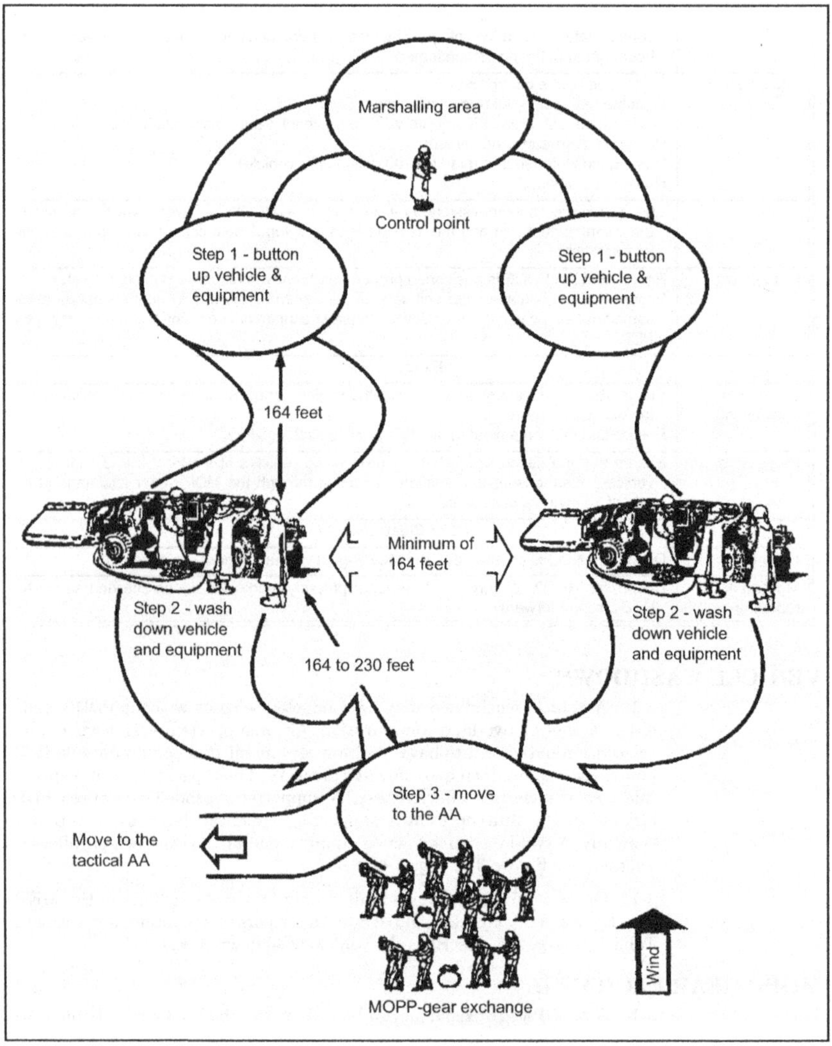

Figure 3-1. Two-Lane Washdown

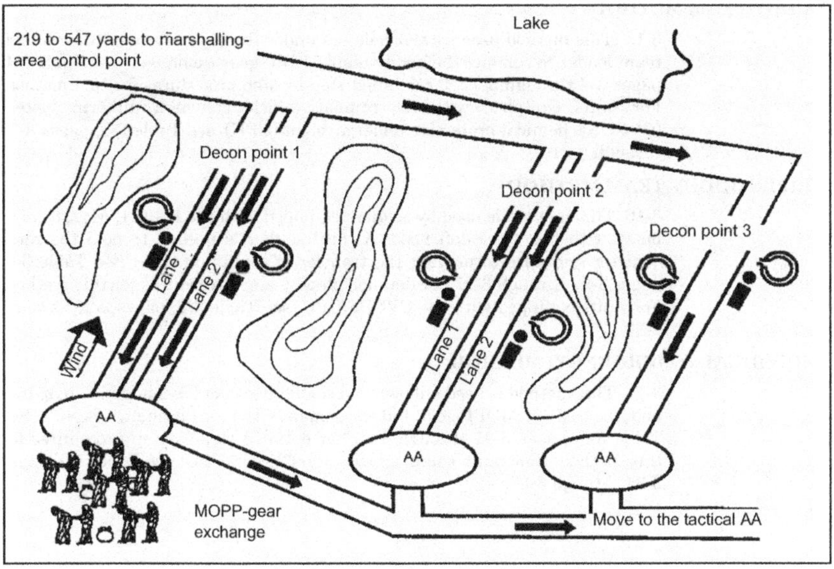

Figure 3-2. Dispersed Operational-Decon Setup

Table 3-3. Vehicle-Washdown Process

Steps	Equipment	Procedures
1. Assemble vehicles. Unit is tactically dispersed. Personnel at the control point direct movement.	Watch	Personnel at the control point supervise preparing vehicles and directing movement out of the AA.
2. Prepare vehicles.	None	Individual/crew closes all access doors, hatches, windows, and other openings. Remove camouflage and cover muzzles. If required, crews (less drivers) move to the MOPP-gear-exchange area. Move to the wash area on order.
3. Perform washdown.	Delivery system (such as the M12 and M17, 65-gallon-per-minute [GPM] pump, fire-fighting equipment, and so forth) that delivers hot, soapy water at 60 to 120 psi. Also sufficient water, fuel, and detergent for vehicles.	Soldiers/personnel wash equipment from top to bottom. Decon crew wears a toxicological apron, protective (TAP) or wet-weather gear over MOPP gear.
4. Move out to the next position.	None	Vehicles move to the MOPP-gear-exchange area (if required) or the next battle position.

BUDDY-TEAM METHOD

3-15. This method uses pairs of soldiers under the supervision of their squad/ team leader to conduct the buddy-team MOPP-gear exchange. See Table 3-4, pages 3-7 through 3-17, for detailed step-by-step procedures for this method. For units equipped with the combat vehicle crewman uniform system (CVCUS)/chemical protective undergarment (CPU), see Table 3-5, pages 3-18 through 3-21.

TRIPLE BUDDY-TEAM METHOD

3-16. This method is used by soldiers equipped with the M40A1, M42, or M43 mask with the quick-doff hood. A third soldier is needed to hold the filter canister and hose to prevent the transfer of contamination. See Table 3-6, pages 3-22 through 3-27, for detailed step-by-step procedures for this method. For units equipped with the CVCUS/CPU, see Table 3-7, pages 3-28 through 3-30.

INDIVIDUAL (EMERGENCY) METHOD

3-17. This method is used only when a soldier does not have buddy to help him and the risk of MOPP-gear failure demands that an exchange occur. See Table 3-8, pages 3-31 through 3-33, for detailed step-by-step procedures for this method. For units equipped with the CVCUS/CPU, see Table 3-9, pages 3-34 through 3-37.

Table 3-4. Steps for the Buddy-Team Method MOPP-Gear Exchange

Steps	Contamination Type	Required Equipment	Procedures
1. Decon gear—removes gross contamination from individual gear (weapons, helmet, load-bearing equipment, and mask carrier).	All contamination	Four long-handled brushes A large piece of plastic (poncho or similar material)	The soldier mixes three parts earth to two parts STB.
	Chemical/ biological	One 30-gallon container, STB dry mix, and one IEDK	The soldier removes and discards the chemical protective helmet cover if worn. He brushes or rubs STB onto his individual gear (helmet and mask carrier) and the hose of the M42 or M43 mask if worn. He gently shakes off any excess STB and sets aside his gear on an uncontaminated surface.
	Radiological	Hot, soapy water	The soldier brushes or wipes radiological contamination from his individual gear. He washes it with hot, soapy water (if available) then sets it aside to dry on an uncontaminated surface (plastic, poncho, or similar material).
2. Prepare to decon—facilitates later removal of overgarment trousers and overboots.	All contamination	Cutting tool	Buddy #1 unfastens the shoulder straps on Buddy #2's hood and pulls them over his shoulder and reattaches them to the Velcro® fasteners. He loosens the drawcord on Buddy #2's hood (see Figure 3-3, page 3-8). He removes the M9 detector paper from Buddy #2's overgarment. He unties the drawcords on the trouser legs of Buddy #2's overgarment. He unzips Buddy #2's trouser legs and rolls a cuff in each trouser leg, ensuring that the cuffs do not come above the tops of his overboots (see Figure 3-4, page 3-8). He unfastens or cuts the fasteners on Buddy #2's overboots. **NOTE: A soldier can do this step by himself or with the help of his buddy. When wearing overboots, unsnap both quick releases on the overboots. The M40 voice amplifier (M7) and the M42A2 detachable microphone cannot be decontaminated and will be disposed of as contaminated waste; however, ensure that these items are contaminated before disposing of them.**
3. Decon mask and hood—removes gross contamination.	Chemical/ biological	Two IEDKs per person	Buddy #1 uses an IEDK to wipe Buddy #2's eye-lens outserts from the top down. **(NOTE: Do not press so hard that you break Buddy #2's face-mask seal.)** He then wipes the rest of Buddy #2's hood from the top of the head to the bottom of the hood (see Figure 3-5, page 3-9). After he has finished wiping Buddy #2's mask, he must wipe his own gloves in preparation for rolling Buddy #2's hood. He starts from the rear and rolls Buddy #2's hood, using 2-inch tucks, until it reaches the center of his head (see Figure 3-6, page 3-9). He rolls the front of Buddy #2's hood tightly under the outlet valve and filter. He ensures that the hood is off of Buddy #2's garment (see Figure 3-7, page 3-10).

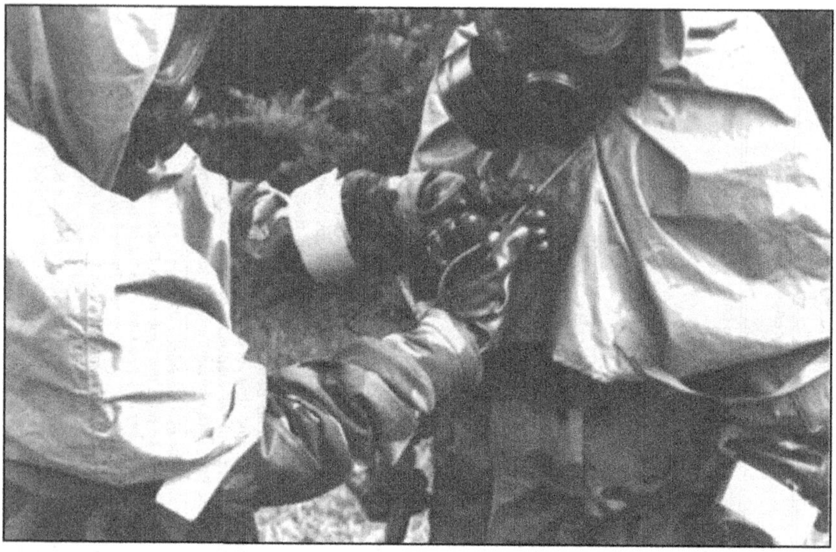

Figure 3-3. Loosening the Drawcord on the Hood

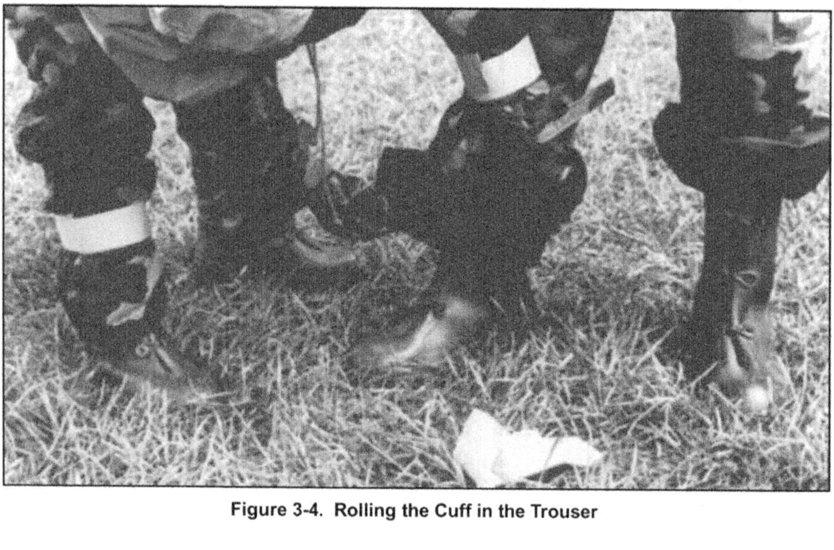

Figure 3-4. Rolling the Cuff in the Trouser

Figure 3-5. Wiping From the Head to the Bottom of the Hood

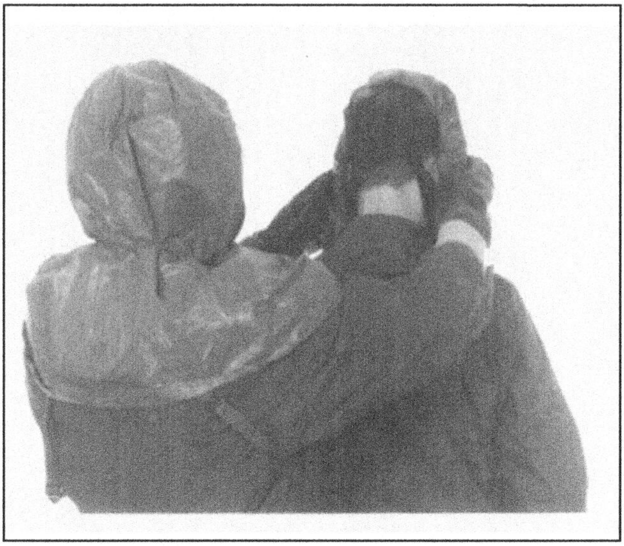

Figure 3-6. Rolling the Hood to the Center of the Head

Figure 3-7. Ensuring That the Hood is off the Garment

Table 3-4. Steps for the Buddy-Team Method MOPP-Gear Exchange (Continued)

Steps	Contamination Type	Required Equipment	Procedures
3. Decon mask and hood (continued).	Radiological	Three containers (about 3-gallon capacity) Two sponges Soapy water Rinse water Paper towels or similar drying material	Buddy #1 wipes Buddy #2's mask and hood with a sponge dipped in hot, soapy water and rinses them with a sponge dipped in clean water. He dries Buddy #2's mask and hood with paper towels or rags. Buddy #2 wipes his own gloves. **NOTE: Cool, soapy water is not as effective for removing contamination, but it can be used if you scrub longer.** If the water supply is limited, use drinking water from your canteen with a wet sponge or cloth. If water is not available, brush and dust off the radioactive dust particles. ***Do not reverse roles.*** Only Buddy #2's hood will be decontaminated and rolled at this time.
4. Remove overgarments and overboots—limits the spread of agents and helps prevent agents from penetrating through to the undergarments or the skin.	All contamination	Two discard containers (for example, plastic bags)	Buddy #1 grasps Buddy #2's overgarment jacket, unsnaps the snaps individually, and unties the drawcord at the bottom of the jacket (see Figure 3-8, page 3-12). He unfastens the Velcro at the wrist of Buddy #2's jacket and then refastens it. He unfastens the Velcro closure over the zippered front of Buddy #2's jacket and unzips the jacket. He grasps Buddy #2's jacket at the shoulders and instructs him to make a fist. He then pulls Buddy #2's jacket down and away from him, ensuring that the black part of the jacket is not touched (see Figure 3-9, page 3-12). He lays Buddy #2's overgarment jacket on the ground, black side up. (It will be used to stand on later.) He carefully unfastens and unzips Buddy #2's trousers. **NOTE: Do not loosen Buddy #2's waist tabs.** He instructs Buddy #2 to loosen his overboots by alternately stepping on each heel and pulling up on his foot (see Figure 3-10, page 3-13). He grasps Buddy #2's trousers and pulls them down to his knees. He instructs Buddy #2 to walk out of his trousers/overboots simultaneously and step onto the black side of the jacket (see Figure 3-11, page 3-13). Buddy #2 should step onto the jacket wearing his mask, battle-dress uniform (BDU), combat boots, and gloves.
5. Remove gloves.	All contamination	Two discard containers (from step 4)	The soldier holds the fingertips of his gloves and partially slides his hand out. When the fingers of both hands are free, he holds his arms away from his body and lets the gloves drop (see Figure 3-12, page 3-14).

Figure 3-8. Unsnapping the Trousers From the Jacket

Figure 3-9. Pulling the Jacket Down and Away

Figure 3-10. Loosening the Overboots

Figure 3-11. Stepping on the Black Lining of the Jacket

Figure 3-12. Removing the Gloves

Table 3-4. Steps for the Buddy-Team Method MOPP-Gear Exchange (Continued)

Steps	Contamination Type	Required Equipment	Procedures
6. Put on overgarment.	All contamination	One set of chemical protective overgarments per person	Buddy #1 opens the package containing the new overgarment without touching the inside of the package. Buddy #2 pulls out the overgarment without touching the outside of the package. He puts on the overgarment and fastens it, leaving the trouser legs open (see Figure 3-13, page 3-16). ***Do not reverse roles.*** Only Buddy #2 will put on the clean overgarment at this time.
7. Put on overboots and gloves.	All contamination	One set of chemical protective overboots per person One set of chemical protective gloves per person (correct size) M9 detector paper	Buddy #1 opens the package of clean overboots without touching the inside of the package. Buddy #2 removes the overboots from the package without touching the outside of the package, puts them on, and fastens his trouser legs. Buddy #1 opens the package of clean gloves without touching the inside of the package. Buddy #2 removes the gloves from the package without touching the outside of the package and puts them on. Buddy #2 puts on the M9 detector paper (see Figure 3-14, page 3-16). ***Do not reverse roles.*** Only Buddy #2 will put on the clean overboots and gloves at this time.

Figure 3-13. Putting on the New Overgarment

Figure 3-14. Putting on the M9 Detector Paper

Table 3-4. Steps for the Buddy-Team Method MOPP-Gear Exchange (Continued)

Steps	Contamination Type	Required Equipment	Procedures
8. Secure hood.	All contamination	One IEDK	Buddy #1 uses an IEDK to wipe Buddy #2's gloves. He unrolls Buddy #2's hood and attaches the straps and tightens the neck cord. Buddy #1 and Buddy #2 reverse roles and repeat steps 2 through 8.
9. Secure gear.	All contamination	One chemical protective helmet cover per person	The soldier secures his individual gear and puts it back on (see Figure 3-15). He puts on a new chemical protective helmet cover and moves to the AA. He uses the buddy system to check the fit of all secured gear.

Figure 3-15. Securing the Hood

Table 3-5. Steps for the CVCUS/CPU (Buddy-Team Method) MOPP-Gear Exchange

Steps	Contamination Type	Required Equipment	Procedures
1. Decon gear—removes gross contamination from individual gear (weapons, helmet, load-bearing equipment, and mask carrier).	All contamination	Four long-handled brushes A large piece of plastic (poncho or similar material)	The soldier mixes three parts earth to two parts STB.
	Chemical/ biological	One 30-gallon container, STB dry mix, and one IEDK	The soldier removes and discards the chemical protective helmet cover if worn. He brushes or rubs the STB onto his individual gear and the hose of the M42 mask if worn. He gently shakes off any excess STB and sets aside his gear on an uncontaminated surface.
	Radiological	Hot, soapy water	The soldier brushes or wipes radiological contamination from his individual gear. He washes it with hot, soapy water (if available) then sets it aside to dry on an uncontaminated surface (plastic, poncho, or similar material).
2. Prepare for decon—facilitates MOPP-gear removal.	All contamination	Cutting tool	Buddy #1 unfastens the shoulder straps on Buddy #2's hood and pulls them over his shoulder and reattaches them to the Velcro fastener. He loosens the drawcord on Buddy #2's hood. He removes the M9 detector paper from Buddy #2's CVCUS. He unzips the ankles on both legs of Buddy #2's CVCUS. He unfastens or cuts the fasteners on Buddy #2's overboots. **NOTE: A soldier can do this step by himself or with the help of his buddy.**
3. Decon mask and hood—removes gross contamination.	Chemical/ biological	Two IEDKs per person	Buddy #1 uses an IEDK to wipe Buddy #2's eye-lens outserts from the top down. **(NOTE: Do not press so hard that you break Buddy #2's face-mask seal.)** He then wipes the rest of Buddy #2's hood from the top of the head to the bottom of the hood (see Figure 3-5, page 3-9). After he has finished wiping Buddy #2's mask, he must wipe his own gloves in preparation for rolling Buddy #2's hood. He starts from the rear and rolls Buddy #2's hood, using 2-inch tucks, until it reaches the center of his head (see Figure 3-6, page 3-9). He rolls the front of Buddy #2's hood tightly under the outlet valve and filter. He ensures that the hood is off of Buddy #2's garment (see Figure 3-7, page 3-10).

Table 3-5. Steps for the CVCUS/CPU (Buddy-Team Method) MOPP-Gear Exchange (Continued)

Steps	Contamination Type	Required Equipment	Procedures
3. Decon mask and hood (continued).	Radiological	Three containers (about 3-gallon capacity) Two sponges Soapy water Rinse water Paper towels or similar drying material	Buddy #1 wipes Buddy #2's mask and hood with a sponge dipped in hot, soapy water and rinses them with a sponge dipped in clean water. He dries Buddy #2's mask and hood with paper towels or rags. Buddy #2 wipes down his own gloves. **NOTE: Cool, soapy water is not as effective for removing contamination, but it can be used if you scrub longer.** If the water supply is limited, use drinking water from your canteen with a wet sponge or cloth. Only Buddy #2's mask and hood will be decontaminated and rolled at this time.
4. Remove CVCUS/ overboots—limits the spread of contamination.	All contamination	Two discard containers (for example, plastic bags)	Buddy #1 opens all the zippers on Buddy #2's CVCUS in the following order: wrist, ankles, and front. While standing behind Buddy #2 and adjacent to the clean area, Buddy #1 grasps the shoulders of Buddy #2's CVCUS and instructs him to make a fist to ensure that his gloves do not come off. He then pulls Buddy #2's CVCUS down below his knees, turning it inside out. He has Buddy #2 to partially remove his overboots by alternately stepping on each heel and pulling up on his foot. While kneeling, Buddy #1 grasps one leg of Buddy #2's CVCUS and his overboot and instructs him to remove his leg and foot simultaneously (see Figure 3-16, page 3-20). As Buddy #2 removes each leg and foot, he steps directly onto the clean area.
5. Remove CPU jacket/gloves.	All contamination	Two discard containers	Buddy #2 holds the fingertips of his gloves and partially slides his hand out. When the fingers of both hands are free, he holds his arms away from his body and lets the gloves drop. Buddy #1 unzips and removes Buddy #2's CPU jacket by grasping it at the shoulders and pulling down, turning it inside out (see Figure 3-17, page 3-20). He places it on the clean area with the clean side up. He removes Buddy #2's protective glove inserts.

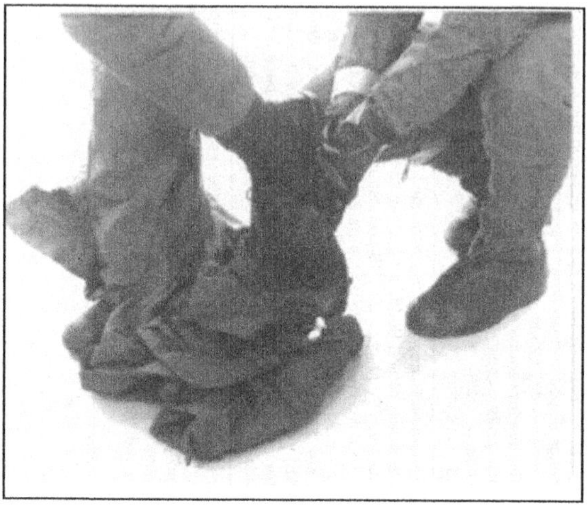

Figure 3-16. Removing Leg and Foot From the Garment

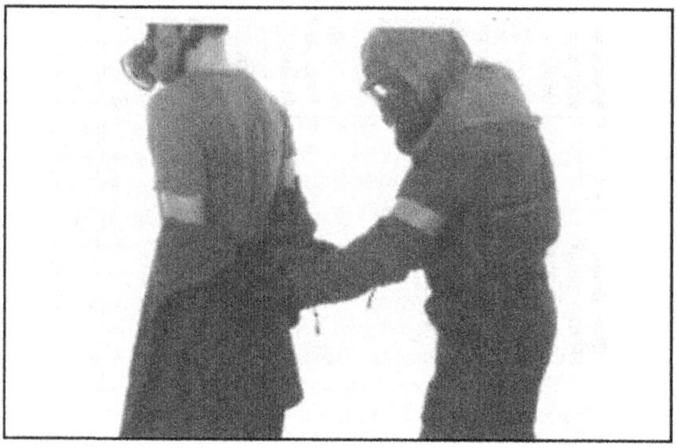

Figure 3-17. Removing the Garment and Turning it Inside Out

Table 3-5. Steps for the CVCUS/CPU (Buddy-Team Method) MOPP-Gear Exchange (Continued)

Steps	Contamination Type	Required Equipment	Procedures
6. Remove combat boots/ CPU pants.	All contamination	Two discard containers	Buddy #1 instructs Buddy #2 to stand on the clean area (but not on the CPU jacket) and unlace his combat boots. Buddy #2 partially removes his boots by alternately stepping on each heel and pulling up on his foot. Buddy #1 grasps one of Buddy #2's boots and instructs him to remove his foot. As Buddy #2 removes his foot, he will step directly onto the clean side of the CPU jacket. Repeat the procedure for the other foot. Next, Buddy #1 removes Buddy #2's CPU drawers by pulling down on the sides and turning them inside out. Buddy #2 is now stripped to his underwear and mask and is ready to put on new clothing. **NOTE: If Buddy #2's feet come in contact with the contaminated pants of the CPU, decon them.**
7. Put on CPU/ CVCUS.	All contamination	One CPU/CVCUS per soldier	Buddy #1 opens the package containing the new undergarment without touching the inside of the package. Buddy #2 pulls out the undergarment without touching the outside of the package. He puts on the new clothing in the following order: CPU drawers, CPU jacket, combat boots, and CVCUS (over the CPU drawers).
8. Put on overboots and gloves.	All contamination	One set of chemical protective overboots per soldier One set of chemical protective gloves per soldier M9 detector paper	Buddy #1 opens the package of clean overboots without touching the inside of the package. Buddy #2 removes the overboots without touching the outside of the package and puts them on. Buddy #1 opens the package of clean gloves without touching the inside. Buddy #2 removes the gloves without touching the outside of the package and puts them on. Buddy #2 puts on the M9 detector paper (see Figure 3-18, page 3-23).
9. Secure hood.	Chemical/ biological	One IEDK	Buddy #1 wipes his rubber gloves with an IEDK. He unrolls Buddy #2's hood and attaches the straps and tightens the neck cord. Buddy #1 and Buddy #2 reverse rolls and repeat steps 2 through 9.
10. Secure gear.	All contamination	One chemical protective helmet cover per person	The soldier secures his *individual gear and puts it back on*. He puts on a new chemical protective helmet cover and moves to the AA. He uses the buddy system to check the fit of all secured gear.

Table 3-6. Steps for the Triple-Buddy Method MOPP-Gear Exchange With the M40A1 and M42

Steps	Contamination Type	Required Equipment	Procedures
1. Decon gear—removes gross contamination from individual gear (weapons, helmet, load-bearing equipment, and mask carrier).	All contamination	Four long-handled brushes A large piece of plastic (poncho or similar material)	The soldier mixes three parts earth to two parts STB.
	Chemical/ biological	One 30-gallon container, STB dry mix, and one IEDK	The soldier removes and discards the chemical protective helmet cover if worn. He brushes or rubs STB onto his individual gear (helmet and mask carrier) and the hose of the M42 or M43 mask if worn. He gently shakes off any excess STB and sets aside his gear on an uncontaminated surface.
	Radiological	Hot, soapy water	The soldier brushes or wipes radiological contamination from his individual gear. He washes it with hot, soapy water (if available) then sets it aside to dry on an uncontaminated surface (plastic, poncho, or similar material).
2. Prepare to decon—facilitates later removal of overgarment trousers and overboots.	All contamination	Cutting tool	Buddy #1 unfastens the shoulder straps on Buddy #2's hood and pulls them over his shoulder and reattaches them to the Velcro fastener. He loosens the drawcord on Buddy #2's hood and ties off the microphone cord to the hose of his mask (see Figure 3-19). He removes the M9 detector paper from Buddy #2's overgarment. He unzips Buddy #2's trouser legs and rolls a cuff in each trouser leg, ensuring that the cuffs do not come above the tops of his overboots. He unfastens or cuts the fasteners on Buddy #2's overboots. **NOTE: A soldier can do this step by himself or with the help of his buddy.**

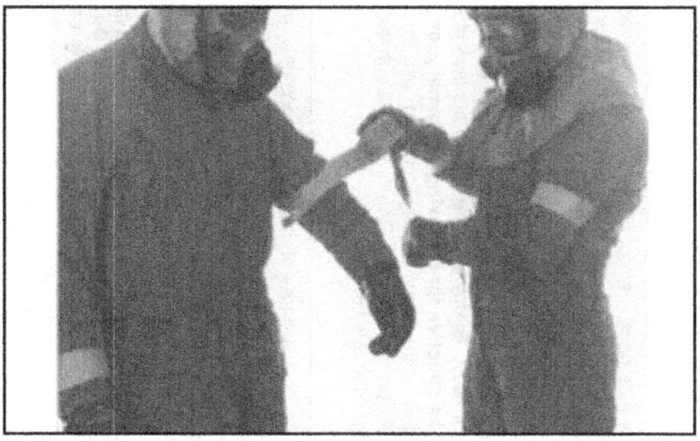

Figure 3-18. Reapplying the M9 Detector Paper

Figure 3-19. Tying the Microphone Cord to the Hose of the Mask

Table 3-6. Steps for the Triple-Buddy Method MOPP-Gear Exchange With the M40A1 and M42 (Continued)

Steps	Contamination Type	Required Equipment	Procedures
3. Decon mask and hood—removes gross contamination.	Chemical/ biological	Two IEDKs per person	Buddy #1 uses an IEDK to wipe Buddy #2's eye-lens outserts from the top down. **(NOTE: Do not press so hard that you break Buddy #2's face-mask seal.)** He then wipes the rest of Buddy #2's hood from the top of the head to the bottom of the hood (see Figure 3-5, page 3-9). After he has finished wiping Buddy #2's mask, he must wipe his own gloves in preparation for rolling Buddy #2's hood. He starts from the rear and rolls Buddy #2's hood, using 2-inch tucks, until it reaches the center of his head (see Figure 3-6, page 3-9). He rolls the front of Buddy #2's hood tightly under the outlet valve and filter. He ensures that the hood is off of Buddy #2's garment (see Figure 3-7, page 3-10).
	Radiological	Three containers (about 3-gallon capacity) Two sponges Soapy water Rinse water Paper towels or similar drying material	Buddy #1 wipes Buddy #2's mask and hood with a sponge dipped in hot, soapy water and rinses them with a sponge dipped in clean water. He dries Buddy #2's mask and hood with paper towels or rags. Buddy #2 wipes down his own gloves. **NOTE: Cool, soapy water is not as effective for removing contamination, but it can be used if you scrub longer.** If the water supply is limited, use drinking water from your canteen with a wet sponge or cloth. Only Buddy #2's mask and hood will be decontaminated and rolled at this time.
4. Remove overgarment overboots—limits the spread of agents and helps prevent agents from penetrating through to the undergarments or the skin.	All contamination	Two discard containers (for example, plastic bags)	Buddy #1 grasps Buddy #2's outside overgarment jacket, unsnaps the snaps individually, and unties the drawcord at the bottom of the jacket. He unfastens the Velcro at the wrist of Buddy #2's jacket and then refastens it. He unfastens the Velcro closure over the zippered front of Buddy #2's jacket and unzips the jacket. Buddy #1 grasps Buddy #2's jacket by the shoulders and instructs him to make a fist. He then pulls the jacket down and away from Buddy #2, ensuring that the black part of the jacket is not touched (see Figure 3-20, page 3-26). He lays Buddy #2's overgarment jacket on the ground, black side up. (It will be used to stand on later.) Buddy #1 carefully unfastens and unzips Buddy #2's trousers. **NOTE: Do not loosen Buddy #2's waist tabs.** He instructs Buddy #2 to break the seals on his overboots by alternately stepping on each heel and pulling up on his foot. He grasps Buddy #2's trousers and pulls them down to his knees. Buddy #1 instructs Buddy #2 to walk out of his trousers/overboots, taking care not to step on the contaminated side of the overgarment. If Buddy #2 is wearing overboots, Buddy #1 removes Buddy #2's trousers first then helps him step out of his overboots onto the black side of the jacket. Buddy #2 should step onto the jacket wearing his mask, BDU, combat boots, and gloves.

Table 3-6. Steps for the Triple-Buddy Method MOPP-Gear Exchange With the M40A1 and M42 (Continued)

Steps	Contamination Type	Required Equipment	Procedures
5. Remove gloves.	All contamination	Discard container (from Step 4)	The soldier holds the fingertips of his gloves and partially slides his hand out. When the fingers of both hands are free, he holds his arms away from his body and lets the gloves drop.
6. Put on overgarment.	All contamination	One set of chemical protective overgarments per person	Buddy #1 opens the package containing the new overgarment without touching the inside of the package. Buddy #2 pulls out the overgarment without touching the outside of the package. He puts on the overgarment and fastens it, leaving the trouser legs open. ***Do not reverse roles.*** Only Buddy #2 will put on the clean overgarments at this time.

Figure 3-20. Removing the Outer Garment

Table 3-6. Steps for the Triple-Buddy Method MOPP-Gear Exchange With the M40A1 and M42 (Continued)

Steps	Contamination Type	Required Equipment	Procedures
7. Put on overboots and gloves.	All contamination	One set of chemical protective overboots per person One set of chemical protective gloves per person M9 detector paper	Buddy #1 opens the package of clean overboots without touching the inside of the package. Buddy #2 removes the overboots without touching the outside of the package, puts them on, and fastens his trouser legs. Buddy #1 opens the package of clean gloves without touching the inside of the package. Buddy #2 removes the gloves without touching the outside of the package and puts them on. Buddy #2 puts on the M9 detector paper. ***Do not reverse roles.*** Only Buddy #2 will put on clean overboots and gloves at this time. **NOTE: The gloves and overboots will have a light powdery coating. This is normal. It is not necessary to remove it. It will not affect the protective qualities.**
8. Secure hood.	All contamination	One IEDK	Buddy #1 wipes his rubber gloves with an IEDK. He unrolls Buddy #2's hood and attaches the straps and tightens the neck cord. He checks the clips and neck cord on Buddy #2's hood. Buddy #1 and Buddy #2 reverse roles and repeat steps 2 through 8.
9. Secure gear.	All contamination	One chemical protective helmet cover per person	The soldier secures his individual gear and puts it back on. He puts on a new chemical protective helmet cover and moves to the AA. He uses the buddy system to check the fit of all secured gear.

Table 3-7. Steps for the CVCUS/CPU (Triple-Buddy Method) MOPP-Gear Exchange With the M40A1, M42, and M43 (Aviation and Armor)

Steps	Contamination Type	Required Equipment	Procedures
1. Decon gear—removes gross contamination from individual gear (weapons, helmet, load-bearing equipment, and mask carrier).	All contamination	Four long-handled brushes A large piece of plastic (poncho or similar material)	The soldier mixes three parts earth to two parts STB.
	Chemical/ biological	One 30-gallon container, STB dry mix, and one IEDK	The soldier removes and discards the chemical protective helmet cover if worn. He brushes or rubs STB onto his individual gear (helmet and mask carrier) and the hose of the M43 mask if worn. He gently shakes off any excess STB and sets aside his gear on an uncontaminated surface.
	Radiological	Hot, soapy water	The soldier brushes or wipes radiological contamination from his individual gear. He washes it with hot, soapy water (if available) and sets it aside to dry on an uncontaminated surface (plastic, poncho, or similar material).
2. Prepare to decon—facilitates later removal of overgarment trousers and overboots.	All contamination	Cutting tool	Buddy #1 unfastens the shoulder straps on Buddy #2's hood, pulls them over his shoulder, and reattaches them to the Velcro fastener. He loosens the drawcord on Buddy #2's hood and ties off the microphone cord to the hose of his mask. He removes the M9 detector paper from Buddy #2's overgarment. He unzips Buddy #2's trouser legs and rolls a cuff in each trouser leg, ensuring that the cuffs do not come above the top of his overboots. He unfastens or cuts the fasteners on Buddy #2's overboots. **NOTE: A soldier can do this step by himself or with the help of his buddy.**

Table 3-7. Steps for the CVCUS/CPU (Triple-Buddy Method) MOPP-Gear Exchange With the M40A1, M42, and M43 (Aviation and Armor) (Continued)

Steps	Contamination Type	Required Equipment	Procedures
3. Decon mask and hood—removes gross contamination.	Chemical/ biological	Two IEDKs per person	If wearing the M42 or M43 mask, use the buddy-team method (three buddies). Buddy #1 uses an IEDK to wipe Buddy #2's hood, mask, canister, and hose of the M42 or M43 mask. **NOTE: Do not press so hard that you break Buddy #2's face-mask seal.** He uses as many wipes as necessary. Buddy #3 wipes off his gloves and holds Buddy #1's hose and canister away from his contaminated overgarment. After Buddy #1 wipes Buddy #2's mask and then his own gloves, he rolls Buddy #2's hood. He starts from the rear and rolls Buddy #2's hood, using 2-inch tucks, until it reaches the center of his head (see Figure 3-6, page 3-9). He rolls the front of Buddy #2's hood tightly under the outlet valve and filter. He ensures that the hood is off of Buddy #2's garment (see Figure 3-7, page 3-10).
	Radiological	Three containers (about 3-gallon capacity) Two sponges Soapy water Rinse water Paper towels or similar drying material	Buddy #1 wipes Buddy #2's mask and hood with a sponge dipped in hot, soapy water and rinses them with a sponge dipped in clean water. He dries Buddy #2's mask and hood with paper towels or rags. Buddy #2 wipes his own gloves. **NOTE: Cool, soapy water is not as effective for removing contamination, but it can be used if you scrub longer.** If the water supply is limited, use drinking water from you canteen with a wet sponge or cloth. If water is not available, brush and dust off the radioactive dust particles. ***Do not reverse roles***. Only Buddy #2's hood will be decontaminated and rolled at this time.
4. Remove CVCUS overboots—limits the spread of contamination.	All contamination	Two discard containers (for example, plastic bags)	Buddy #1 opens all the zippers on Buddy #2's CVCUS in the following order: wrist, ankles, and front. While standing behind Buddy #2 and adjacent to the clean area, Buddy #1 grasps the shoulders of Buddy #2's CVCUS and instructs him to make a fist to ensure that his gloves do not come off. He then pulls Buddy #2's CVCUS down below his knees, turning it inside out. He has Buddy #2 to partially remove his overboots by alternately stepping on each heel and pulling up on his foot. While kneeling, Buddy #1 grasps one leg of Buddy #2's CVCUS and his overboot and instructs him to remove his leg and foot simultaneously. As Buddy #2 removes each leg and foot, he steps directly onto the clean area. Buddy #1 decons his gloves again.

Table 3-7. Steps for the CVCUS/CPU (Triple-Buddy Method) MOPP-Gear Exchange With the M40A1, M42, and M43 (Aviation and Armor) (Continued)

Steps	Contamination Type	Required Equipment	Procedures
5. Remove CPU jacket/ gloves.	All contamination	Two discard containers	Buddy #2 holds the fingertips of his gloves and partially slides his hand out. When the fingers of both hands are free, he holds his arms away from his body and lets the gloves drop. Buddy #1 unzips and removes Buddy #2's CPU jacket by grasping it at the shoulders and pulling down, turning it inside out. He places it on the clean area with the clean side up. He removes Buddy #2's protective glove inserts.
6. Remove combat boots/ CPU pants.	All contamination	Two discard containers	Buddy #1 instructs Buddy #2 to stand on the clean area (but not on the CPU jacket) and unlace his combat boots. Buddy #2 partially removes his boots by alternately stepping on each heel and pulling up on his foot. Buddy #1 grasps one of Buddy #2's boots and instructs him to remove his foot. As Buddy #2 removes his foot, he steps directly onto the clean side of the CPU jacket. Repeat the procedure on the other foot. Next, Buddy #1 removes Buddy #2's CPU drawers by pulling down on the sides and turning them inside out. Buddy #2 is now stripped to his underwear and mask and is ready to put on new clothing. **NOTE: If Buddy #2's feet come in contact with the contaminated pants of the CPU, decon them.**
7. Put on CPU/ CVCUS.	All contamination	One CPU/CVCUS per person	Buddy #1 opens the package containing the new undergarment without touching the inside. Buddy #2 pulls out the undergarment without touching the outside of the package. He puts on the new clothing in the following order: CPU drawers, CPU jacket, combat boots, and CVCUS (over the CPU drawers).
8. Put on overboots and gloves.	All contamination	One set of chemical protective over-boots per person One set of chemical protective gloves per person M9 detector paper	Buddy #1 opens the package of clean overboots without touching the inside of the package. Buddy #2 removes the overboots without touching the outside of the package and puts them on. Buddy #1 opens the package of clean gloves without touching the inside of the package. Buddy #2 removes the gloves without touching the outside of the package and puts them on. Buddy #2 puts on the M9 detector paper.
9. Secure hood.	Chemical/ biological	One IEDK	Buddy #1 wipes his rubber gloves with an IEDK. He unrolls Buddy #2 hood and attaches the straps and tightens the neck cord. He checks the snaps and neck cord on Buddy #2's hood to ensure that they are closed. Buddy #1 and Buddy #2 reverse rolls and repeat steps 2 through 9.
10. Secure gear.	All contamination	One chemical protective helmet cover per person	The soldier secures his individual gear and puts it back on. He puts on a new chemical protective helmet cover and moves to the AA. He uses the buddy system to check the fit of all secured gear.

Table 3-8. Steps for the Individual (Emergency Method) MOPP-Gear Exchange

Steps	Contamination Type	Required Equipment	Procedures
1. Decon gear—removes gross contamination from individual gear (weapon, helmet, load-bearing equipment, and mask carrier).	Chemical/ biological	One IEDK	The soldier uses M8 detector paper to determine the areas of gross contamination and field-expedient absorbents (sand, dirt, or rags) to remove gross liquid contamination. He should take special care to avoid touching these areas during overgarment removal. He uses an IEDK to decon his individual gear (helmet, load-carrying equipment (LCE), weapon, and mask carrier).
	Radiological	Hot, soapy water	The soldier brushes or wipes radiological contamination from his individual gear. He washes it with hot, soapy water (if available) then sets it aside to dry on an uncontaminated surface (plastic, poncho, or similar material).
2. Prepare to decon—facilities removal of overgarment trousers and overboots.	All contamination	Cutting tool	The soldier unfastens the shoulder straps on his hood, pulls them over his shoulders, and reattaches them. He loosens the drawcord on the hood of his protective mask. **NOTE: If wearing the M43 protective mask, tie off the microphone cord to the hose of the mask.** He removes the M9 detector paper from his overgarment and unties/cuts the drawcords on the trouser legs of his overgarment. He unzips the trouser legs and rolls a cuff in each trouser leg, ensuring that the cuffs do not come above the top of his overboots. He unfastens or cuts the fasteners on the overboots.
3. Decon mask and hood—removes gross contamination from the mask and hood.	Chemical/ biological	One IEDK	The soldier wipes the eye-lens outserts on his mask from the top down. He wipes his mask and gloves and rolls his hood. He grasps the straps of the hood and lifts the hood off his shoulders and partially over his head until most of the back of his head is exposed. He rolls the hood, starting at the chin, and works around the entire mask until the rolled hood will stay up and off of his shoulders. He tucks the straps and neck cord into the roll. He rolls the hood tightly against his mask with-out pulling the hood off the back of his head. He tucks the tail between the upper part of the canister and the mask. (Tie the tail over and under the hose for the M42 mask.) The soldier removes the applicator mitt from the package with his nondominant hand. Making a "V", he wipes down his hand, paying particular attention to the areas between his fingers. Once he has thoroughly wiped down his dominant hand, he must insert it into the applicator mitt and thoroughly wipe down his other hand. He gently pats the voicemitter with black powder until it is covered.

Table 3-8. Steps for the Individual (Emergency Method) MOPP-Gear Exchange (Continued)

Steps	Contamination Type	Required Equipment	Procedures
3. Decon mask and hood (continued).			The soldier starts at the top of his hood and wipes down and away, patting until the surface of the hood is covered by the black powder. He rewipes his gloves, starting with his nondominant hand. The soldier lifts the hood off of his shoulders by grasping the shoulder straps in one hand and placing the other hand on top of the head and pulling the hood over his head until the elastic band is over his knuckles and most of the back of his head is exposed. He should not expose his ears or pull the hood completely over his face or mask. He tucks his shoulder straps, underarm straps, and the rolled portion of his hood under the elastic band. Follow these procedures when using the M40A1 mask with the quick-doff hood: The soldier removes the underarm straps from the front of his hood and places them over his shoulders. He refastens them on the front of his hood. **NOTE: Anytime you decon your mask or roll your hood, place two fingers on the voicemitter of the mask to prevent accidental breakage of the seal.**
	Radiological	Three containers (about 3-gallon capacity) Two sponges Soapy water Rinse water Paper towels or similar drying material	The soldier wipes his mask and hood with a sponge dipped in hot, soapy water and rinses them with a sponge dipped in clean water. He dries the mask and hood with paper towels or rags. **NOTE: Cool, soapy water is not as effective for removing contamination, but it can be used if you scrub longer.** If water is not available, soldier should brush and dust off the radioactive dust particles.
4. Remove overgarment/ overboots—prevents agent from penetrating through to the undergarments or the skin.	All contamination	Two discard containers (recommend plastic bags). Chemical protective suit with protective gloves and overboots	The soldier grasps his overgarment jacket and unsnaps the snaps individually. He unties the drawcord at the bottom of the jacket. He unfastens the Velcro at the waist and then refastens it. He unfastens the Velcro closure over the zippered front of the jacket and unzips the jacket. He grasps the front of the jacket and pulls the jacket back until it is off his shoulders. He puts his arms behind his back and works his arms out of the sleeves. He should not let the outside of the jacket touch his body. When the jacket is off, he lays it on the ground with the black side up. He unfastens and unzips his trousers. He should not loosen his waist tabs. He loosens his overboots by alternately stepping on each heel and pulling up on his foot. He grasps his trousers and pushes them down to his knees. He walks out of his trousers/ overboots simultaneously and steps onto the black side of the jacket. He steps onto the jacket wearing his mask, BDU, combat boots, and gloves.

Table 3-8. Steps for the Individual (Emergency Method) MOPP-Gear Exchange (Continued)

Steps	Contamination Type	Required Equipment	Procedures
5. Remove gloves.	All contamination	Two discard containers (from step 4) One IEDK	The soldier wipes around the edges of the packages containing the new items (gloves, overgarments, and overboots) with an IEDK. He opens the new packages. He holds the fingertips of his gloves and partially slides his hands out. When the fingers of both hands are free, he holds his arms away from his body and lets his gloves drop.
6. Put on overgarment.	All contamination	One set of chemical protective overgar-ments per person One IEDK	The soldier removes the overgarment from its package without touching the outside. He puts on the overgarment and fastens it, leaving the trouser legs open until he puts on the new overboots.
7. Put on overboots and gloves.	All contamination	One set of chemical protective overboots per person One set of chemical protective gloves per person	The soldier removes the overboots (one at a time) from their package without touching the outside and puts them on. He removes the gloves from their package without touching the outside, puts them on, and fastens his trouser legs. **NOTE: The gloves and overboots will have a light powdery coating. This is normal. Do not remove it. It will not affect the protective qualities.**
8. Secure hood.	All contamination	One chemical protective helmet cover per person	The soldier secures his individual gear and puts it back on. He puts on a new chemical protective helmet cover and moves to the AA.

Table 3-9. Steps for the CVCUS/CPU, Individual (Emergency Method) MOPP-Gear Exchange

Steps	Contamination Type	Required Equipment	Procedures
1. Decon gear—removes gross contamination from individual gear (weapon, helmet, load-bearing equipment, and mask carrier).	Chemical/ biological	One IEDK	The soldier uses M8 detector paper to determine the areas of gross contamination and field-expedient absorbents (sand, dirt, or rags) to remove gross liquid contamination. He should take special care to avoid touching these areas during overgarment removal. He uses an IEDK to decon his individual gear (helmet LCE, weapon, and mask carrier).
	Radiological	Hot, soapy water	The soldier brushes or wipes radiological contamination from his individual gear. He washes it with hot, soapy water (if available) then sets it aside to dry on an uncontaminated surface (plastic, poncho, or similar material).
2. Prepare to decon—facilities removal of overgarment trousers and overboots.	All contamination	Cutting tool	The soldier unfastens the shoulder straps on his hood, pulls them over his shoulders, and reattaches them to the front of the hood. He loosens the drawcord on the hood of his protective mask. **NOTE: If wearing the M43 protective mask, tie off the microphone cord to the hose of the mask.** He removes the M9 detector paper from his overgarment and unties/cuts the drawcords on the trouser legs of his overgarment. He unzips the trouser legs and rolls a cuff in each trouser leg, ensuring that the cuffs do not come above the top of his overboots. He unfastens or cuts the fasteners on the overboots.
3. Decon mask and hood—removes gross contamination from the mask and hood.	Chemical/ biological	One IEDK	The soldier wipes the eye-lens outserts on his mask from the top down. He wipes his mask and gloves and rolls his hood. He grasps the straps of the hood and lifts the hood off his shoulders and partially over his head until most of the back of his head is exposed. He rolls the hood, starting at the chin, and works around the entire mask until the rolled hood will stay up and off of his shoulders. He tucks the straps and neck cord into the roll. He rolls the hood tightly against his mask without pulling the hood off the back of his head. He tucks the tail between the upper part of the canister and the mask. (Tie the tail over and under the hose for the M42 mask.) The soldier removes the applicator mitt from the package with his nondominant hand. Making a "V", he wipes down his hand, paying particular attention to the areas between his fingers. Once he has thoroughly wiped down his dominant hand, he must insert it into the applicator mitt and thoroughly wipe down his other hand.

Table 3-9. Steps for the CVCUS/CPU, Individual (Emergency Method) MOPP-Gear Exchange (Continued)

Steps	Contamination Type	Required Equipment	Procedures
3. Decon mask and hood (continued).			The soldier gently pats the voicemitter with black powder until it has been covered. He starts at the top of his hood and wipes down and away, patting until the surface of the hood is covered by the black powder. He rewipes his gloves, starting with his nondominant hand. The soldier lifts the hood off of his shoulders by grasping the shoulder straps in one hand and placing the other hand on top of the head and pulling the hood over his head until the elastic band is over his knuckles and most of the back of his head is exposed. He should not expose his ears or pull the hood completely over his face or mask. He tucks his shoulder straps, underarm straps, and the rolled portion of his hood under the elastic band. Follow these procedures when using the M40A1 mask with the quick-doff hood: The soldier removes the underarm straps from the front of his hood and places them over his shoulders. He refastens them on the front of his hood. **NOTE: Anytime you decon your mask or roll your hood, place two fingers on the voicemitter of the mask to prevent accidental breakage of the seal.**
	Radiological	Three containers (about 3-gallon capacity) Two sponges Soapy water Rinse water Paper towels or similar drying material	The soldier wipes his mask and hood with a sponge dipped in hot, soapy water and rinses them with a sponge dipped in clean water. He dries the mask and hood with paper towels or rags. **NOTE: Cool, soapy water is not as effective for removing contamination, but it can be used if you scrub longer.** If water is not available, the soldier should brush and dust off the radioactive dust particles.
4. Remove CVCUS/ overboots.	All contamination	Two discard containers (plastic bags)	The soldier opens all the zippers on his CVCUS in the following order: wrists, ankles, and front. While standing adjacent to the clean area, he grasps the front of his CVCUS and pulls it off of his shoulders. He makes a fist to ensure that his gloves do not come off when removing his CVCUS. He puts his arms behind his back and works his arms out of the sleeves. He loosens his overboots by alternately stepping on each heel and pulling up on his foot. He pulls his CVCUS down below his knees, turning it inside out. He should not let the outside of his CVCUS touch his body. He walks out of his CVCUS/overboots simultaneously and steps onto the clean area.

Table 3-9. Steps for the CVCUS/CPU, Individual (Emergency Method) MOPP-Gear Exchange (Continued)

Steps	Contamination Type	Required Equipment	Procedures
5. Remove jacket/ gloves.	All contamination	Two discard containers	The soldier wipes his gloves and around the edges of the packages containing the new items (gloves, undergarments, and overboots) with an IEDK. He opens the new packages. He holds the fingertips of the gloves and partially slides his hands out. When the fingers of both hands are free, he holds his arms away from his body and lets the gloves drop. The soldier opens all Velcro wrist fasteners and unzips his CPU jacket. He removes the CPU jacket by grasping it at the shoulders and pulling down, turning it inside out. He places the jacket on a clean area with the clean side up. He removes the protective glove inserts.
6. Remove combat boot/CPU pants.	All contamination	Two discard containers	The soldier unlaces his combat boots, removes them, and steps directly onto his CPU jacket. He takes off his CPU drawers by placing his hands beneath the waistband and removing them, turning them inside out. The soldier is now stripped to his underwear and mask and is ready to put on his new clothing. The soldier follows these procedures if wearing the BDU. He stands on the clean area (but not on the CPU jacket) and unlaces his combat boots. He partially removes his boots by alternately stepping on each heel and pulling up on his foot. He removes each trouser leg and combat boot simultaneously and steps directly onto the clean side of the CPU jacket.
7. Put on CPU.	All contamination	One set of chemical protective undergarments per soldier	The soldier removes the undergarment from the package without touching the outside. He puts on the undergarment in the following order: CPU drawers, CPU jacket, combat boots, and CVCUS (over the CPU drawers).
8. Put on overboots and gloves.	Chemical/ biological	One set of chemical protective overboots per soldier One set of chemical protective gloves per soldier M9 detector paper	The soldier removes the overboots from their package without touching the outside and puts them on. He removes the gloves from their package without touching the outside and puts them on. He puts on the M9 detector paper.

Table 3-9. Steps for the CVCUS/CPU, Individual (Emergency Method) MOPP-Gear Exchange (Continued)

Steps	Contamination Type	Required Equipment	Procedures
9. Secure hood.	Chemical/ biological	One IEDK	The soldier wipes his rubber gloves with an IEDK. He unrolls his hood, attaches the straps, and tightens the neck cord.
10. Secure gear.	All contamination	One chemical protective helmet cover per soldier	The soldier secures his individual gear and puts it back on. He puts on a new chemical protective helmet cover and moves to the AA.

Chapter 4

Thorough Decon

Thorough-decon operations reduce and sometimes eliminate contamination. They restore combat power by removing nearly all contamination from unit and individual equipment so that troops can operate it safely for extended periods at reduced MOPP levels. Since a small risk of contamination remains, periodic checks with the CAM, the M8/M9 detector paper, or the M256A1 detector kit must be performed after each operation. A thorough decon can be done as part of reconstitution, or it can be done to support operations throughout the battlefield. After a thorough decon, the unit moves out of the decon site into a tactical AA. The unit, while in this tactical assembly, may undergo reconstitution or prepare for future operations. Combat-service-support (CSS) elements replenish combat stocks, refit equipment, and replace personnel and equipment, as required. The contaminated unit, with some assistance from a chemical unit, does the DTD. A supporting chemical unit does the DED or DAD. The planning considerations that are required to conduct thorough-decon operations and the methods that various chemical units (decon and smoke/decon) use to conduct DTD/DED are discussed in this chapter. The exact layout of a thorough-decon site is determined by METT-TC.

PLANNING

4-1. A thorough decon is the most effective type of decon but is the most resource intensive. A combination of weather conditions, agent used, mission requirements, time, troops, and supplies available may make weathering the decon option of choice. Commanders decide which type of decon is required based on the recommendation of the chemical personnel.

4-2. Generally, thorough-decon operations are conducted beyond the range of enemy direct-fire systems. If a contaminated unit requires a thorough decon as part of reconstitution operations, the decon site is established near the reconstitution area or the brigade's rear area. Usually, company-sized units are reconstituted in their brigade's rear area while battalion-level units are reconstituted in their division's rear area. Organizations larger than a battalion will be reconstituted in the corps's rear area. Further information about reconstitution can be found in FMs 100-5 and 100-9.

4-3. All echelons prepare for thorough-decon operations as part of the overall planning process. From the commander's general guidance, the chemical staff can begin to develop the decon plan. Coordination with higher HQ is required to determine the availability of engineer support for site preparation and closure. Engineers can provide support for sump preparation, drainage ditches, and sumps. Coordinate with the civil-affairs office for host-nation support (personnel, equipment, and supplies) and for environmental requirements and restrictions.

4-4. The chemical staff selects possible decon sites throughout the unit's area of operation (AO) based on such factors as decon type, terrain, mission, threat, road network, and availability of water. After the decon sites are selected, linkup points are chosen to support each site. A site may have more than one linkup point.

4-5. Since decon assets are limited, the commander must establish a priority of decon support. The priority of support lists the units in the order they will receive decon support. This can change from phase to phase during an operation. The chemical staff develops the priority of support based on an understanding of the commander's intent.

4-6. To give priority of support to the lead task force during the assault phase may not be the best choice since any contaminated elements will most likely not stop for decon until after the assault is completed. The commander should establish a priority of work that specifies the order in which equipment will be decontaminated. For example, a priority of work may be in this order: engineer equipment, artillery pieces, main battle tank, and long-haul vehicles. A limiting factor is the availability of water. A typical vehicle requires 500 gallons of water during the DED. The actual amount of water required varies by the vehicle and its contamination level. The supported unit's chemical staff must develop a water resupply plan for thorough-decon operations.

4-7. A water resupply plan can be as simple as selecting a series of linkup points along a route where the chemical unit can link up with a bulk water truck. More complex water resupply plans include caching water throughout the AO, coordinating for the movement of water bladders by aircraft, and identifying water sources in the unit's AO. The use of nonpotable, salt, and brackish waters should be considered.

PREPARATION

4-8. Subordinate units review their higher HQ decon plan. If additional decon sites or assets are required, request them from the higher HQ. The supporting decon-platoon leader conducts a leader's recon of the designated decon sites. Since the decon sites are initially selected by map recon, the actual site may not be suitable. If the selected site cannot support decon operations, the chemical unit leader attempts to find another site close to the original site and notifies the supported unit and his higher HQ of the change.

4-9. The decon-platoon unit leader determines the support requirements that are needed to operate the decon site. Engineers are required to dig sumps, improve access to the site, and dig ditches for runoff. Air defense (AD) and military police (MP) are required to provide security. Medical support is also required. Depending on the operational situation and the threat, some DED sites may be prepared in advance. Routes must be established and published in the OPORD, leading the contaminated unit to the linkup point.

NOTE: Units must be informed not to proceed directly to the decon site.

EXECUTION

4-10. The actual decon operation begins once a unit requests decon support. The unit leader must assess his situation and accurately report it to his HQ. This situation report should include the type and extent of contamination, the current location, his assessment of the unit's ability to perform the current mission while contaminated, and the time he expects the unit to become combat ineffective. The contaminated unit should perform immediate-decon techniques to increase its survivability and to limit the spread of contamination. An operational decon should also be considered and performed, as appropriate. The request for decon support must contain several essential elements of information to assist the chemical staff and the decon-platoon leader in coordinating the decon operation. These essential elements of information are the—

- Designation of the contaminated unit.
- Location of the contaminated unit.
- Time the unit became contaminated.
- Earliest possible time the unit can move/begin decon.
- Type of contamination.
- Number and type of vehicles contaminated.
- Special requirements (patient-decon station, recovery assets, unit decon team, and so forth).
- Supported unit's frequency and call sign to facilitate initial linkup.

4-11. The supporting chemical unit is given a warning order (WO) to conduct a decon operation. Subsequent WOs provide more detailed information. After requesting decon support, the supported unit issues an OPORD/fragmentary order (FRAGO) to the chemical unit.

4-12. The supported unit must coordinate the movement of the contaminated unit to the linkup point. If the contaminated unit is out of the supported unit's AO, the move must be coordinated with the higher HQ.

4-13. Once the chemical unit is set up, the decon-platoon leader moves to the linkup point. Other elements may provide assets to support the decon operations, such as medical, engineer, AD, and MP. These elements linkup with the chemical unit before the arrival of the contaminated unit (see Table 4-1, page 4-4).

4-14. The supported unit is responsible for site security and overall control. The decon-platoon leader is responsible for operating the predecon area and processing vehicles.

4-15. After linkup is achieved and all support assets are in position, the actual decon operation begins. The decon-platoon leader, with help from the commander of the contaminated unit or the supported unit's command and control cell, supervises the decon operation. All equipment not necessary for decon operations should be located in a clear area to limit the spread of contamination.

Table 4-1. Thorough-Decon Support Matrix

Tasks	Division CMO	Division	Brigade	DISCOM	FSB	Chemical Unit	Contaminated Unit	Battalion
Preparation-Phase Tasks								
Request	S						P	S
Coordination	S		S					P
Site selection		S OR S				P		
Advance-party linkup						S	P	S
Site setup				S	S	P		
Execution-Phase Tasks								
Site control/security						S	P	
Predecon actions						P	S	
Processing						P	S	
Site-Clearance-Phase Tasks								
Cleanup						P	P	
Marking and reporting						P		

S - supporting responsibility
P - primary responsibility

4-16. A thorough-decon site consists of the following areas: predecon staging area, DED and DTD areas, and postdecon AA (see Figure 4-1). The chemical unit leader selects these areas based on operational guidance, road network, available cover and concealment, and water supply. The contaminated unit uses the predecon staging area to ready itself. The postdecon AA is the location where the vehicles and soldiers exiting the DED and DTD areas are linked up before moving from the decon site.

PREDECON STAGING AREA

4-17. The contaminated unit moves to the predecon staging area about 274 to 547 yards downwind of the thorough-decon site. The contaminated unit performs predecon actions to include segregating vehicles by checking for contamination.

4-18. For chemical contamination, use the CAM and the M8/M9 detector paper. The contamination level can be estimated by comparing the M8/M9 detector paper to Figure 4-2. When monitoring vehicles for contamination, there should be about 33 to 49 feet between each vehicle to prevent false positive readings with the CAM. For radiological contamination, use the AN/PDR-77 or AN/VDR-2 radiac detector. If the vehicle only has isolated areas of contamination, use the M11 or M13 to decon those areas.

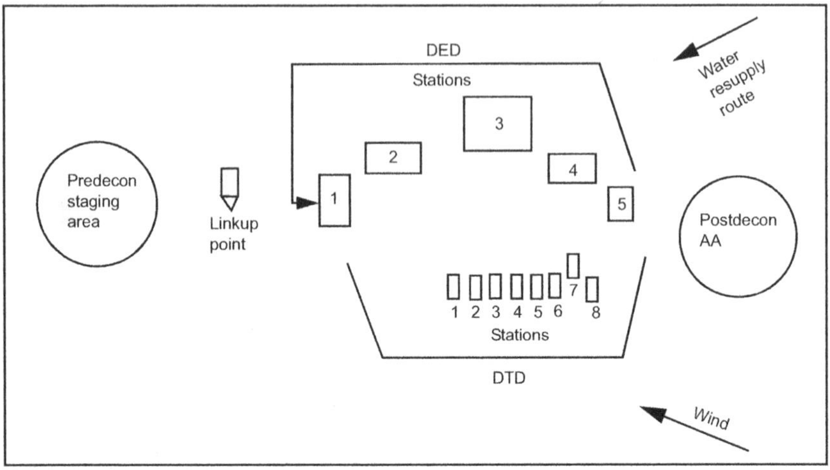

Figure 4-1. Thorough-Decon Site

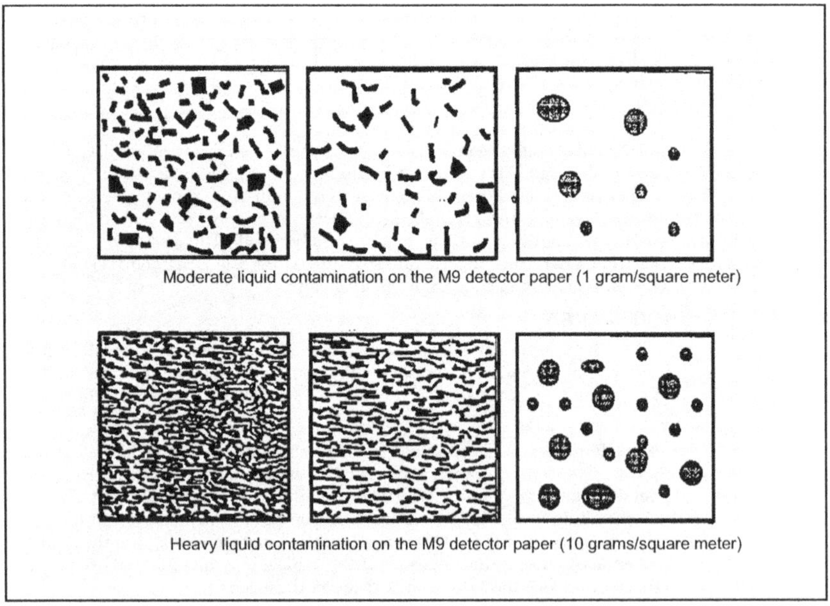

Figure 4-2. Contamination Levels

Vehicle Crews

4-19. The vehicle crews, except for the drivers, dismount. As the crews dismount, they remove all contaminated equipment, to include electro-optical systems from the top and sides of the vehicles. Once the crews have exited the vehicle, they will not reenter. This prevents contamination from being spread into the vehicle interior.

Vehicle Preparation

4-20. Using the pioneer tools from the contaminated vehicles, the crews remove all heavy mud and debris. They concentrate on the vehicle's undercarriage, which is the most likely place for contamination to collect and the hardest place to decon. Once the crews are finished with the pioneer tools, they are placed back on the vehicle. Initial removal of the mud and debris makes it more likely that the decon wash will remove any remaining contamination.

4-21. Seat covers (when applicable), canvas items, camouflage netting, and any other materials that can absorb liquid contamination are removed. These items create a potential transfer hazard and are not easily decontaminated. Left untreated, absorbed chemical agents will desorb after being decontaminated and will create a vapor hazard. Remove items that cannot be decontaminated by the standard methods used in the DTD, and place them at the collection point. Chemical-unit personnel provide advice concerning the decon or disposal of these items.

4-22. Design vehicle-loading plans to minimize the amount of equipment carried on the outside of the vehicle that cannot be readily decontaminated. Whenever possible, NBC covers should be used when a chemical attack is expected (see FM 3-4). All NBC covers are removed and disposed of as contaminated waste during the predecon actions.

DED AND DTD AREAS

4-23. In coordination with the chemical unit leader operating the decon site, the contaminated unit begins sending contaminated vehicles by priority for processing. Communication is maintained between the predecon staging area and the chemical unit leaders. All assistant vehicle drivers are the first individuals sent through the DTD to ensure that there is a driver exchange at station 3 (see paragraph 4-94).

POSTDECON AA

4-24. The chemical unit leader selects the general location for the postdecon AA. It must be big enough to hold the entire unit undergoing the thorough decon with proper cover and concealment. The postdecon AA is located about 0.62-mile upwind from the DED and DTD areas. The unit assembles in the postdecon AA after completing DTD and DED operations. The unit occupies the postdecon AA before moving to a reconstitution location or a tactical AA to prepare for future operations.

DETAILED TROOP DECON

4-25. The contaminated unit or its higher HQ is responsible for setting up, operating, manning, and closing the DTD area at the thorough-decon site. The chemical unit leader determines the general location of the DTD within the decon site and provides technical advice on setting up, operating, and closing the DTD area. The supervisor of the DTD must establish a work/rest cycle.

4-26. There are eight stations for a DTD. Spacing between the stations is 10 to 16 feet (see Figure 4-3, page 4-8). Recapitulation of personnel and equipment for a DTD is provided at Table 4-2, page 4-9.

STATION 1 - INDIVIDUAL-GEAR DECON

4-27. At this station, contamination is removed from individual gear (LCE, mask carrier, helmet, and weapon) to a negligible risk level (see Chapter 1 of this manual).

Preparation

4-28. At this station, the following equipment and supplies are needed:

- Four 30-gallon containers for each two decon lanes used.
- 2 long-handled brushes for each two decon lanes used.
- Sufficient STB slurry mix for chemical contamination.
- Hot, soapy water and a sponge for radiological contamination.
- Rinse water.
- 2 ponchos or plastic sheets.
- Engineer tape.

4-29. Dig a sump that is 6 feet long, 6 feet wide, and 4 feet deep (minimum). Place three 30-gallon containers near the sump for ease of changing. Fill two containers with an STB slurry mix. Fill the other two containers with clean water for rinsing and place it about 3 feet forward of the STB cans. Place two long-handled scrub brushes at each can of STB slurry.

4-30. To prepare the slurry, mix 100 pounds of STB with 20 gallons of hot water. The chemical unit provides the hot water. Change the mixture after 20 soldiers have decontaminated their gear. The rinse water should be changed after every 10 soldiers or when it appears dirty (place the waste mixture in the sump).

4-31. Place a poncho or a plastic sheet on the ground at the checkpoint. Divide the poncho or the sheet in half using engineer tape. This is the contamination control line. The checkpoint will be a minimum of 10 feet from all other stations in order to get a true reading on the detection equipment. Place the following chemical detection/identification equipment at the checkpoint for an average company-sized unit:

- 1 CAM.
- 8 books of M8 detector paper.
- 4 M256A1 detector kits.
- 100 plastic trash bags.
- 1 AN/VDR-2 for radiological contamination.

4-32. Three soldiers are required to operate this station. One soldier supervises the decon of the individual gear and takes the decontaminated equipment to the checkpoint. He also prepares a new slurry mixture as necessary. One soldier remains at the checkpoint and checks all the gear using the detection equipment to ensure that it is decontaminated. One soldier transports the decontaminated gear to the reissue point.

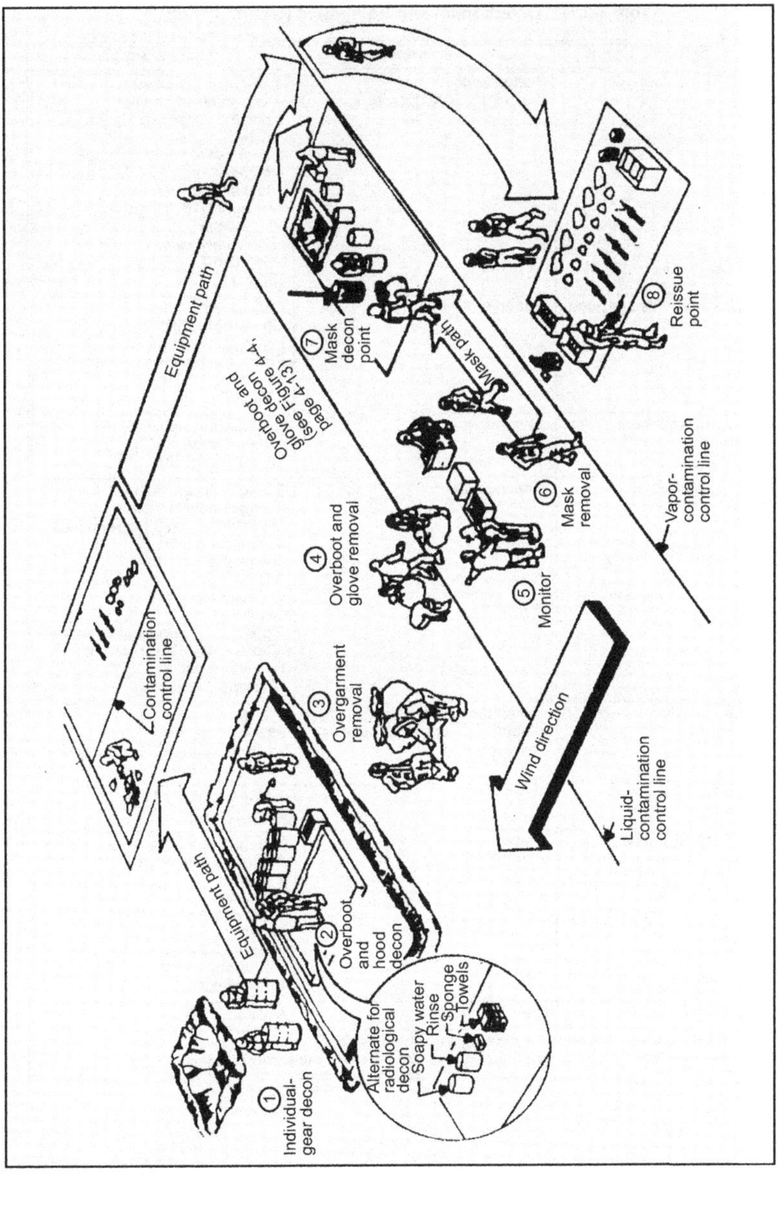

Figure 4-3. DTD Layout

Table 4-2. DTD Personnel and Equipment Recapitulation

Stations	Personnel	Equipment/Supplies
Station 1 - individual-gear decon	2 attendants 1 monitor (CAM operator)	4 30-gallon containers 2 long-handled brushes 2 ponchos or plastic sheets 1 CAM 8 books of M8 detector paper 4 M256A1 detector kits 100 plastic trash bags Sufficient STB slurry mix Rinse water *Hot, soapy water *AN/VDR-2 *Sponge
Station 2 - overboots and hood decon	1 attendant	Cutting tool One SKD or IEDK per person Large plastic sheet Plastic trash bags (as required) 10 drums of STB 1 shovel
Station 3 - overgarment removal	1 attendant	2 30-gallon containers 100 plastic trash bags 10 boxes of SDKs
Station 4 - overboot and glove removal	3 attendants	8 30-gallon containers 2 long-handled brushes M8 detector paper (as required) 1 immersion heater 100 plastic trash bags 1 CAM Engineer tape 1 cutting tool 2 ponchos or large plastic sheets 10 percent STB/HTH solution Hot, soapy water Cold rinse water
Station 5 - monitor	1 attendant (CAM operator) 1 aidman (or combat lifesaver)	1 CAM First aid supplies 5 books of M8 detector papers 1 case of SDKs
Station 6 - mask removal	2 attendants	1 M8A1 or M22 ACAA Engineer tape
Station 7 - mask decon point	2 attendants 1 monitor	1 30-gallon container 4 3-gallon containers 1 CAM 2 sponges One case of paper towels 1 immersion heater Mask sanitizing solution Hot, soapy water Rinse water
Station 8 - reissue point	Unit NBC NCO/supply sergeant	Mask PLL
* Denotes the equipment needed for radiological contamination. **NOTE: Assume that an operational decon was done before the thorough decon.**		

Execution

4-33. The soldier decontaminates his gear by washing and scrubbing it for 6 minutes in a decontaminant container with hot, soapy water or an STB slurry mix. If he is wearing the M42 mask, he should use hot, soapy water and a sponge or an STB slurry mix to decon the hose and canister.

4-34. The soldier dips his gear into the clean water and rinses it for 4 minutes then hands it to the attendant and proceeds to the next station. The attendant takes the gear to the equipment checkpoint and places the decontaminated gear on the "dirty" side of the contamination control line and returns to the containers to pick up more gear. The monitor at the checkpoint checks the gear using the appropriate detection device and the procedures associated with that device. If the residual contamination exceeds negligible risks, recycle the gear and decon it again (see Chapter 1 of this manual). If the gear passes the check, place it on the clean side of the contamination control line. The attendant will carry the equipment to the reissue point.

4-35. Depending on the time available, more extensive washing and checking procedures may be used. The longer the gear is washed or left out in the air after washing, the lower the contamination level. The gear may be put in closed areas or plastic bags and checked for hazardous vapors with the M256A1 detector kit or the CAM. The CAM only detects the G-series nerve-agent vapors and the H-series blister-agent vapors.

Risk

4-36. If this step is not done properly, contamination may remain on the equipment. The resulting vapor hazard could cause casualties to unmasked personnel, particularly in closed areas (vehicle interiors) or heavily wooded areas where air circulation is poor.

STATION 2 - OVERBOOT AND HOOD DECON

4-37. At this station, gross contamination on overboots, trouser legs, mask, and hood is neutralized. If ample hoods are available at the reissue point, the hood should be cut away.

Preparation

4-38. Prepare a shuffle pit by digging a shallow pit about 3 feet long, 3 feet wide, and 6 inches deep. Fill the shuffle pit with an STB dry mix or an STB slurry, depending on water availability. Prepare the STB dry mix by mixing three parts of earth to two parts of STB. Prepare the STB slurry in the same manner as at station 1 (see paragraph 4-30). Add more STB to the mix after 10 soldiers have processed through the shuffle pit. The chemical unit will provide 10 drums of STB for every company-sized unit that goes through the station.

4-39. At this station, the following equipment and supplies are needed for the mask and the shuffle pit: 1 cutting tool for each decon lane used, 1 SDK or IEDK per person, plastic trash bags (as required), 10 drums of STB, and 1 shovel (to mix).

4-40. One soldier is required to operate this station. The attendant directs and "observes" the soldiers as they decon their overboots and hoods.

Execution

4-41. The soldier walks into the shuffle pit and spreads his legs apart (double-shoulder width), bends at the waist, and uses his hands to thoroughly rub the STB dry mix or the STB slurry on his overboots and lower trousers legs. He should take special care to rub the rear of his overboots. He should also remove any excess decontaminant from his gloves.

4-42. If a replacement hood is available, remove the hood as follows. Buddy #1 cuts the shoulder straps and drawcord on Buddy #2's hood. Buddy #1 pulls Buddy #2's hood inside out over the front of the mask, being careful not to touch the exposed neck or head. Buddy #1 gathers Buddy #2's hood in one hand and using a cutting tool, cuts away the hood as close as possible to the eye-lens outsert, voicemitter, and inlet-valve covers. Make sure nothing is left dangling below the bottom of the mask.

4-43. If a replacement hood is not available, Buddy #1 decontaminates and rolls Buddy #2's hood in the same manner as for a MOPP-gear exchange. When the task is completed, Buddy #1 and Buddy #2 reverse roles.

NOTE: Soldiers should check their overboots, rubber gloves, and overgarment for damage. Any rips, tears, or punctures in these items should be reported to the monitor at station 5. This allows the monitor at station 5 to check the soldiers for chemical-agent symptoms and their clothing for possible contamination.

Risk

4-44. If this step is not done properly, contamination can be transferred to the combat boots and the head and neck area from the hood.

STATION 3 - OVERGARMENT REMOVAL

4-45. At this station, contaminated overgarments are removed before the agent penetrates the overgarment material and touches the undergarments or the skin.

Preparation

4-46. At this station, the following equipment and supplies are needed:

- Two 30-gallon containers for each two decon lanes used.
- 100 plastic trash bags (or about one per soldier).
- 10 boxes of SDKs.

4-47. One soldier is required to operate this station. He directs and "monitors" the soldiers as they remove their overgarments in the same manner as a MOPP-gear exchange.

Execution

4-48. The attendant assists the soldier in removing his overgarment. The attendant cuts and removes the M9 detector paper from around the soldier's wrist. He unfastens the Velcro closure over the jacket zipper, waist cord, and wrist Velcro straps on the soldier's jacket. He unfastens the back snaps and instructs the soldier to make a fist. He then pulls the soldier's jacket down and away from him.

4-49. The attendant cuts and removes the M9 detector paper from the soldier's trousers. He unfastens the Velcro straps and zippers on the cuffs of the soldier's trousers. He also unfastens the front waist snaps and unzips the front zipper. He has the soldier lift one leg and point that foot down and bend slightly at the knees for stability. The attendant grasps the cuff of the elevated foot with a hand on each side and pulls the cuff in an alternating, jerking motion until the soldier can step out of the trouser leg. The process is repeated on the other leg.

NOTE: The attendant ensures that the soldier steps wide enough so as not to rub his clean leg against the contaminated boot and overgarment.

Risk

4-50. If this step is not done properly, the agent may be transferred to the undergarment or the skin.

STATION 4 - OVERBOOT AND GLOVE REMOVAL

4-51. At this station, contaminated overboots and gloves are removed to limit the spread of contamination. The overboots and gloves may also be decontaminated for reissue (if serviceable).

Preparation

4-52. At this station, the following equipment and supplies are needed for every two lanes used:

- Engineer tape.
- 1 cutting tool for each lane used.
- 100 plastic trash bags (or about one per soldier).

4-53. If the overboots and gloves are not being decontaminated, two 30-gallon containers are needed.

4-54. If the overboots and gloves are being decontaminated, the following are needed:

- Eight 30-gallon containers.
- 2 long-handled brushes.
- 2 ponchos or large plastic sheets.
- 1 CAM.
- 1 AN/VDR-2 for radiological contamination.
- 4 M256A1 detection kits.
- 1 immersion heater.
- 100 plastic trash bags.
- M8 detector paper, as required.
- 10 percent STB/HTH solution.
- Hot, soapy water.
- Cold rinse water.

4-55. If replacement overboots and gloves are available, set up the station as follows (see Figure 4-4). Establish a liquid-contamination control line and set

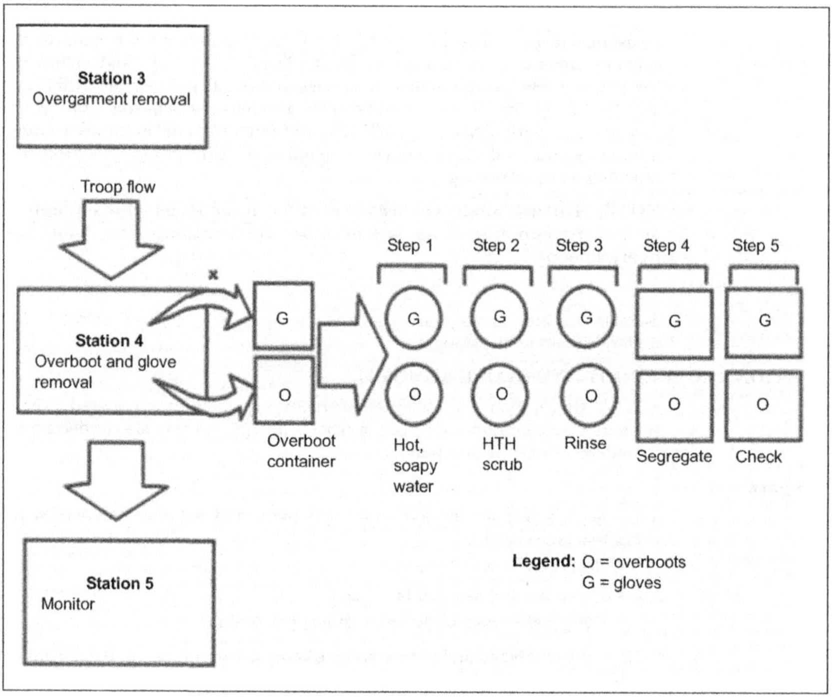

Figure 4-4. Station 4 Layout

two 30-gallon containers 1 foot back from the line. Soldiers support themselves using the containers and discard their overgarments into the containers. An attendant directs and monitors the soldiers as they remove their overboots and gloves in the same manner as a MOPP-gear exchange; however, the soldiers step over the control line instead of onto a jacket.

4-56. If replacement overboots and gloves are not available, set up the station as follows: Fill two 30-gallon containers with hot, soapy water and two 30-gallon containers with a 10 percent STB/HTH solution, placing two scrub brushes near the containers. Then fill two 30-gallon containers with cold rinse water. One attendant supervises and assists the soldiers wearing the M42 mask. The other two attendants decon the overboots and the gloves; one processes the overboots, while the other processes the gloves.

NOTE: Replace the water in the container once 20 items have been processed. When available, the decon platoon will assist with the water requirements.

4-57. For both situations, use engineer tape to mark the liquid-contamination control line on the ground. Place the cutting tools, two containers, and plastic bags on the "dirty" side of the liquid-contamination control line. The liquid-contamination control line separates the "dirty" and "clean" areas. No liquid agent should be tracked on the ground beyond the liquid-contamination control line.

Execution

4-58. The liquid-contamination control line separates the "dirty" and "clean" areas. The attendant unfastens or cuts the elastic closures on the soldier's overboots. The soldier faces the liquid-contamination control line and steps back from it about 12 inches. The attendant steps on the back of the soldier's overboot and instructs him to lift his heel and work his foot out of the overboot and step across the liquid-contamination control line. Repeat the process on the other foot. If the overboot cannot be removed by this process, the attendant cuts it off and discards it in the designated container. The soldier holds the fingertips of his gloves and partially slides his hands out. The attendant remove the soldier's gloves.

4-59. If the soldier is wearing the M42 protective mask, the attendant from station 6 carries the soldier's filter canister until it is removed.

NOTE: The attendant at station 4 performs his duty from the "dirty" side of the liquid-contamination control line.

4-60. The overboots and gloves are decontaminated using the following steps:

NOTE: Check all items for holes, tears, and punctures and discard any item with this defect. Do not decon any item that is unserviceable.

- Step 1. Submerge the gloves and overboots in their respective container of hot, soapy water. Some of the contamination is removed during this step. When the overboots and gloves are removed from the container, ensure that no water remains inside them.
- Step 2. Submerge the gloves and overboots in their respective containers of STB/HTH solution. Thoroughly scrub the items until no visible contamination remains. After scrubbing, submerge each item once more before moving to the rinse container.
- Step 3. Thoroughly rinse the scrubbed items, making sure that they are rinsed inside as well as out.
- Step 4. Place usable items on a poncho or a plastic sheet to air dry and weather.
- Step 5. Place usable items in plastic trash bags along with an M256A1 detector kit. If the detector kit shows contamination remaining, the attendants can recycle the items or discard them. However, if the kit show no contamination, the items can be reused.

Risk

4-61. If the overboots are not properly removed, the combat boots may become contaminated and contamination may be spread to clean areas. If the gloves are not properly removed, the undergarments and skin may become contaminated.

STATION 5 - MONITOR

4-62. At this station, contamination on personnel is identified, spot-decon capabilities are provided, and medical aid is provided, as required.

Preparation

4-63. At this station, the following equipment and supplies are needed:

- First aid supplies.
- 1 CAM.
- 1 AN/VDR-2 for radiological contamination.
- 5 books of M8 detector paper per 100 soldiers.
- 1 case of SDKs.

4-64. An aidman should be present to treat any soldiers suffering from chemical-agent symptoms. If an aidman is unavailable, a combat lifesaver should be present.

Execution

4-65. The attendant checks the soldiers for contamination using the CAM. Liquid agents can be detected with M8 detector paper. Small quantities of agent vapor can be detected with the CAM. Symptoms of agent poisoning are the most obvious indication of skin contamination. At this station, the medic checks each soldier for symptoms of agent poisoning and treats, as required. Soldiers should report any damage to their MOPP gear that was identified at stations 2, 3, and 4. The attendant can decon any areas identified as contaminated with an SDK. Soldiers are remonitored after decon. It is possible that all liquid-chemical contamination is absorbed into the clothing. If so, M8/M9 detector paper will indicate negative, even though there is a hazard.

Risk

4-66. If this station is omitted, the soldier could become a casualty. After this station, the soldier is not checked for contamination or decontamined again. Commanders can choose to conduct more extensive contamination checks here if medical assistance and time are available. This decreases the risk of casualties.

STATION 6 - MASK REMOVAL

4-67. At this station, the mask is removed without contaminating the soldier. The mask is taken to a mask decon point, limiting agent transfer at the station.

Preparation

4-68. At this station, the M8A1 or M22 automatic chemical-agent alarm (ACAA) and engineer tape are needed. Two soldiers are needed to operate this station. They remove and carry the masks to station 7.

Execution

4-69. If the hood is still attached to the mask, the attendant pulls the hood over the front of the mask, grabs the mask by the voicemitter cover, and pulls the mask off the soldier. The soldier holds his breath as the mask is removed. If the mask has optical inserts, the attendant holds the mask open so that the soldier can remove the inserts without touching the outside of the mask. The soldier walks upwind 16 feet, crosses the vapor-contamination control line, and then resumes breathing. The attendant brings the mask to station 7.

4-70. If the wind direction remains constant, no chemical vapor hazard is expected beyond the vapor-contamination control line. Position the M8A1 or the M22 ACAA upwind of the station to warn of vapor hazards. The soldier getting decontaminated moves straight ahead while his mask, which may still give off vapors, is held on the vapor dirty side of the line and taken to station 7 where it is decontaminated.

Risk

4-71. If the step is not done properly, the soldier could breathe toxic vapors. There is a high probability that the vapor hazard is still present on the mask and the hood. The soldier must not touch the outside of the mask because it could contaminate his bare hands. The soldier proceeds to station 8.

STATION 7 - MASK DECON POINT

4-72. At this station, all the contamination is removed from the mask.

NOTE: Once 20 items have been processed in the wash containers, replace the water. Once 10 item have been processed in the rinse water, replace it. Place the contaminated waste into the sump.

Preparation

4-73. At this station, the following materials and equipment are needed:

- 4 containers (about a 3-gallon capacity).
- 1 CAM (for chemical only).
- Hot, soapy water.
- Rinse water.
- Mask sanitizing solution.
- Immersion heater.
- One 30-gallon container.
- 2 sponges.
- 1 case of paper towels per company.

4-74. Dig a sump that is 4 feet long, 4 feet wide, and 4 feet deep in which to discard used filters and canisters. Place the equipment and materials as shown in Figure 4-3, page 4-8. Three soldiers are needed to operate this station. Two soldiers strip, wash, rinse, sanitize, and dry masks. The other soldier checks the masks and carriers them to the reissue point.

Execution

4-75. Remove the eye-lens outserts and the hood if the hood was not cut off at station 2. Remove and discard the filters or canisters. Put the items into the properly marked containers. Wash the mask, hood, eye-lens outserts, and the hoses on the M42 and M43 masks in hot, soapy water. Rinse these items in clean water, dip them into the sanitizing solution, agitate them for 5 minutes, and then rinse them again in clean water. Add one tube of mask sanitizing solution (calcium hypochlorite) to each quart of water. Wipe the masks with rags until they are almost dry. Discard each container of mask sanitizing solution into a sump after every 10 masks. The attendant checks the masks for contamination with a CAM. If the masks are still contaminated, the attendant recycles them for more decon and then decons his rubber gloves. If the masks are not contaminated, the attendant takes the masks to the reissue point. The attendant must take care not to contaminate the reissue point or himself.

Risk

4-76. If this step is not done properly, the soldiers may become contaminated when the masks are reissued to them at the reissue point. Even though the step is done correctly, there is still a possible danger when many masks are stacked together. Small amounts of residual vapor from each mask can become potentially dangerous.

NOTE: Runners between stations 7 and 8 are in MOPP2 and are prepared to go into MOPP4.

STATION 8 - REISSUE POINT

4-77. At this station, the mask with its components is provided to the soldier for reassembly.

Preparation

4-78. At this station, the protective-mask prescribed load-list (PLL) parts are needed. The unit NBC NCO/supply sergeant sets up the reissue point to provide the soldiers with replacement parts for all types of protective masks and assist in mask maintenance.

Execution

4-79. The mask with its components is reissued to the soldier who assembles it in the AA. The unit chemical NCO affixes canisters to the cleaned M42 and M43 hoses. The soldier picks up individual gear and moves to the postdecon AA.

Risk

4-80. If this step is not done properly, the soldier may be inadequately equipped for future operations.

NOTE: If time is not available, the NBC NCO will have replacement chemical suits, overboots, and gloves at this station for reissue. If time is available, personnel will receive this equipment at the postdecon AA.

RESUPPLY SUPPORT RESPONSIBILITIES

4-81. The supporting chemical unit (decon or smoke/decon) must supply the majority of the equipment and supplies that are required to operate the DTD. This does not relieve the supported unit of the requirement to maintain adequate supplies to conduct DTD procedures.

4-82. A chemical unit's operational control continues to receive support from its parent organization. The supported unit provides the chemical unit with replacements, supplies, and material at the end of the DED. This shortens the time the chemical unit is nonmission capable following the DED. The supported unit must coordinate for the supplies and equipment necessary to operate the DTD.

4-83. Typical nonchemical units do not have more than two CAMs per company-sized unit. However, the DTD requires a minimum of three CAMs. The supported HQ may need to provide additional CAMs. The division support command (DISCOM) sustains the division's chemical units. A chemical unit attached to a maneuver force identifies the resources that are needed to accomplish the mission and passes these requirements to the maneuver staff's Supply Officer (S4)/Logistics Staff at Corps and Division (G4). The S4/G4 then coordinates with the supporting CSS element to fill the requirement. The supply units that operate the Class II points issue the protective masks and overgarments. They normally package protective clothing as complete sets of MOPP equipment to support a predetermined number of soldiers. Decon supplies may be issued in preconfigured "push" packages.

DETAILED EQUIPMENT DECON

4-84. Chemical units (decon and smoke/decon platoons) are responsible for setting up, operating, and closing the DED portion of the thorough-decon operation. The DED for chemical and biological contamination is comprised of five stations. For radiological contamination, the DED uses all but station 2. Stations are normally 98 to 164 feet apart; however, spacing is METT-TC dependent.

DECON STATIONS

4-85. The actions that are taken at each of the stations are as follows:

Station 1 - Primary Wash

4-86. At this station, the gross contamination and dirt are removed from the vehicle. The vehicle is sprayed for 2 to 3 minutes with hot, soapy water. The vehicle is then scrubbed to help remove caked-on dirt. The mechanical action of scrubbing also helps remove thickened chemical agents. Although the undersurfaces are difficult to reach, try to remove as much dirt as possible. After scrubbing the vehicle, spray it again for 2 to 3 minutes to remove loosened dirt and contamination. This station uses about 250 gallons of water per vehicle. Larger vehicles with large quantities of dirt use more water. The runoff from this station is contaminated and must be treated as hazardous. The contaminated runoff must be controlled; if available, engineer support may be used. This station requires high water-pressure systems (M12A1 power-driven decon apparatus [PDDA] and M17 Lightweight Decontaminating System [LDS]) rather than low water-volume systems (65-GPM pumps).

NOTE: Thirty-five cubic feet of space per 250 gallons of liquid runoff should be used when calculating the size for the drainage pump.

4-87. The effectiveness of the wash depends on the type of wash (hot, soapy water; hot water; cold water; or steam). The relative effectiveness rankings for selected surfaces are listed in Table 4-3, page 4-20. Hot, soapy water is water heated to about 120° to 140° F to which a detergent has been added to reduce its surface tension. The detergent removes the agent by emulsification, which is followed by the mechanical displacement of the suspension. Hot water alone is less effective than hot, soapy water. Because of the high temperature, some agents are best removed by steam through vaporization. Finally, for some chemical agents, cold water exhibits better solvent characteristics.

Station 2 - DS2 Application

4-88. At this station, a decontaminant is applied to the entire vehicle. The vehicle is divided into four parts, and a member of the scrubbing team is assigned that part of the vehicle. This limits the workload of each member of the scrubbing team and avoids duplication of work. DS2 is applied starting at the top of the vehicle and working towards the undercarriage. Every effort is made to apply DS2 to the undercarriage, especially if the vehicle has crossed a contaminated area. The mop is the least tiring method of applying DS2. Using a mop to apply DS2 creates a large amount of spillage. However, continual use of the M13 decontaminating apparatus, portable (DAP) requires the scrubbing team to exert more energy than with using the mop. The M13 DAP can be used to apply DS2 to hard-to-reach areas. Planning factors for DS2 application are provided at Table 4-4, page 4-20.

4-89. Before starting the decon operation, the scrubbing team pours 5-gallon cans of DS2 into 30-gallon containers if mops are going to be used instead of M13 DAPs. Each member of the scrubbing team wears a TAP apron or wet-weather gear to protect himself from the DS2.

4-90. Water adversely affects the DS2's ability to react with chemical agents. If there is excess water remaining on the vehicle from station 1, the scrubbing team can—

- Wait for the majority of the water to evaporate.
- Remove the excess water.
- Increase the amount of DS2 applied.

4-91. There must be sufficient DS2 on the item being decontaminated for complete neutralization to occur. The DS2-to-agent ratio needs to be 55 to 1 for H agents and 25 to 1 for G agents. For a vehicle the size of an M1A1 tank, this corresponds to 15 and 7 gallons, respectively.

Station 3 - Contact Time/Interior Decon

4-92. At this station, the DS2 is allowed to completely neutralize the chemical agent and the interior of the vehicle is decontaminated. Vehicles are moved to a concealed position. Vehicles will remain in station 3 for no less than 30 minutes. DS2 reacts with most chemical agents within 5 minutes. However, by allowing the DS2 to remain on the contaminated surface for 30 minutes, the amount of agent that will later desorb (off gas) will be significantly reduced. When there is a 30-minute contact time, there will be no desorption after decon operations for most chemical agents. However, studies indicate that distilled-mustard (HD) vapors desorb after decon even if DS2 is allowed to remain for 30 minutes.

4-93. While the vehicle is held at this station for the DS2 to completely react, the driver inspects the interior of the vehicle for liquid contamination. He is given M8 detector paper to check for chemical contamination. If he identifies chemical contamination, he is given decon supplies to decon the interior of the vehicle. The best decon solution for use in the interior of vehicles is a 5 percent solution of HTH or STB. The driver wipes all reasonably accessible

Table 4-3. Effectiveness of Types of Wash

Agents/Surfaces	Relative Effectiveness[1]
TGD on alkyd-painted metal	Hot water and/or steam; hot, soapy water; cold, soapy water
TGD on CARC-painted metal	Hot, soapy water; hot water and/or steam; cold water
THD on alkyd-painted metal	Hot water and/or steam; hot, soapy water; cold water
THD on CARC-painted metal	Hot, soapy water; hot water and/or steam; cold water
HD on alkyd-painted metal	Hot, soapy water; hot water and or steam; cold water
HD on CARC-painted metal	Hot, soapy water and/or steam; cold water; hot water
VX on alkyd-painted metal	Steam; cold water; hot water and/or hot, soapy water
TGD on canvas/webbing	Steam; hot water and/or hot, soapy water; cold water
GD on canvas/webbing	Steam; hot, soapy water and/or hot water; cold water
THD on canvas/webbing	Steam; hot, soapy water; cold water; hot water
HD on canvas/webbing	Steam; hot water; hot, soapy water; cold water
VX on canvas/webbing	Steam; hot water and/or hot, soapy water; cold water
Overall	Steam; hot water and/or hot, soapy water; cold water

[1]The types of wash are listed in the order of effectiveness, starting with the most effective to the least effective.

Table 4-4. Planning Factors for DS2 Application

Planning Categories	Tank	Truck	APC
Gallons of DS2 applied	15	8	7
Minutes to apply solution	35	29	18

surfaces with a rag or sponge soaked in the HTH or STB solution. He should not attempt to decon areas where there is little likelihood of contamination (electrical assemblies, area beneath the turret floor, and so forth).

CAUTION

Do not mix HTH or STB with DS2. If mixed, a violent reaction will occur.

4-94. Once the interior decon is completed, the driver places a cover over the seat and floor of the vehicle. Then he dismounts from the vehicle and proceeds to the start of the DTD. The assistant driver, having completed the DTD, mounts the vehicle and moves it to the next station. Drivers must exercise caution when entering or exiting the vehicle. A DS2-coated surface is slippery, and the DS2 may react with chemical protective footwear. Personnel should avoid stepping in DS2 and tracking it into the vehicle.

4-95. For radiological contamination, use an AN/PDR-77 or AN/VDR-2 radiac detector to determine the extent and location of contamination inside the vehicle. If there is contamination, determine the intensity of the contamination inside the vehicle. If the contamination has an intensity greater than 0.33cGy, the interior of the vehicle must be decontaminated. Use a sponge to wipe the interior of the vehicle.

Station 4 - Rinse

4-96. At this station, the DS2 is removed from the vehicle. The vehicle is sprayed with water from top to bottom. Take care to rinse the undercarriage. This station uses about 200 gallons of water per vehicle. Failure to remove all DS2 from the vehicle may cause a false-positive reading at station 5. If high water-pressure systems (M12A1 PDDA and M17 LDS) are not available, large-volume water pumps (65-GPM pumps) should be used at this station. The driver removes plastic or other material (if present) covering the seats and floor and disposes of it as hazardous waste. See Table 4-5 for the planning factors for the rinse station.

Table 4-5. Planning Factors for the Rinse Station

Planning Categories	M1 Tank	M2 BFV	M113 APC	M109A PALADIN	HEMTT	5-Ton Truck	HMMWV
M12A1 PDDA rinse (gallons applied)	325	325	203	325	180	158	90
M12A1 PDDA rinse (minutes applied)	12	12	9	12	8	7	4
M17 LDS rinse[1] (gallons applied)	57	57	38	57	30	42	23
M17 LDS rinse[1] (minutes applied)	14	14	10	14	12	11	6
[1] Rinse is done with a spray wand.							

Station 5 - Check

4-97. At this station, the vehicle is checked to see if it has a negligible contamination level or if it still has significant contamination remaining. Detection procedures will vary depending on the type of contamination. If significant contamination is found on the vehicle, the vehicle will be recycled to station 2 for chemical contamination or station 1 for radiological contamination. An assistant driver takes the vehicle to the AA.

4-98. **Chemical.** The CAM is used to check for the presence of vapor from residual liquid contamination. A one bar or lower reading on the CAM indicates a negligible contamination level. Once the CAM indicates the presence of vapor contamination, M8 detector paper is used to verify the presence of liquid contamination. If it is suspected that both the CAM and M8 detector paper are producing a false positive, use an M256A1 detector kit to confirm or deny the presence of contamination. See Table 4-6, page 4-22, for a list of common interferences that can cause false-positive readings on the CAM. If the vehicle has significant contamination remaining, recycle it. The commander may modify the recycle criteria based on mission requirements.

4-99. There will be desorption of chemical agents from the surfaces after decon. The desorption of vapors on surfaces painted with the CARC will stop sooner then those surfaces painted with alkyd. Consider this when checking decontaminated items for overall decon effectiveness.

Table 4-6. Common Interferences of the CAM

Interferences	G-Bar Response	H-Bar Response
DS2	Low	
Insect repellent	Low to very high	
Brake fluid	High to very high	Very high
General-purpose cleaner	High	
Burning kerosene		High
Breath mints	High	
Gasoline vapors	Low	Low
Burning grass	Low to high	Low
Burning gas	Low	
Green smoke	Low	Low to high
Break-free oil	Low	
Ammonia	Very high	

4-100. **Radiological.** Use the AN/PDR-77 or AN/VDR-2 to determine if any contamination remains. If there is contamination remaining, determine the intensity of the contamination inside and outside of the vehicle. If the contamination has an intensity greater than 0.33cGy/hr, the vehicle is recycled to station 1.

RECYCLE CRITERIA

4-101. The commander, with the chemical unit leader's help, establishes the recycle criteria before starting decon operations. The recycle criteria determines which vehicles will return to station 1 after contamination is detected at station 5. If the unit has sufficient time and resources, any vehicle having more contamination than the acceptable level should be recycled. However, time and resources are usually limited, and not all vehicles can be recycled. The recycle criteria is based on the weathering effects.

RECONSTITUTION CRITERIA

4-102. The operational- and thorough-decon operations that are performed on vehicles or major equipment will be recorded on DA Form 2404 (Army) or NAVMC 10245 (Marine). This information will become a permanent record for these vehicles or major equipment. The entries that are required on these forms are the type of decon performed, decontaminant used, date/time group completed, location of decon site, and type of monitoring equipment used to verify decon completeness. These forms will remain a permanent part of the unit's, vehicle's, and equipment's records. This information will assist with reconstitution operations once the conflict or war has ended.

DED LAYOUTS

4-103. Dual-purpose and decon platoons establish thorough decon sites differently because of organization and equipment differences. The optimum layout for each type of platoon is described in the following paragraphs (see

Figures 4-5 through 4-8, pages 4-24 through 4-27). This layout provides the maximum output for platoons at 100 percent personnel and equipment. The equipment and personnel requirements for the optimum layout are identified for both the chemical unit (decon platoon) and supported unit (augmenter) (see Tables 4-7 through 4-10, pages 4-28 through 4-31).

4-104. Since it is likely that all platoons will not be at 100 percent strength, alternate layouts are discussed. For each alternate layout, chemical-unit equipment requirements are identified. However, personnel requirements are not identified by chemical unit and supported unit. The total number of personnel that are needed is identified, with a minimum number of chemical unit personnel.

ALTERNATE-LAYOUT PLANNING CONSIDERATIONS

4-105. Chemical unit leaders use METT-TC to determine the best possible DED layout to execute their mission. When determining alternate DED layouts, use the following guidelines:

- The ability to spray hot, soapy water or steam under pressure must be retained at station 1.
- The largest number of people is required at station 2.
- Experienced and qualified CAM operators are required at station 5.
- Water does not have to be hot to rinse off the DS2; however, the lower the water pressure the more water that is required for the rinse.

DECON PLATOON

4-106. The optimum DED layout for an M12A1 PDDA-equipped platoon requires the use of all authorized equipment and personnel (see Figure 4-5, page 4-24). While this DED layout is manpower and equipment intensive, it provides for the rapid decon of vehicles and equipment (eight vehicles processed per hour). This layout uses dual lanes at stations 1, 4, and 5 to process two vehicles at one time. Since the most time- and labor-intensive work takes place at station 3, this station is designed to process three vehicles at a time. The processing rate of this layout will be affected by any work/rest cycle (see Table 4-11, page 4-32).

SMOKE/DECON PLATOON

4-107. Smoke/decon platoons equipped with the M17 LDS will set up the DED differently then the decon units equipped with the M12A1 PDDA. While the optimum DED layout is manpower and equipment intensive, it provides for the rapid decon of vehicles and equipment (eight vehicles processed per hour). The layout uses a dual lane at stations 1, 4, and 5 to process two vehicles at a time. Since the most time- and labor-intensive work takes place at station 3, this station is designed to process three vehicles at a time. The processing rate of this layout will be affected by any work/rest cycle (see Table 4-11, page 4-32).

4-108. It may not be possible, for a variety of reasons, for an M17 LDS-equipped decon platoon to use the optimum DED layout. Figure 4-7, page 4-26, shows an example of an optimum DED layout. Limited personnel or equipment will affect the DED layout. Figure 4-8, page 4-27, shows an example of an alternate DED layout.

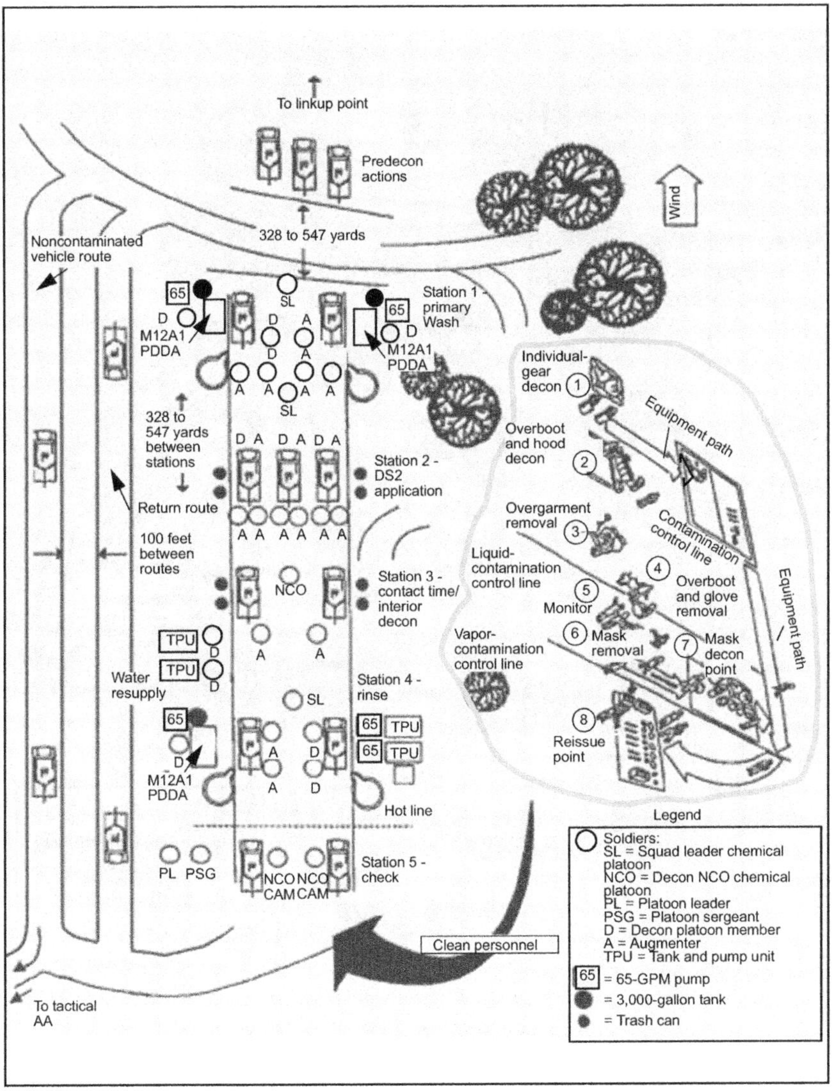

Figure 4-5. Optimum DED Layout for the M12A1 PDDA-Equipped Unit

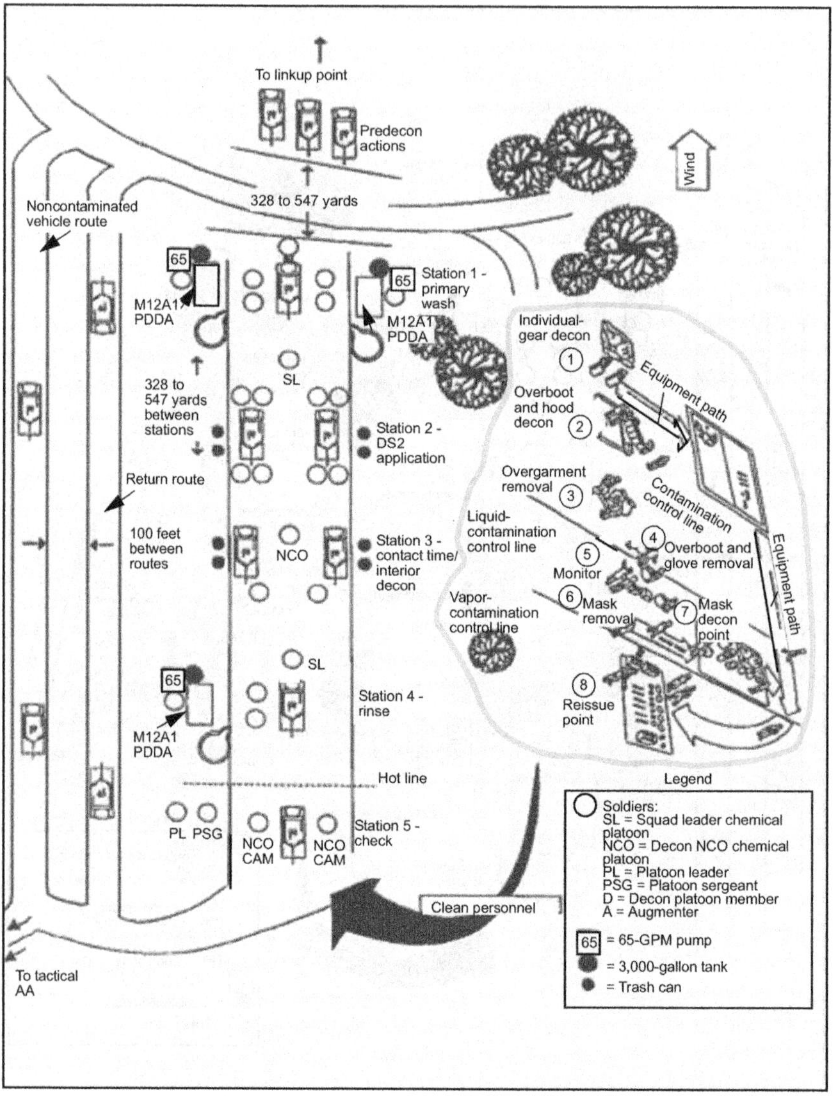

Figure 4-6. Alternate DED Layout for the M12A1 PDDA-Equipped Unit

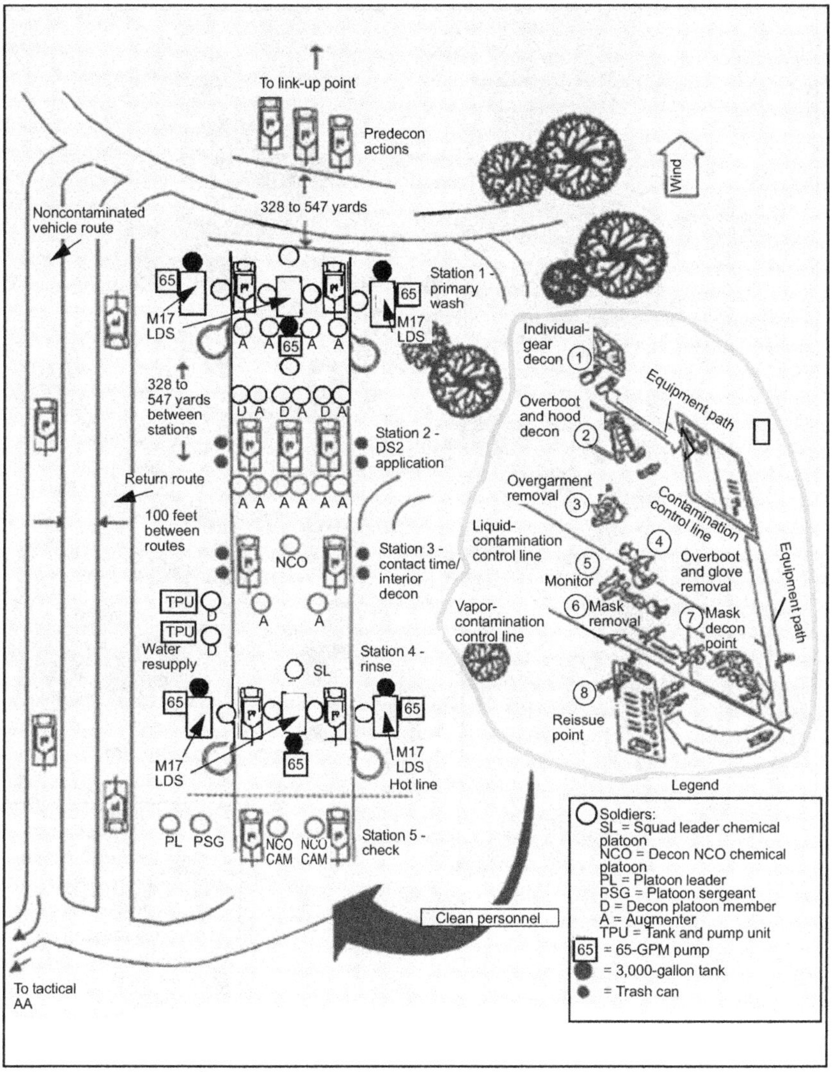

Figure 4-7. Optimum DED Layout for the M17 LDS-Equipped Unit

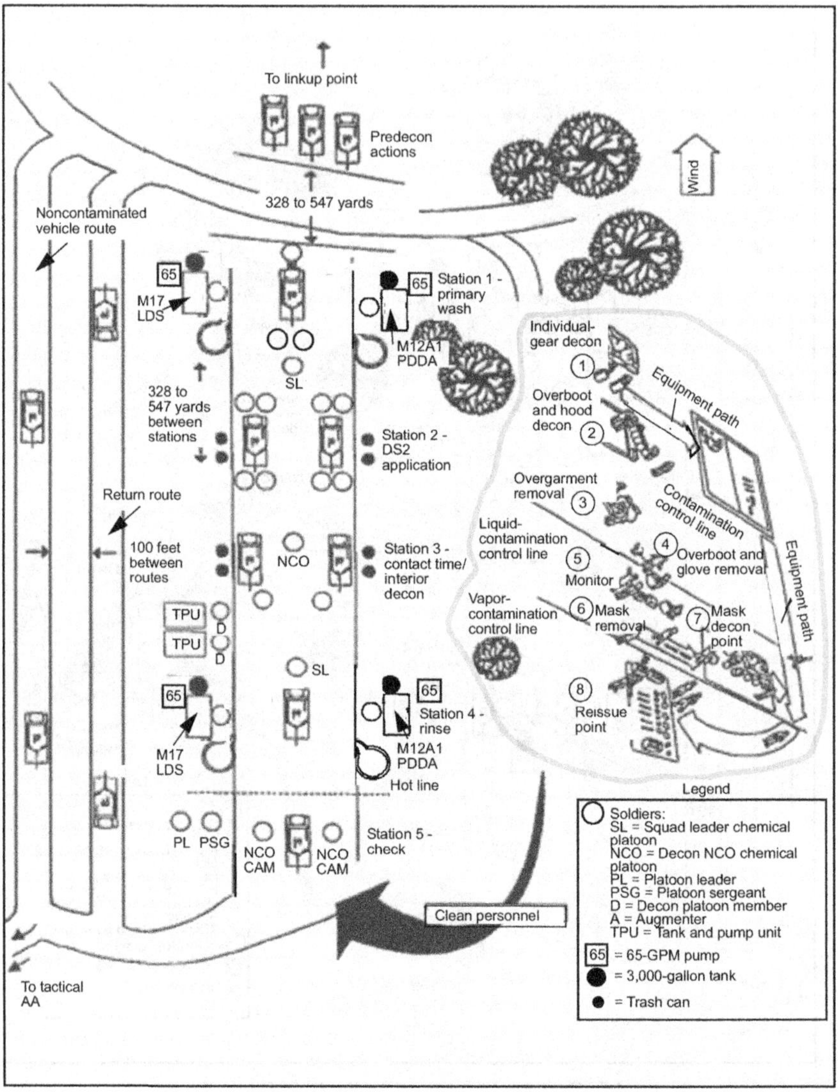

Figure 4-8. Alternate DED Layout for the M17 LDS-Equipped Unit

Table 4-7. Personnel and Equipment Requirements for the Optimum DED Layout of the M12A1 PDDA-Equipped Unit

Stations	Personnel		Equipment/Supplies
	Decon Platoon	Augmenter	
Station 1 - primary wash	1 squad leader 2 PDDE operators 4 sprayers	4 scrubbers	2 M12A1 PDDAs 2 3,000-gallon tanks 2 65-GPM pumps 6 long-handled brushes 8 TAP aprons Liquid detergent
Station 2 - DS2 application	1 squad leader 3 appliers	9 appliers	18 long-handled brushes 9 mops with extra mop heads 3 30-gallon containers 9 M13 DAPs Sufficient DS2
Station 3 - contact time/ interior decon	1 NCO	2 interior decon assis-tants	2 AN/VDR-2 3 TAP aprons 6 30-gallon containers 10 books of M8 detector paper 30 sponges 8 M256A1 detector kits 50 trash bags 1 clipboard with a pen 1 stopwatch
Station 4 - rinse	1 squad leader 1 PDDE operator 2 pump operators	2 sprayers	1 M12A1 PDDA 1 3,000-gallon tank 3 65-GPM pumps 2 TPUs 2 TAP aprons
Station 5 - check	2 NCO/CAM opera-tors		2 CAMs 10 M256A1 detector kits 20 books of M8 detector paper 2 AN/VDR-2 2 M8A1 or M22 ACAAs
C^2	1 PL 1 PSG		1 HMMWV with a radio 3 NBC marking kits
Total personnel	20	17	

WARNING

Chemical-unit leaders must consider the impact of the work/rest cycle on their unit's ability to process vehicles through the DED. Failure to initiate a work/rest cycle could result in heat casualties and mission failure.

4-109. It may not be possible, for a variety of reasons, for an M12A1 PDDA-equipped decon platoon to use the optimum DED layout. Limited personnel or equipment will affect the DED layout. Figure 4-6, page 4-25, shows an example of an alternate DED layout.

Table 4-8. Personnel and Equipment Requirements for the Alternate DED Layout of the M12A1 PDDA-Equipped Unit

Stations	Personnel	Equipment/Supplies
Station 1 - primary wash	1 NCOIC* 2 PDDE operators* 4 sprayers 2 scrubbers	2 M12A1 PDDAs 2 3,000-gallon tanks 2 65-GPM pumps 4 long-handled brushes 6 TAP aprons Liquid detergent
Station 2 - DS2 application	1 NCOIC* 8 appliers	14 long-handled brushes 8 mops with extra mop heads 3 30-gallon containers 8 M13 DAPs Sufficient DS2
Station 3 - contact time/ interior decon	1 NCO* 2 interior decon assistants	2 AN/VDR-2 3 TAP aprons 6 30-gallon containers 10 books of M8 detector paper 30 sponges 8 M256A1 detector kits 50 trash bags 1 clipboard with a pen 1 stopwatch
Station 4 - rinse	1 NCOIC* 1 PDDE operator* 2 sprayers	1 M12A1 PDDA 1 3,000-gallon tank 3 65-GPM pumps 2 TPUs 2 TAP aprons
Station 5 - check	2 NCO/CAM operators*	2 CAMs 10 M256A1 detector kits 20 books of M8 detector paper 2 AN/VDR-2 2 M8A1 or M22 ACAAs
C^2	1 PL* 1 PSG*	1 HMMWV with a radio 3 NBC marking kits
Total personnel	29 (Minimum of 11 soldiers from the decon platoon)	
*** These individuals should be from the decon platoon.**		

THOROUGH-DECON-SITE CLOSURE

4-110. Once all vehicles and soldiers from the contaminated unit have processed through the thorough-decon site, the site can be closed. Coordination with the supported unit is necessary before actually closing the site to ensure that all contaminated elements have been processed.

4-111. The chemical unit (decon or smoke/decon platoons) closes the DED first. Once the DED is closed, the chemical unit processes through the DTD. After the chemical unit has processed through the DTD, the DTD is closed. At that point, the chemical unit marks the area as a contaminated area and reports its exact location to the supported unit, using an NBC-5 report. The exact procedures for closing the DED and DTD are described as follows:

Table 4-9. Personnel and Equipment Requirements for the Optimum DED Layout of the M17 LDS-Equipped Unit

Stations	Personnel		Equipment/Supplies
	Decon Platoon	Augmenter	
Station 1 - primary wash	1 squad leader 4 sprayers 2 scrubbers	2 scrubbers	3 M12A1 PDDAs 3 1,500-gallon tanks 3 65-GPM pumps 6 long-handled brushes 8 TAP aprons Liquid detergent
Station 2 - DS2 application	1 squad leader 3 appliers	9 appliers	18 long-handled brushes 9 mops with extra mop heads 3 30-gallon containers 9 M13 DAPs Sufficient DS2
Station 3 - contact time/ interior decon	1 NCO 2 interior decon assis- tants		2 AN/VDR-2 3 TAP aprons 6 30-gallon containers 10 books of M8 detector paper 30 sponges 8 M256A1 detector kits 50 trash bags 1 clipboard with a pen 1 stopwatch
Station 4 - rinse	1 squad leader 4 sprayers		3 M17 LDSs 3 3,000-gallon tanks 3 65-GPM pumps 2 TPUs 4 TAP aprons
Station 5 - check	2 NCO/CAM opera- tors		2 CAMs 10 M256A1 detector kits 20 books of M8 detector paper 2 AN/VDR-2 2 M8A1 or M22 ACAAs
C^2	1 PL 1 PSG		1 HMMWV with a radio 3 NBC marking kits
Total personnel	23	11	

CLOSING THE DED AREA

4-112. The DED is closed in sequence, starting at station 1. All vehicles, equipment, and nonexpendable supplies are inspected for contamination. If contamination is found on an item, it is decontaminated. The actions at each station are described as follows:

Station 1 - Primary Wash

4-113. Spray all vehicles and equipment with hot, soapy water to remove any contamination that could have been transferred during the primary wash operations. Drain the water billets or fabric tanks of water. Inspect all equipment and vehicles for contamination using the appropriate detection equipment.

Table 4-10. Personnel and Equipment Requirements for the Alternate DED Layout of the M17 LDS-Equipped Unit

Stations	Personnel	Equipment/Supplies
Station 1 - primary wash	1 NCOIC* 4 sprayers 4 scrubbers	3 M17 LDSs 3 3,000-gallon tanks 3 65-GPM pumps 4 long-handled brushes 8 TAP aprons Liquid detergent
Station 2 - DS2 application	1 NCOIC* 8 appliers	14 long-handled brushes 8 mops with extra mop heads 3 30-gallon containers 8 M13 DAPs Sufficient DS2
Station 3 - contact time/ interior decon	1 NCO* 2 interior decon assistants	2 AN/VDR-2 3 TAP aprons 6 30-gallon containers 10 books of M8 detector paper 30 sponges 8 M256A1 detector kits 50 trash bags 1 clipboard with a pen 1 stopwatch
Station 4 - rinse	1 NCOIC* 4 sprayers	2 M17 LDS 2 3,000-gallon tanks 2 65-GPM pumps 4 TAP aprons
Station 5 - check	2 NCO/CAM operators*	2 CAMs 10 M256A1 detector kits 20 books of M8 detector paper 2 AN/VDR-2 2 M8A1 or M22 ACAAs
C^2	1 PL* 1 PSG*	1 HMMWV with a radio 3 NBC marking kits
Water resupply	2 drivers*	2 TPUs
Total personnel	32 (Minimum of 10 soldiers from the smoke/decon platoon)	
*** These individuals should be from the smoke/decon platoon.**		

Station 2 - DS2 Application

4-114. Throw the mops and brushes that were used in applying the DS2 into a sump or bury them. Load the unused cans of the DS2 and the M13 DAPs on a vehicle.

Station 3 - Contact Time/Interior Decon

4-115. Inspect any unused supplies and equipment for contamination. If no contamination is detected, load the equipment and supplies on a vehicle. Throw all contaminated supplies into the nearest sump. In calculating sump requirements, 35 cubic feet of space is required for 250 gallons of liquid drain off. Once a sump is 2/3 full, prepare a new sump.

Table 4-11. Work/Rest Table

Temperatures	Stations and Workload				
	Station 1 – Primary Wash	Station 2- DS2 Application	Station 3 – Contact Time/Interior Decon	Station 4 - Rinse	Station 5 - Check
	Physical Exertion				
	Moderate	Heavy	Moderate	Moderate	Light
Cool (<68°F)	60 work 15 rest	30 work 30 rest	60 work 15 rest	60 work 15 rest	60 work 15 rest
Warm (68° - 74°F)	45 work 45 rest	20 work 20 rest	45 work 45 rest	45 work 45 rest	50 work 50 rest
Hot (74° - 84°F)	30 work 60 rest	15 work 30 rest	30 work 60 rest	30 work 60 rest	40 work 80 rest
Very Hot (>84°F)	20 work 60 rest	10 work 30 rest	20 work 60 rest	20 work 60 rest	25 work 75 rest

NOTE: When operating in temperatures above 75°F, consider the ability of the soldiers to accomplish the mission. Once the soldiers have reached their maximum workload for heat stress, they cannot recover quickly enough to accomplish the decon mission. A viable option is to postpone the decon operation until a cooler part of the day or evening. This will reduce the heat-stress load on the soldiers and increase the probability of mission success. For further information, see FM 3-4.

Station 4 – Rinse

4-116. Spray all vehicles and equipment with hot, soapy water to remove any contamination that could have been transferred during the rinse operations. Drain the water billets or fabric tanks of water. Inspect all equipment and vehicles for contamination using the appropriate detection equipment. If no contamination is detected, load the equipment on the vehicles. Spread one can of STB dry mix into each of the sumps and then cover them. Ensure that DS2 is not present in the sumps. Post NBC hazard markers near the covered sumps. **(NOTE: DS2 destroys chemical agents; however, some of the by-products created from it are toxic.)**

Station 5 – Check

4-117. Ensure that all equipment has been checked for contamination. If it is not contaminated, load it on a vehicle. If it is contaminated, decon it according to the appropriate technical manual (TM). Throw any contaminated supplies into the nearest sump. Move all vehicles upwind of station 5 and inspect them again for contamination. If any contamination is detected, the crew will use either the M11 or M13 DAP to decon the identified areas. Once the vehicles are staged, all the soldiers will proceed to the DTD.

CLOSING THE DTD AREA

4-118. Once all the soldiers from the DED have processed through the DTD, it may be closed. After the last person has exited the DTD—

- Pick up all the used supplies from station 7 and put them in station 7's sump. Take up the contamination control line. If engineer tape was used, dispose of it in station 7's sump.
- Move all usable supplies and equipment from all the stations to station 1. Discard all unusable supplies from stations 3, 4, and 5 in station 1's sump.
- Decon all supplies and equipment collected at station 1 using the decon and rinse water at station 1. Empty the decon and rinse water from the station into the sump and decontaminant the containers.
- Mark the entire decon area. Remove your overgarment using the MOPP-gear-exchange technique and dispose of it in the sump at station 1.
- Move any equipment used to fill the sump upwind of the decon area. Decon your rubber gloves and move all equipment and supplies in station 1 upwind of the decon area. Keep this equipment and supplies separate from that used to fill the sump.

THOROUGH DECON UNDER UNUSUAL CONDITIONS

4-119. Decon operations are not restricted to daylight hours. The enemy may employ chemical and biological weapons at night, since weather conditions are usually more favorable for their employment at that time. Consequently, NBC contamination encountered at night may require decon before daylight.

4-120. Conducting decon operations at night is a challenge. "White light" cannot be used without possibly revealing your location. However, decon personnel must have illumination to perform such essential decon tasks as spraying water, applying decontaminants, using detection equipment, and taking off the MOPP gear.

4-121. Use M8/M9 detector paper to check for chemical contamination and to see how well the chemical decon was performed. If contamination is present, the paper color changes to red and cannot be seen if red-filtered light is used. The CAM and M256A1 detector kit can be used to identify and detect chemical agents but will require a light source in order to obtain a reading.

4-122. Several vehicles awaiting or undergoing decon present a significant infrared "signature" due to the hot engines. To enter, exit, and move within the decon site is difficult under blackout conditions.

4-123. Night-decon operations are difficult but can be accomplished. Whenever possible, conduct thorough-decon operations in built-up areas. Use a building for DTD operations and a warehouse for DED operations. This allows for the use of white light. Soldiers can see what they are decontaminating, see where they are going, and read the color changes of the M8/M9 detector paper.

4-124. There will be a need for additional soldiers to act as ground guides in the DED area. The NCOICs of stations 1, 2, and 4 will inspect each vehicle with a flashlight or chemical light before the vehicle proceeds to the next station. There should be at least two flashlights at each station in both the DED and DTD areas.

Chapter 5

Special Decon Operations

The decon of terrain or structures allows personnel to increase their stay time in an area and facilitates their passage through an area. Large-scale terrain decon requires extensive amounts of equipment, material, and time. Thus, terrain decon is limited to areas of critical importance. Contaminated areas and structures that can be avoided do not need to be decontaminated except by weathering. If they must be used, decon may be necessary. When decon is not practical, use protection to deal with the hazard. Most terrain and structural surfaces (excluding unpainted metal and glass) absorb the agent. Therefore, the agent may be transferred to clothing by physical contact. The agent may also be transferred through vaporization and subsequent inhalation or diffusion into the eyes and the clothing. The decon of equipment containing vulnerable components presents certain challenges. Electronics and optics are especially vulnerable to damage if not carefully decontaminated. Some materials, such as canvas, absorb chemical agents and may not be decontaminated and reused.

TERRAIN DECON

5-1. A terrain decon provides more long-term rather than short-term benefits. Generally, you cannot decon occupied terrain well enough to allow you to reduce your MOPP level right away. At best, all you can do is speed up the weathering process. This may allow you to reduce your MOPP level a little sooner, for example, in 2 days instead of 4 or 5. Terrain decon can be so expensive and so ineffective that you should consider relocating.

5-2. If you decide on a terrain decon, limit the amount you do. Build a causeway or breach a path through the contamination for entry/exit into a major facility. Restrict your efforts to areas where you must work or move. No matter what you do, you will probably have to remain in MOPP; but you can reduce the danger of picking up liquid contamination and spreading it to other areas.

5-3. The most effective techniques for a terrain decon are the natural methods (weathering or aging). Using standard decontaminants will quickly exhaust your resources (see Appendix B). Usually the most practical method is to scrap off contaminated surfaces. Fire is a terrain-decon method; however, consideration must be given to the potential downwind vapor hazard. If possible, a contaminated area should be evacuated until time and weather remove the hazard. An STB slurry can be used for vital areas. To determine the persistency of chemicals at various temperatures, refer to FM 3-9.

5-4. A terrain decon may be accomplished using several methods. These methods are listed in the order of decreasing effectiveness. Although they reduce the contact hazard, none will prevent hazardous vapors desorbing from contaminated surfaces. Therefore, appropriate protective measures must still be taken.

WEATHERING

5-5. Weathering is the simplest and easiest form of decon.

Chemical and Biological Contamination

5-6. Warm, windy weather can significantly reduce terrain contamination. In some cases, this occurs in a few hours or it may take a few days. Many variables affect the persistency of chemical and biological hazards, so it is impossible to accurately predict how long it takes such contamination to weather. Sunlight is especially effective against most biological agents.

Radiological Contamination

5-7. Although the term "weathering" is used to describe the decon process, weather has little effect on radiological hazards. Heavy rain and wind may remove some contamination, but only time will reduce the radiation emanating from the contamination.

REMOVING/COVERING

5-8. Removing or covering contamination does not destroy it, but it does keep the hazards away from you temporarily.

Chemical and Biological Contamination

5-9. Contamination can be removed from paths and unpaved roadways by scraping off a layer of earth using heavy earthmoving equipment (bulldozers and road graders). Contaminated paths and paved surfaces can be covered with a 4-inch layer of earth, roofing paper, plastic sheets, and wood mats. This is a temporary measure because the agents may penetrate the covering. When the contamination penetrates the covering or when the covering is removed, the hazards will reappear. The coverings may also extend the life of the contamination hazard by reducing its exposure to air and sunlight.

Radiological Contamination

5-10. Radiological contamination must be covered by thick layers of dense material, like earth. Three inches of earth will decrease radiation dose rates by half because of the shielding provided by the soil. However, 12 inches is more effective. The job will be easier with earthmoving equipment, but the equipment and operators probably will have to undergo decon.

NEUTRALIZING

5-11. The STB dry mix or STB slurry may be used effectively against chemical or biological contamination, but not against radiological contamination. Trained chemical troops and equipment are required when using the STB slurry, which is costly both in time and material.

Chemical and Biological Contamination

5-12. The STB dry mix is spread on solid surfaces and raked into soft surfaces like sand or earth.

5-13. The STB slurry is applied to terrain by using a spray hose that is attached to an M12A1 decon apparatus. Operators wear TAP aprons over their MOPP gear and sit on the front fenders of a vehicle to spray the STB slurry. The speed of the vehicle should be adjusted according to the terrain; this is normally 2 to 4 miles per hour. The operators spray a fan pattern 3 to 5 feet ahead of the vehicle. The pattern overlaps on the center and extends past both sides of the vehicle. The STB slurry forms a seal over the surface and must be renewed periodically due to deterioration from traffic or weathering. One load of STB slurry will decon a concrete surface about 328 feet long by 33 feet wide. Other surfaces may require more decontaminants. It takes about an hour to load and mix one load of SBT slurry.

Radiological Contamination

5-14. The STB dry mix and STB slurry have no effect on radiological contamination or its hazards.

BURNING

5-15. Burning works well against chemically or biologically contaminated vegetation; but it is of no value against radiological contamination.

Chemical and Biological Contamination

5-16. Fuel may be used to burn grass or short undergrowth. Burning also works on dirt surfaces. Soak the area with diesel fuel, kerosene, or fuel oil and ignite remotely. Do not use gasoline; it burns too quickly. Burning will cause vapor hazards downwind, and protective measures will have to be used by downwind units. Area commanders must warn these units of the vapor hazards.

Radiological Contamination

5-17. Radiologically contaminated surfaces should not be burned. Burning will not destroy radiological contamination or its hazards. It may spread contamination if radioactive particles become suspended in smoke spread by the wind.

EXPLODING

5-18. Explosives can be used to blow up areas that are chemically and biologically contaminated; however they are of no value against radiological contamination.

Chemical and Biological Contamination

5-19. Detonating cord, bangalore torpedoes, or mine-clearing line charges may be used with the STB dry mix to clear small paths through a contaminated area. The two methods that are used follow:

- First method. Remove the holding band and loosen the lid of the bleach drum. Set the drum with the lid in place upside down over about 3 feet of looped detonation cord. Remove the drum and leave the bleach piled on the lid as shown in Figure 5-1, A, page 5-4. Lids may be fired in a series with a single strand of detonation cord connecting the loops.

- Second method. Remove the holding band and loosen the lid of the bleach drum. Set the drum in a shallow hole as shown in Figure 5-1, B. Make sure the lid is loose. Use 5 feet of looped detonating cord (7 feet in soft ground) for each drum. The drums may be fired individually or in a series. For gross contamination, additional firings may be necessary.

5-20. For either emplacement method, the suggested spacing for charges is 33 feet apart for a 100-foot-front contaminated area. The charges should be placed along the upwind edge of the contaminated area (see Figure 5-1, C).

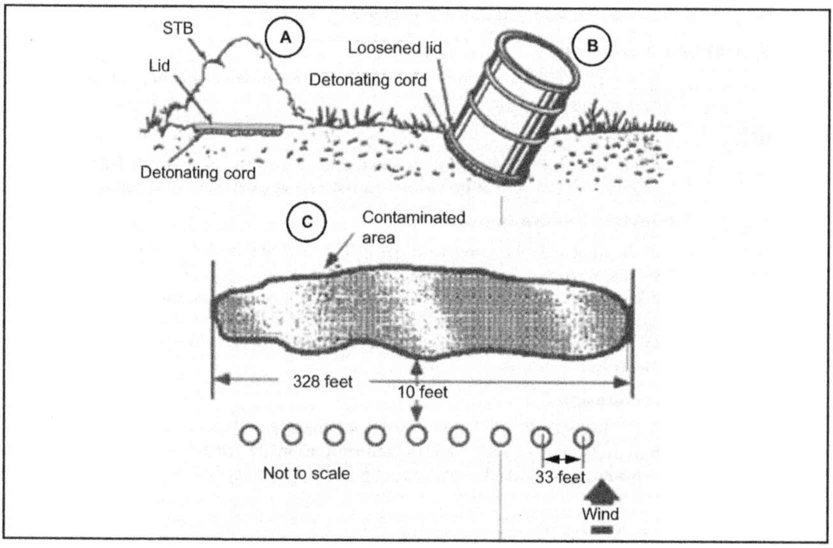

Figure 5-1. Two Methods of Emplacing Dry STB for Distribution by Detonation

Radiological Contamination

5-21. Explosives will not destroy radiological contamination or its hazards.

FLUSHING

5-22. Flushing with large quantities of water removes contamination. A pressurized stream of hot, soapy water delivered by the PDDA, M17 LDS, or fire trucks can remove significant amounts of an agent. Scrubbing removes even more.

Chemical and Biological Contamination

5-23. Flushing is not very effective against some kinds of agents, especially thickened agents. However, flushing may speed up the weathering process.

Radiological Contamination

5-24. Flushing is very effective for removing loose radioactive contamination such as fallout. Rain out, however, can coat surfaces with a film that resists flushing. Such films must be scrubbed. Any contamination removed by flushing and scrubbing will remain radioactive, so control the runoff.

CLEARING

5-25. To clear passageways, remove the contaminated layers covering the terrain.

Chemical and Biological Contamination

5-26. Most liquid chemical agents will not penetrate the soil more than 2 inches, and biological agents penetrate even less. Earthmoving equipment or hand tools can be used to scrap aside contamination. By scraping off 2 to 4 inches of earth, a passageway can be created that is free of transfer hazards. However, contamination on either side of the passageway will continue to present a vapor hazard.

Radiological Contamination

5-27. Fallout does not penetrate the top layer of soil unless it is followed by rain. It can be scraped aside. Move contaminated soil as far away as possible; the piles tend to concentrate radiation. Immediately below a nuclear blast area, the soil may be radioactive to a depth of 5 feet. In most cases, decon of this much earth is not practical.

FIXED-SITE DECON

5-28. Personnel and equipment need to be decontaminated to reduce the hazard and the spread of contamination, limit casualties, and minimize operational degradation. The goal is to restore the unit to an acceptable level of operational effectiveness. The inability of a fixed site to move away from contamination makes decon more critical. Decon is necessary when protective items cannot be worn in performing the mission or when long-term occupancy is anticipated.

5-29. Decon capabilities may be designed and built into a facility and used if the facility becomes contaminated. Mobile decon equipment may also be available at a fixed site to decon equipment, roads, and buildings. Interior decon may be accomplished with large commercial heaters. Loading docks, entries and exits, and building exteriors can be decontaminated with more conventional methods, such as using an STB dry mix or STB slurry, soap and water, or an M13 DAP. Commanders should identify all systems that are capable of contributing to the decon effort (for example, fire trucks, steam cleaners, and water pumps).

5-30. Decon equipment can be retrofitted to accommodate the process in each area. For instance, rest and relief shelters need rapid, personnel decon systems, and supply operations need a decon system that can handle moderate size pieces of equipment, but at a high volume. Light maintenance needs a system that can decon equipment sensitive to harsh decontaminants or water, and heavy maintenance needs a system that can clean large equipment.

5-31. The commander should designate and train teams that can perform decon for fixed-site operations. The following are fixed-site decon techniques that the teams could apply:

AIRFIELD

5-32. Decon the area required to sustain the mission (see Figure 5-2). Figure 5-3 illustrates the STB application on the entrance/exit of the airfield hangers. Note that the STB application includes 10 feet on each side of the runway between the entrance and the exit. A transfer hazard will remain a problem until all parts of the airfield have weathered. Buildings' or hangers' entrance/exit and doors must be decontaminated accordingly. Figure 5-3 illustrates a field-expedient method that could be employed at any fixed site to cover the major entry/exit toward the site.

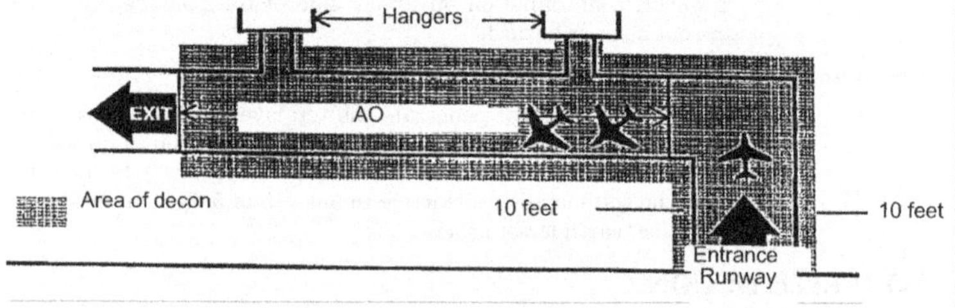

Figure 5-2. Airfield-Area Decon

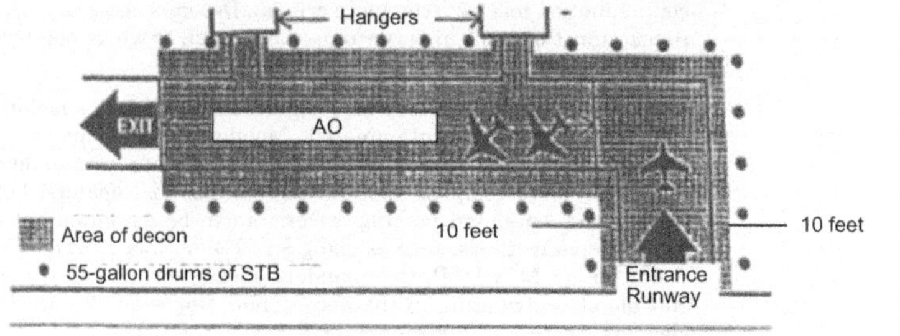

Figure 5-3. Field-Expedient Airfield Decon

HELIPAD

5-33. To decon a helipad site, spray the STB slurry on the entrance/exit approach or the path (see Figure 5-4). Since airborne STB particles can damage sensitive aircraft components, the decontaminated area should be washed with water before landing helicopters.

POMCUS/MOTOR PARK

5-34. At a prepositioning of material configured to unit sets (POMCUS)/motor-park site, decon entrance/exit approaches to limit contamination. Exposed vehicles in a POMCUS/motor park will become contaminated but most likely there will not be enough personnel to drive the vehicles to a decon station. Decon parked equipment by spraying it with an STB slurry or with hot, soapy water (see Figure 5-5). The M12A1 PDDA crew drives around the parked vehicles and performs vehicle washdown to accelerate the weathering process.

NOTE: For more detailed guidance on the decon procedures of fixed sites, ports, and airfields, refer to FM 3-4-1.

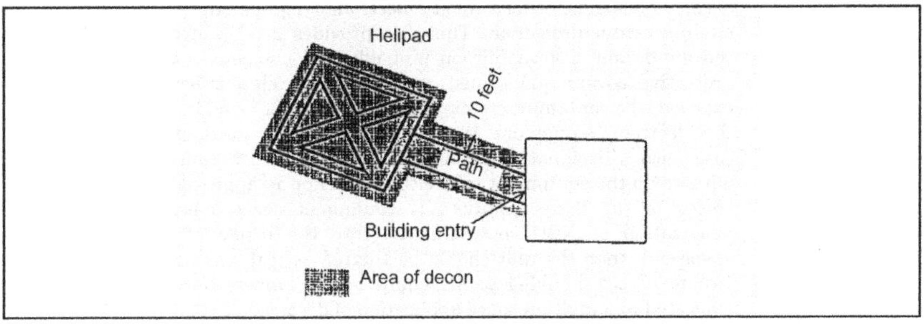

Figure 5-4. Helipad-Area Decon

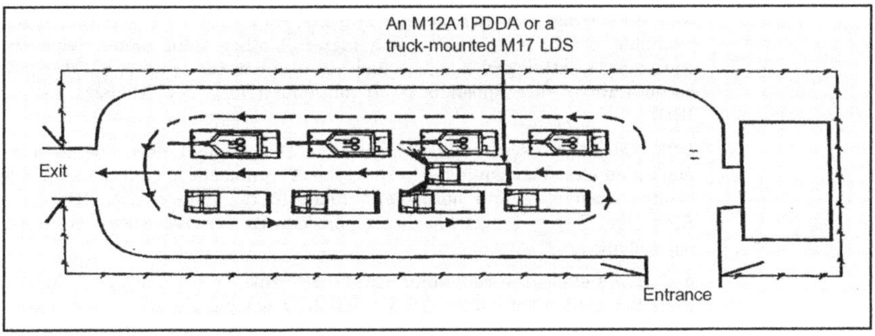

Figure 5-5. POMCUS/Motor-Park-Site Decon

VULNERABLE EQUIPMENT DECON

5-35. Most military equipment has not been critically assessed for its ability to withstand decon operations. As more testing is done, specific decon instructions will be included in the TM for all types of equipment. Some equipment is extremely vulnerable to damage when subjected to decon operations. Electronics and optic equipment are especially vulnerable. Some materials (such as canvas and rubber) tend to absorb chemical agents, and decontaminating absorbent surfaces is extremely difficult, if not impossible. Most decontaminants are highly corrosive and cannot be used on certain materials. They may also render ammunition unserviceable. Do not subject vulnerable equipment to unnecessary decon operations. Radiac meters may measure low levels of radiation that are not dangerous. The commander's operational exposure guidance (OEG) determines the maximum permissible radiation dose rate (usually 0.33 cGy per hour). If this dose rate is exceeded, you must decon.

ELECTRONICS

5-36. Nonsealed electronic-equipment circuitry can be damaged by corrosive decon materials. Most field electronic equipment is watertight for environmental protection. This also provides good protection against NBC contamination. Contamination probably will not penetrate gasket-equipped protective covers and sealed components on electronic equipment; but if exposed, the contaminants may be present on the outside of cases containing the electronic equipment. Wipe down the outside portions of the equipment case with a designated decontaminant. After decontaminating the outside, wipe down the equipment with either water or an approved solvent to remove traces of the decon solution. If equipment seals appear damaged or if penetration of NBC contamination into the inside of the equipment is suspected, then the unit should be treated as if it was nonsealed. Under no circumstances should electronic equipment be immersed in a decon solution or subjected to a high-pressure application of decon solutions.

5-37. For chemical contamination, wipe all metal electronic cases with DS2. If a chemical agent is identified as HD, allow 30 minutes contact time. Next, wipe the case down thoroughly with a damp cloth, dry it off, and wipe it down with a designated oil (for biological agents, wait 10 minutes). If DS2 is not available or the case is of some material other than metal, wipe the equipment's exterior with a cloth and hot, soapy water or use a nonstandard decontaminant (see Appendix B). If contamination is not extensive, use an IEDK.

5-38. Corrosive decontaminants, such as DS2, should never be used on nonsealed electronic equipment. This type of equipment is often found inside shelter assemblies and helicopters. Refer to the appropriate TMs or to Appendix C for acceptable decon procedures for nonsealed electronic equipment.

5-39. For radiological contamination, brush, wipe, or vacuum contamination from the equipment. Since contamination is not destroyed but just moved from one place to another, control the runoff and treat it is as a hazardous substance.

OPTICS

5-40. Optic systems are extremely vulnerable to decon materials that might scratch or adversely affect the lenses. Wipe optic systems with a soft, nonabrasive material such as lens-cleaning tissue, cotton wadding, or soft cloth dipped in hot, soapy water. Wipe the optic system with decontaminants. Do not immerse it.

5-41. Hot, soapy water is the preferred decontaminant for chemical and biological contamination. When DS2 is applied to coated lenses, it causes some damage. Likewise, exposing lenses to an STB slurry removes some of the magnesium from the coating. Small optic surfaces can be decontaminated with the IEDK.

5-42. Radiological contamination should be blown off with a stream of air or wiped off with hot, soapy water. Rinse by wiping with a sponge dipped in clean water.

AMMUNITION

5-43. Decon contaminated ammunition with cool, soapy water. Apply the soapy water with the PDDA or with brushes, mops, rags, or brooms. Cool, soapy water is the preferred decontaminant for all types of contamination.

WARNING

Do not use DS2 or an STB powder on ammunition. It removes critical markings from the ammunition. The STB powder may cause a fire when it comes in contact with a blister agent or DS2. It may also corrode and render the ammunition unserviceable. Do not use nonstandard decontaminants that are corrosive. They also may remove critical markings from the ammunition.

CANVAS ITEMS

5-44. These items include load-carrying equipment and web gear. It may be necessary to burn or bury them if they are heavily contaminated. Either an STB dry mix or an STB slurry may be used to decon them. The STB slurry is more effective. In many cases, weathering may be the preferred decon technique because scrubbing canvas frequently imbeds the contamination further and worsens the situation. If the item must be decontaminated, boil it for 1 hour in soapy water. This is the preferred method for chemical and biological decon. Radiological contamination can be removed by brushing it off, then washing. It may also be vacuumed off a surface.

5-45. Salvage of contaminated canvas requires a great amount of logistic and manpower requirements. Other options for disposing of contaminated canvas should be considered against METT-TC. NBC covers protect vulnerable items, but these covers must be buried or destroyed after use.

FOOD AND WATER

5-46. Most units cannot test the purity of food and water even if they can decon it. Food and water decon usually takes special handling by trained technicians. If your water becomes contaminated, contact medical personnel. Water-treatment specialists test and treat the water in an NBC environment. Direct contaminated-food questions to supporting veterinary personnel.

5-47. Food and water should be kept in sealed containers and protective packaging. Leave food in wrappers. Keep wrapped food in boxes, tins, or other containers. Store these containers in closed refrigerators or vans. See FM 3-3 for further discussion on avoiding contamination.

5-48. Contaminated food should be placed into three groups for handling.

Group I

5-49. Group I consists of canned and unopened items that have been exposed only to the vapors of an agent. Generally, the items in this group will be safe for issue to personnel after a brief period of outdoor airing to remove clinging vapors.

Group II

5-50. Group II consists of canned and unopened items that have been contaminated with a liquid chemical agent. Attempts to decon porous packaging materials, such as cardboard or wood, are likely to be unsuccessful and may result in spreading the contamination. To handle such items, strip off the outer contaminated coverings and examine the inner layer to determine if agent penetration has occurred. If it has, continue stripping off layers until an uncontaminated layer is reached. Items packaged for military operations are usually packed in boxes, so this procedure is feasible. When the inner uncontaminated package is reached, place it in group I. If the agent has penetrated to the food, place it in group III.

Group III

5-51. Group III consists of unpackaged or poorly packaged items that have been exposed to an agent in either a vapor or liquid form. Foodstuffs in this group should be decontaminated only when absolutely necessary.

5-52. If you do not know if the food is contaminated, do not use it. Do not decon items in group III without the help of veterinary personnel.

CHEMICAL MUNITIONS DISPOSAL AND DECON

5-53. In combat, chemical munitions may be encountered. Captured enemy chemical munitions may have sustained damage or have deteriorated to where they are leaking. Leaking munitions must be decontaminated, evacuated, and disposed of safely by specialized units with properly trained soldiers performing this task. All soldiers must use protective clothing and masks when there is a possibility of exposure to a toxic chemical agent. The MOPP level will be specified by the unit commander. Use detection equipment (CAM, M256A1 detector kit, M8A1 or M22 ACAA, and M8/M9 detector paper) to monitor chemical munitions.

5-54. Unexploded leaking munitions should be treated in the same manner as conventional ordnance. A spot report must be submitted to an explosive ordnance disposal (EOD) unit.

5-55. When disposing of and decontaminating chemical munitions, see Army regulation (AR) 385-10 and TM 10-277 for additional safety guidance.

HANDLING

5-56. Chemical projectiles should not be handled. A good rule of thumb is to treat them as you would a fused high-explosive (HE) round. Security and protective measures must match those given to conventional munitions. Notify an EOD unit for disposal and decon of chemical munitions. EOD units have the responsibility for sealing, packaging, and disposing of chemical munitions.

CONTAMINATION CONTROL

5-57. Whenever a chemical projectile is damaged and starts to leak, the round presents the same hazards to personnel in the area as an enemy chemical attack. Only munitions that have not completely detonated or have been severely damaged may exude liquid agent. It is the vapor hazard created when the chemical leakage occurs that poses the greatest threat to personnel. When a leaking chemical munition is discovered, react as if you are under chemical attack.

DEPLETED URANIUM (DU) DECON

5-58. DU is an extremely dense metal used in munitions to penetrate heavy armor or as a protective shielding (armor packages). DU is also used as equipment components. All components containing radioactive isotopes are listed in TB 43-0116.

5-59. DU exposure and incidents may occur anytime there is damage to the DU armor package, a vehicle is hit with DU munitions, DU munitions are damaged, or equipment components containing DU are damaged. The DU armor package can be damaged during vehicle maneuvers, on-board fires, maintenance activities, or ballistic impacts. DU munition problems may occur during storage, transportation, combat, testing, or manufacturing. DU contamination may be present on the ground in areas where equipment was destroyed or damaged.

5-60. DU contamination may include DU oxides (dust), contaminated shrapnel, munition components, or armor components. DU emits primarily alpha particles; however, beta, gamma, and x-ray ionizing radiation are also emitted. DU contamination can be inhaled, ingested, or injected. DU contamination does not pose an immediate health risk. Consequently, contamination should be removed from personnel or vehicle surfaces when directed by the unit commander based on METT-TC.

5-61. Visual signs that DU contamination is present include heavy, dull-black dust or small round holes. DU contamination can only be verified with a radiac meter. An AN/VDR-2 or AN/PDR-77 with an alpha or beta probe (flat-pancake surface) is used to detect and measure DU contamination. These probes are within the radiological protection officer (RPO) kit.

5-62. When working on or within DU contaminated equipment, soldiers should wear their protective mask. Depending on temperature, protective-clothing availability, DU contamination levels, and tasks to be performed, soldiers can wear the BDO or coveralls or they can roll down their sleeves and blouse their trousers as directed by the unit's chemical or medical personnel.

5-63. General decon procedures follow:

- Use a radiac meter to determine if DU contamination is present.
- Provide protection, including appropriate clothing, for workers as directed by the unit's chemical or medical personnel.
- Identify what is to be decontaminated.
- Obtain necessary equipment and materials (Appendix D).

- Brush, wash, or wipe off contamination with a damp cloth. Use a high-efficiency particulate air filter (HEPA) vacuum cleaner, if available.
- Work from the outside of the contaminated area to the inside.
- Cover fixed contamination with tape, paint, paper, plastic, or other disposable material.
- Use the standard double-bag-and-tag process for hazardous waste. The only contaminated waste generated by DU will be the vacuum-cleaner bags after use on multiple vehicles.
- Continue the decontaminating process until contamination levels are reduced to the appropriate level (see AR 11-9).
- Decon personnel and equipment according to Chapter 4 of this manual.

CONTAMINATED REMAINS DECON

5-64. Mortuary affairs personnel establish and operate the mortuary affairs decontamination collection point (MADCP). When an NBC event has occurred on the battlefield, there is a very high probability that many deceased personnel were exposed to contaminating agents. If the situation does not lend itself to the determination of a hazard on an individual basis, all remains within the affected area will be treated as if contaminated. If the theater surgeon or his staff determines that biological agents have been employed, all remains will be treated as if contaminated.

5-65. Whenever the threat of NBC warfare exists in a theater of operations, the unified commander will direct the Joint Mortuary Affairs Office (JMAO) to be prepared to handle contaminated remains. The JMAO will serve as the theater central point of coordination for the operation. Joint operations of contaminated remains are described in Joint Publication 4-06.

5-66. Personnel support is required after completing the evacuation mission to the MADCP. A DTD takes about 1 hour. The MADCP site will require a complete detailed decon by a chemical decon unit.

RADIOLOGICAL DECON

5-67. Radiological contamination may occur in the form of one element. In the following paragraphs, the decon of six specific, commonly found radioactive elements is discussed. The discussion is applicable not only to these elements but also to other elements having similar chemical properties.

CESIUM

5-68. The common radioisotope of cesium is cesium-137. It emits beta and gamma radiation, decaying to stable barium-137. Cesium-137 is widely used in gamma sources. It occurs in these sources as cesium-chloride pellets. Cesium chloride is a soluble salt. The contamination from a sealed-source leak absorbs water, becomes damp, and creeps. Contamination from a sealed cesium source is best decontaminated by wet procedures unless the contamination is on a porous surface, in which case wet procedures should be preceded by vacuuming. Cesium is known to adsorb from a solution onto glass surfaces. Decontaminating a cesium liquid-contaminated surface is best accomplished by wetting the surface, absorbing the solution with a rag or other absorbent material, and rinsing the area several times with water. If the contamination persists, use a detergent solution and scrub with a brush. A cesium-contaminated solution that has been standing for some time is best decontaminated by absorbing any remaining liquid, treating the surface several times with water (allowing the water to stand on the surface for about 1 minute each time), and then absorbing the liquid from the surface. If the contamination remains, further treatment depends on the surface. Metallic surfaces are treated with strong mineral or oxidizing acids. Waxed surfaces are removed. If contamination still persists, abrasives or other removal techniques are used.

COBALT

5-69. The common radioisotope of cobalt is cobalt-60, a beta gamma emitter. Metallic cobalt-60 is commonly used in sealed gamma sources. Particles of cobalt dust adhering to small articles are readily removed by ultrasonic cleaners or by dipping the article in a dilute solution of nitric, hydrochloric, or sulfuric acid. Cobalt-dust contamination that exists over a large area is best removed by vacuuming. Sealed cobalt sources may leak as a result of electrolytic action between the cobalt and the container. The result is often a soluble cobalt salt, which creeps and spreads. This is best decontaminated with a detergent or an ethylene diamine tetra acetic acid (EDTA) solution, followed by a treatment with mineral acids. Contamination from solutions containing cobalt may be treated with water.

PLUTONIUM

5-70. The most common isotope in which plutonium may be present as a contaminant is plutonium-239, an alpha emitter. This isotope is present in the AN/UDM-6 calibration source. Plutonium contamination may be the result of a nuclear weapon's accident, in which case, the plutonium will be scattered as a metal or oxide in a dust form. Both forms of plutonium are insoluble. Aging of plutonium-239 contamination is impractical since it has a 24,000-year half-life. Plutonium contamination that covers a small area is best decontaminated by vacuuming. If contamination remains, the area should be washed with a detergent solution. Any contamination that remains can be sealed in a protective coating of paint, varnish, or plastic. Plutonium oxide or metal dust spread over a large area, such as a field, is best decontaminated by removing the top layer of soil and disposing of it as radioactive waste. Personnel should wear respiratory protection when decontaminating or moving the soil.

STRONTIUM

5-71. The most common radioisotope of a strontium is strontium-90, a beta emitter. The daughter particle of strontium-90 is yttrium-90, which is also a beta emitter. Strontium-90-yttrium-90 is commonly used in sealed beta sources such as the M6 source. Generally, it is present as chlorine or carbonate. The chlorine is hygroscopic; it absorbs water and creeps out of the container. This contamination is best decontaminated by vacuuming, followed by a treatment with water, a complexing-agent solution, and a mineral acid, in that order. Contamination resulting from a dilution containing strontium is best decontaminated by absorbing the solution and washing the area with a detergent solution. If strontium contamination persists, the top layer of the surface should be removed by abrasives or other removal procedures and a sealing coat should be placed over the surface.

TRITIUM

5-72. Tritium is the radioisotope of hydrogen and is a weak beta emitter. If it is released to an area as a gas, the best decon method is to flush the area with air. Since inhalation of tritium can present an internal hazard, personnel entering an area containing tritium gas should wear an appropriate self-contained type of breathing apparatus. Objects in an area exposed to tritium for a length of time may absorb the gas and should be disposed of, if possible. They may be degassed, under a vacuum, by flushing with helium or hydrogen. A surface which monitors clean may become contaminated again in a matter of hours by percolation. There is no practical way of removing tritium oxide (T20) from water due to its similarity to natural water.

URANIUM

5-73. The most probable source of uranium contamination is a nuclear weapon accident in which the fissionable uranium is spread as a metal or oxide dust. The common isotopes of uranium contamination are uranium-235 and uranium-238. This metal or oxide is insoluble and is best removed from a contaminated surface by brushing or vacuuming, followed by a treatment with mineral acids or oxidizing acids and then sealed. Large-area uranium contamination is best decontaminated by removing the top layer of the surface or by sealing it.

Chapter 6

Effects of Environmental Conditions

US forces may find themselves anywhere in the world and subjected to the conditions in that region. Weather and terrain conditions will dictate how to fight and what equipment is most effective to support the fighting force under NBC conditions. The conditions in cold- and arctic-weather areas, warmed areas, urban areas, mountains, jungles, and deserts impact how decon operations are performed.

COLD- AND ARCTIC-WEATHER AREAS

6-1. These regions comprise about 45 percent of the North American continent and about 65 percent of the Eurasian land mass. They are characterized by extreme cold weather and deep snow during winter months. Spring breakup and mud restrict mobility. Whiteout and grayout cause loss of depth perception, making vehicle movement hazardous. Ice fog (clouds of ice crystals) covers troops, vehicles, bivouac areas, and permanent facilities, marking their location. Chemical agents become more persistent in low temperatures (see Table 6-1, page 6-2). As a result, some normally nonpersistent chemicals, such as sarin (GB), become persistent; this nerve agent could remain a transfer hazard for up to 30 days in arctic climates. Normally, persistent agents become more persistent. Although frozen agents do not present a significant problem in solid state, they become hazards when they warm up. Decon is a problem because low temperatures also reduce the effect of decontaminants.

6-2. The freezing points of decontaminants are as follows:

- Water, 32°F.
- STB slurry with antiset, 0°F.
- DS2, -25°F.
- SKD, -50°F.
- IEDK, -50°F.

DECONTAMINANTS

6-3. When temperatures reach 32°F, water can no longer be used as a decon. In such situations, use DS2. DS2 is effective down to -25°F. For temperatures below -25°F, use STB or HTH as a dry mix (two parts of STB to three parts of earth or snow) for decon. Apply the dry mix by shoveling it on contaminated surfaces or by filling sandbags with the mix and dusting it on the surfaces. Remove the dry mix by brushing, scraping, or using uncontaminated earth or snow to "wash" it off. Other decontaminating methods using nonstandard solvents and fuels may be used, but observe fire safety, protect personnel from corrosives, and take precautions against super-cooling effects. Because of their low freezing points, solvents such as aviation fuel (JP8), diesel fuel, and kerosene may be used to physically remove contamination. These solvents only flush the agent from the surfaces. They generally do not neutralize agents nor do they eliminate agents that soak into surfaces. Nonstandard solvents generally are very flammable and must be handled with care. See Appendix B for specific nonstandard solvents.

Table 6-1. Freezing Points of Selected Chemical Agents

Agents	Symbols	Contact Hazard	Vapor Hazard	Freezing Point
		Nerve		
Tabun	GA	Extreme	Low-moderate	+23°F
Sarin	GB	Extreme	Extreme	-69°F
Soman	GD	Extreme	Probable	-44°F
	GF	Extreme	Probable	-22°F
	VX	Extreme	Negligible	-60°F
	VR-55	Extreme	Probable	Unknown
	TGD	Extreme	Probable	Depends on the percent of thickener
		Blister		
Distilled mustard	HD	Extreme	Negligible	+57°F
Mustand-lewisite	HL	Extreme	Low	-14°F (Pure) -44°F (Plant purity)
Nitrogen mustard	HN-1	Extreme	Low	-29°F
Nitrogen mustard	HN-2	Extreme	Low	-76°F
Lewisite	L	Extreme	Negligible	0°F
Nitrogen mustard	HN-3	Extreme	Low	+25°F
Phosgene oxime	CX	Extreme	Low	+95°F
		Blood		
Hydrogen cyanide	AC	Low	Extreme	+8°F
Cyanogen chloride	CK	Low	Extreme	+20°F
Arsine	SA	Low	Extreme	-177°F
		Choking		
Phosgene	CG	Slight	Extreme	-198°F
Diphosgene	DP	Slight	Extreme	-71°F

EQUIPMENT-DECON DEVICES

6-4. Discussed in the following paragraphs are the different types of equipment decon devices:

M11 Decon Apparatus

6-5. This apparatus is designed to dispense DS2 at temperatures as low as -15°F. DS2 is effective down to -25°F, but it becomes thicker as the temperature drops. Normally, one nitrogen cylinder is used to charge the M11. In cold weather, a second nitrogen cylinder may be used if inadequate pressure is developed with one. Never use more than two nitrogen cylinders.

M13 Decon Apparatus

6-6. This apparatus is designed to dispense DS2 at temperatures as low as -15°F. As the temperature drops, the flow of decontaminants through the hose and wand becomes degraded.

M17 Lightweight Decon System and M12A1 Decon Apparatus

6-7. Both apparatuses have problems in the cold similar to any other water pump. Normal engine "cold-soaking" problems that cause starting difficulties and brittle metal parts have been observed. Also, internal pumps and lines will crack and break when water freezes within these systems. To avoid freeze-up problems, use the shutdown procedures described in TMs 3-4230-228-10 and 3-4230-209-20&P.

DETECTION AND WARNING DEVICES

6-8. In the following paragraphs, the different types of detection and warning devices are discussed:

M8/M9 Detector Paper

6-9. M8/M9 detector paper is limited in the cold because only agents in a liquid form can be detected. If the substance is thickened or solidified, collect a sample with a stick or scraper and wipe it onto a sheet of M8/M9 detector paper. Place the sample on a heated surface, such as an operating vehicle or a power generator, to thaw the suspected agent. Because of off gassing, do not do this in a heated vehicle or tent. Put a box or can over the paper to concentrate any vapors. Put an M256A1 detector kit inside the box or can. Heat the suspected agent enough to enable detection and allow agent identification.

M256A1 Detector Kit

6-10. As temperatures drop toward 32°F, the M256A1 detector kit takes longer to give a positive or negative indication. When temperatures reach freezing, carry the liquid components in the breast pockets of the chemical protective clothing. It may also be necessary to put the test sampler back into the shipping bag after sampling and place it on a heated surface to hasten the reaction. The kit will not work at temperatures below -25°F.

Chemical-Agent Monitor

6-11. The CAM will have difficulty in detecting agents at low temperatures. Contaminated personnel entering a building or a collective protection shelter will desorb agents as they warm up. Use the CAM in a vestibule or an air lock where the temperature has been raised to that of the building's or shelter's interior. Likewise, check aircraft and vehicles requiring maintenance as they pass through a warmed-up site before entry into a hanger or maintenance bay.

M8A1 Automatic Chemical-Agent Alarm

6-12. When operating an M8A1 ACAA at temperatures below 20°F, use the M253 winterization kit. Warm-up time with this kit is 50 minutes.

EQUIPMENT-DECON STATIONS

6-13. Place decon sites in built-up areas and near road junctions and intersections of forest lanes to facilitate access. Divide the decon sites and approach routes into sectors, and assign units to keep them open in spite of mud or heavy snow. With present technology, equipment decon problems in the field are difficult to overcome in an arctic climate. Commanders should consider fighting dirty in arctic regions. Fresh units can be rotated into the field so that dirty units can be moved back to built-up areas to decon.

6-14. Decon apparatuses and water trucks may have to be deployed with empty tanks instead of full ones as in temperate climates. To prevent freezing, it may be necessary to preheat water when loading the water trucks and the tanks of the decon apparatuses and keep it heated until it is used. These vehicles may have to be enclosed and warmed so that the engines will start. Decon apparatuses and water trucks must be drained immediately after use to prevent damage from freezing. Commanders may choose to conduct minimum decon until they can get back to fixed facilities and conduct a more thorough decon.

6-15. Vehicles and personnel covered with contaminated snow must be decontaminated before the snow has a chance to melt and freeze. Such snow forms layers of ice that make contamination difficult to remove. Radioactive fallout that is mixed with snowfall must be removed as soon as possible. Use tree branches (if available) to remove contaminated snow.

6-16. Snow can be used to cover contamination; however, the snow can blow away or the contamination can resurface when tracked vehicles, troop movement, or digging disturbs it. Snow cover provides some protection if left undisturbed, but this protection is too uncertain to rely on.

WARMED AREAS

6-17. Preventing contamination from entering warmed areas is a challenging problem. For example, if soldiers get frozen agents on their clothing, it will be hard to detect because low temperatures have slowed its effects. However, if the temperature warms or if the soldiers enter a warmed area, the agents become dangerous. Because of this, it may be necessary to set up a warming station for each warmed area to isolate the agents. Otherwise, all occupants may be subjected to hazardous liquids and vapors.

6-18. Additional soldiers and equipment are needed for these warming stations. Therefore, in cold-weather operations, the decon and detection priority must be heated support facilities. FM 3-4 discusses individual and collective protection in detail.

URBAN AREAS

6-19. The need for decon operations in urban areas may also exist, and it may be easier to support. Water sources are generally available and commercial chemicals may be used as decontaminants. Site security may be simplified because of limited observation and poor fields of fire. Decon operations in urban areas will not differ significantly from similar operations in the field; however, keep the following considerations in mind:

STREETS AND STRUCTURES

6-20. When buildings are contaminated with persistent chemicals, their value for cover, concealment, and shelter is reduced. Wood and concrete tend to absorb liquid agents, and they may give off toxic vapors for days or weeks. Building decon is very difficult and requires large quantities of decontaminants. The hazard can be reduced by covering the contamination with plastic sheets, STB slurry, sodium silicate, or other substances that cover or absorb the agent. Even though a particular part of a building is not intended for occupation, it may still need to be decontaminated to prevent the contamination from spreading.

6-21. Streets and sidewalks also absorb liquid agents, then give off toxic vapors when heated by the sun. These surfaces may need to be decontaminated several times to reduce hazards. Streets, sidewalks, or other porous surfaces are best decontaminated by weathering if the time and the situation permit.

CIVILIANS

6-22. Try to locate decon operations away from civilians because they may injure themselves or interfere with the work. Civilians must not be forced to prepare decon sites or to help with decon operations. However, volunteers who have been cleared by the Assistant Chief of Staff, G5 (Civil Affairs) (G5) may be used.

SUPPLIES AND AMMUNITION

6-23. Store ammunition, as well as other supplies, under cover to protect it from NBC contamination. FM 3-4 explains contamination covers in detail.

SANITATION SYSTEMS

6-24. Urban areas may have sophisticated sanitation systems. When those systems are destroyed, sanitary conditions become far worse than those in areas where sanitation systems have never existed. Sanitation systems must be maintained to avoid epidemics and to avoid overloading biological-decon capabilities. Contaminated water and residue must be controlled so that they will not create a hazard to the civilian population or interfere with sanitation systems. Engineer support may be necessary to construct controlled runoff areas.

MOUNTAINS

6-25. Excluding the extremely high, alpine-type mountains, most mountain systems are characterized by heavy woods or jungle, compartments and ridge systems, limited routes, and highly variable weather conditions. All these factors will affect decon operations.

MOBILITY

6-26. The terrain and the disruption of existing routes may dictate that mountain operations be decentralized. Additional water-carrying vehicles may be needed to support these operations. Forces must be organized to be self-sufficient and should have decon elements attached to them initially. Decon platoons are attached to brigades and decon squads are attached to battalions. When decon elements operate independently, their parent chemical companies may not be able to provide support. Decon elements must be equipped with greater hauling capabilities, especially for water. Administrative and logistical support to the decon element will have to come from the support unit.

WIND

6-27. Mountains have changing weather and constant winds that promote weathering at a faster rate. In high mountains, calm winds in the ridges and passes are seldom; whereas, in protected valleys, strong winds are rare. Normally, wind velocity increases with altitude. Generally, downwind contamination will be greater because high winds are more frequent. However, variable winds will make predictions unreliable.

TEMPERATURE

6-28. Cool or cold temperatures have an adverse effect on decon operations. With every 328-foot gain in elevation, the temperature will normally drop 3^{0} to 5^{0}F. The drier the air, the more pronounced this drop will be. At high altitudes, there may be differences of 40^{0} to 50^{0}F between the temperature in the sun and that in the shade. Consequently, the temperature rises swiftly after sunrise and drops quickly after sunset. Low temperatures that normally exist at extremely high altitudes may demand decon procedures and precautions similar to those used in cold-weather operations. Daytime operations can be scheduled to avoid some low-temperature problems.

SUNLIGHT

6-29. Sunlight destroys most biological agents. Above the timberline, there is little protection from the sun, so the effectiveness of a biological attack is reduced.

JUNGLE

6-30. Jungle climates vary with the location. Close to the equator, all seasons are nearly alike, with rains throughout the year. Farther from the equator, especially in India and Southeast Asia, jungles have distinct wet (monsoon) and dry seasons. Both zones have high temperatures (averaging 78^{0} to 98^{0}F), heavy rainfall (as much as 400 inches annually), and high humidity (90 percent) throughout the year.

6-31. Transportation is extremely difficult due to natural obstacles; therefore, decon operations will be decentralized as much as possible. Because of high humidity and temperatures, MOPP gear should not be worn for long periods. Since decon crews will be more susceptible to heat stress, staffs should plan for frequent crew rotations and provide enough decon personnel to conduct decon operations.

6-32. Contamination will be retained temporarily in the jungle canopy, reducing the immediate hazard. Later, rains will wash these particles to the ground and concentrate them in low areas. These areas are likely to become contamination "hot spots." Decon sites should be put on high ground during the rainy season.

6-33. Tropical rainstorms will flood decon sites unless the sites are adequately drained. Ground that appears firm may become impassable when it rains. Rain and intense heat will speed the deterioration of many supplies. Do not break the seals on packing material until it is absolutely necessary.

6-34. Rain and high humidity (wetness) degrade the protective abilities of the overgarment. Take care to ensure the maximum length of time of the overgarment's usefulness. Additional quantities of overgarments must be maintained to replace those that have gotten wet. Rain helps the decon process by washing away contamination on exposed surfaces. Rain can also hydrolyze some agents. However, runoff may contaminate the soil.

TEMPERATURE AND HUMIDITY

6-35. When temperatures rise to the 85^{0} to 100^{0}F range, troops can only continue medium or heavy workloads by reducing their MOPP level. More soldiers will be needed to make up for extra rest periods and more frequent rotation periods during decon operations.

6-36. Frequently oil exposed metal parts and grease wheel bearings to protect them from moisture. Since petroleum, oils, and lubricants (POL) tend to absorb chemical agents, additional decon may be needed. Measures taken to protect electronic communications equipment from moisture will reduce the need for decon since these measures provide good NBC protection. Seal radios with silicone compound and pack them in waterproof containers when they are not in use (remove batteries). Protect handsets and microphones with plastic bags and check the bags frequently for moisture buildup.

6-37. The jungle's high humidity and temperature increase the need for maintenance. Contact teams must operate independently as they move from site to site doing repairs. There may be no decon specialists to help them.

6-38. Biological agents thrive in the heat, humidity, and shade, which are characteristics of the jungle. Therefore, weathering is not a practical means of biological decon.

6-39. Solid decontaminants, such as STB powder, tend to cake and decompose at a faster rate than in temperate climates. Caking is no problem, but the decomposition eventually makes STB powder ineffective.

TIME

6-40. Many tasks in the jungle take more time than in other environments. More time for movement and security must be allowed, which may leave less time for planning and preparation. In the jungle, more decon personnel will be required to perform decon operations.

SECURITY

6-41. Decon stations are lucrative targets; therefore, security must be a prime consideration. Relocating a decon site may require a small security force, thus additional personnel will be required.

RESUPPLY

6-42. Because of the terrain, aerial resupply usually will be common practice. Keep this in mind when determining the amounts of decontaminants to keep on hand since they will compete with ammunition, POL, medical supplies, food, and water for priority of delivery.

PERSISTENCY

6-43. Sunlight and wind are reduced within thick jungles due to the vegetation. As a result, chemical and biological agents are less likely to disperse and will remain a hazard for an extended time.

DESERT

6-44. Desert regions are characterized by extreme temperatures with ranges varying between 30° to 130°F over a 24-hour period. These regions have long drought periods, which are interrupted by sudden rains that bring flash floods. There are shortages of suitable groundwater. Large areas suitable for tracked-vehicle maneuvers may sometime have impassable ravines; wet, spongy grounds; and sand areas.

WATER

6-45. The principal problem for decon in the desert is the lack of water. The use of STB slurry and DS2 will burden the logistical system because of the water required for mixing and rinsing. Nonwater-based decontaminants (natural solvents) may be required (see Appendix B of this manual). Contamination avoidance becomes increasingly important in desert operations because of limited water sources. Therefore, maximum use should be made of field-expedient covers, especially at night when an attack by chemical or biological agents is most likely.

HEAT STRESS

6-46. Heat stress is a critical problem for soldiers working in the desert environment in full MOPP. Operating decon stations in daytime temperatures may require short periods of work followed by long periods of rest. Performing operations at night to avoid heat stress may create light discipline problems.

WEATHERING

6-47. Weathering is a viable decon option. High daytime temperatures can increase evaporation of liquid contamination. As a result, vapor concentrations will be high but may not last long. Refer to FM 3-4 for MOPP opening/unmasking guidance to estimate the chemical hazards resulting from enemy attacks. If liquid contamination soaks into soft, porous soil (such as loose sand), evaporation is not as quick. Strong winds also increase the evaporation rate. Low temperatures during the night have a reverse effect and tend to increase the persistency of chemical and biological contamination. The sandblasting effect of sandstorms may remove contamination from surfaces facing the storm. Desert sunlight and high temperatures will destroy many chemical and biological agents without additional decon measures.

BEARINGS AND OTHER CRITICAL MOVING PARTS

6-48. Bearings and other critical moving parts need extra lubrication in the desert. This complicates decon because lubricants tend to absorb chemical agents. After a sandstorm, maintenance must be conducted regardless of the last scheduled maintenance. Therefore, perform decon first and if time is critical, decon only those surfaces that will be touched during maintenance. This will not eliminate vapor hazards.

Chapter 7

Aviation

Contamination in the battle space causes a unique problem for aviation. Since an aircraft uses its high mobility to maneuver across the entire battle space, avoiding contamination can be very difficult. Once an aircraft is contaminated, it is very difficult to decon. Therefore, aviation units need to prepare effective standing operating procedures (SOPs) that emphasize contamination avoidance, protection, and methods to decon each type of aircraft using an immediate, an operational, and a thorough decon.

CONTAMINATION AVOIDANCE

7-1. Contamination avoidance is critical due to the difficulty in decon. Aircrews must be aware of NBC hazards. For instance, an aircraft that hovers or lands in a contaminated area could splash contamination onto itself. Likewise, contaminated passengers or ground crews pose a danger of transferring contaminants into the aircraft where they are difficult to remove. Aircraft could also possibly pick up some contamination by flying through an invisible vapor cloud. Hazards that will only incapacitate ground personnel for a short time can be fatal in the air.

7-2. Some things that can be done to limit contamination are as follows:

- Know what areas are contaminated and avoid these, if possible.
- Pick landing zones that will have a reduced splash effect if an aircraft must land in a contaminated area.
- Limit the spread of contamination into the aircraft from the outside. For example, ground crews at the forward arming and refueling point (FARP) could conduct arming and refueling without requiring the aircrew to exit the aircraft.
- Conduct inspections without touching or shaking items. The contaminated crews should perform this task. Many inspection points can be inspected visually.
- Increase the use of covers when not flying. Use engine covers, flyaway gear, and hatches. If possible, provide overhead cover for parked aircraft.
- Limit the number of aircraft that must operate in a contaminated area or use aircraft already contaminated.
- Line the troop compartment with plastic (a field expedient way to limit the spread of contamination) when carrying contaminated personnel or casualties. A plastic curtain can be fastened between the troop compartment and the flight compartment with tape or Velcro to limit contamination transfer. The aircraft's heater can be used with

the curtain to create an overpressure in the pilot's compartment. This will limit vapors from entering the compartment.

- Apply M9 detector paper to the landing gear of the aircraft. FARP personnel should always check the M9 detector paper before servicing the aircraft. Another piece of M9 detector paper can be placed on the windscreen where the aircrew can see it.

CAUTION

M9 detector paper will show false positives if exposed to petroleum products.

DECONTAMINANTS

7-3. Only approved cleaning compounds may be used to decon aircraft. Caustic decontaminants such as DS2, STB, bleaches, and sodium hypochlorite are not considered safe. STB corrodes metal components and the aircraft's skin, and DS2 corrodes rubber, plastic, and acrylic-plastic windows.

7-4. Soap and water, kerosene, JP8, and diesel fuels are approved as decontaminants on selected parts of the aircraft. JP8 is effective in removing some agents from the aircraft's skin and components; however, it does not neutralize the agents. Personnel must use care when handling JP8. When using a cloth soaked in JP8, personnel must avoid wiping internal components near the exhaust. If water is available, use it to rinse off the JP8. Many portions of the aircraft are delicate and cannot stand high-pressure water or extreme hot air.

7-5. Sodium carbonate is a fair decontaminant against G nerve agents, but it is not effective against V nerve and blister agents. The chemical reaction of the V nerve agent and sodium carbonate is extremely slow and produces a product that is very toxic. Most of the field-expedient decontaminants are corrosive and could cause damage to the aircraft's skin or acrylic-plastic windows.

AIRCRAFT DECON

7-6. When an aircraft becomes contaminated, the aircrew must fly in MOPP gear. However, wearing MOPP gear degrades the aircrew's ability to perform its jobs over time.

7-7. If air and ground crews are careful when operating in a contaminated environment, contamination can be contained on the exterior of the aircraft. The increased airflow over the aircraft's skin increases the rate of evaporation. Still, some agents will remain and continue to be a hazard. Thickened agents evaporate more slowly and may remain a hazard even after prolonged flights. If the interior is contaminated, flying the aircraft with the doors open can help reduce the hazard. The heat from the running engine decontaminates the interior of the engine compartment. However, more active methods of decon may be needed. Use the immediate-, operational-, or thorough-decon method with modifications to perform an aircrew spot decon, an aircraft washdown, and the DAD.

DECON TYPES

7-8. Table 7-1 shows the types of aircraft decon. The crew of a contaminated aircraft will use the techniques described in previous chapters to decon. Unit personnel conduct the aircrew spot decon and aircraft washdown, but a chemical decon unit usually conducts the DAD.

Aircrew Spot Decon

7-9. After personal wipe down, aviators may conduct a spot decon on their aircraft (see Table 7-1). The aircrew spot decon provides the same benefits as the operator's spray down. It limits the transfer and spread of contamination by decontaminating the surfaces that must be touched during operations. Spot decon applies to the first six functional activities. The goal is to limit the transfer and spread of contamination while sustaining flight operations.

Table 7-1. Aircraft Decon Types

Decon Types	When, Why, Who Performs Decon	Areas to Decon	Types of Decon Agents
Aircrew spot decon	The crew decontaminates after the immediate decon to allow continued operations	Essential operating surfaces on the aircraft	IEDK, JP8[1]
Aircraft washdown	The unit decon PDDE crew or decon unit performs decon within 6-24 hours to reduce contact hazards[2]	The entire exterior surface of the aircraft[3]	Hot, soapy water
DAD	The decon units decon when the mission allows to reduce the hazard to negligible risk levels	The entire exterior surface and selected interior surfaces of the aircraft	10 percent sodium-carbonate solution (10 pounds of sodium carbonate to 12 gallons of water)

[1]JP8 should not be used inside the aircraft.

[2]An aircraft washdown is most effective if conducted within 1 hour of contamination.

[3]An aircrew spot decon is performed to reduce contact hazards inside the aircraft. Do not spray water inside the aircraft.

7-10. Air or ground crews may perform the aircrew spot decon. Wash exterior surfaces with decontaminants to flush off contamination. Fuel, soap, and water are most commonly used. Exact procedures and areas to decon are identified in the aircraft's maintenance manual. See Table 7-2, pages 7-4 through 7-6, for guidance on the aircrew spot decon.

Aircraft Washdown

7-11. An aircraft washdown site consists of a single station; with other decon sites, there may be several stations. Before arriving at the site, the aircrew should fly its aircraft at high speeds to help evaporate exterior contamination. If the aircraft's interior has been contaminated, open the doors; however, before arriving at the washdown site, close the doors. At the washdown site, the aircraft will land and reduce its speed to an idle. If the aircrew is contaminated, it should exit the aircraft and perform a MOPP-gear exchange.

Table 7-2. Aircraft Spot Decon

Actions and Locations	Areas to Decon	Personnel Responsible	Decontaminants	Procedures	Remarks
Refueling at FARP	Fuel ports, hatches, and all areas that FARP personnel touch	POL handler	Diesel fuel; JP8; or hot, soapy water	Wipe fuel ports and hatches with a sponge dipped in the decontaminant. Do not allow the decontaminant to enter the fuel system. Control runoff because the agent will not be neutralized. This method simply flushes contamination from the surface.	FARP personnel should conduct an operational decon before servicing aircraft if they are contaminated.
Arming at FARP	Armament system	Ammo handler	JP8 or hot, soapy water	Wipe the armament system with a sponge dipped in the decontaminant. Control runoff because the agent will not be neutralized. This method simply flushes contamination from the surface.	**CAUTION** **Take care to prevent certain areas of the armament system from being exposed to the decontaminants.** The TM on the armament system should be checked to know what areas to avoid.
Entering and exiting aircraft anywhere	Door handles, steps, ladders, handholds, and all other areas that the aircraft crew is likely to touch	Crew members and FARP personnel	JP8 or hot, soapy water	Apply the decontaminant. Take care to prevent the spread of liquid contamination from the outside of the aircraft to the inside. This includes controlling the contaminated runoff. This method simply flushes contamination from the surface. Precautions might include having the crew chief do the decon, which reduces the chance of contamination spreading into the pilot and copilot compartment. All crew members should diligently practice NBC avoidance.	Procedures should be developed for each type of aircraft. Before entering the aircraft, use an IEDK on boots and gloves. FARP personnel can decon the area for exiting aircrew members.

Table 7-2. Aircraft Spot Decon (Continued)

Actions and Locations	Areas to Decon	Personnel Responsible	Decontaminants	Procedures	Remarks
Preflight and postflight checks anywhere. Maintenance inspections at aviation intermediate maintenance (AVIM) facilities	Areas that must be touched as part of the inspection	Aircrew	Diesel fuel; JP8; or hot, soapy water on exterior surfaces. Use hot air on interiors or areas that are not compatible with decontaminants or liquids.	Wipe areas that are required to be touched for preflight and postflight checks with a sponge dipped in the decontaminant. Wash gloves in a decontaminant before touching uncontaminated surfaces. Decon gloves with an IEDK after the inspection is completed. Control runoff because the agent will not be neutralized. This method simply flushes contamination from the surface.	Overheated air should not be used directly on instrumentation. Crews may want to wear wet-weather gear to keep most of the contamination off overgarments.
Repair and recovery anywhere	Parts or areas that will need to be decontaminated depending on the situation	Battle-damage repair team	Diesel fuel; JP8; or hot, soapy water on exterior surfaces. Use hot air on interiors or areas that are not compatible with liquid decontaminants.	Decon only those parts that need to be touched during repair.	Contaminated aircraft should not be brought into an uncontaminated area.
Cannibalization anywhere on the battle space	Parts or assemblies being cannibalized	Maintenance personnel (someone who knows what assembly or parts are needed)	DS2; diesel fuel; 5 percent chlorine solution; or hot, soapy water on exterior parts or surfaces. Use hot air on interior parts that are not compatible with liquids or corrosive decontaminants.	Decon only those parts or assemblies that need to be touched during cannibalization.	Caustic decontaminants should be used only on those areas that have been removed from the aircraft. Assemblies must be rinsed thoroughly before replacing.

Table 7-2. Aircraft Spot Decon (Continued)

Actions and Locations	Areas to Decon	Personnel Responsible	Decontaminants	Procedures	Remarks
Overhaul at AVIM facilities	All areas and equipment required to be worked on during the overhaul	Maintenance personnel and/or a chemical decon platoon from division or corps	DS2; diesel fuel; 5 percent chlorine solution; or hot, soapy water. Surfaces that are easily destroyed by liquid or corrosion should be decontaminated using hot air.	Wash with diesel fuel and then hot, soapy water. Then rinse. Check for contamination with the M8 detector paper or the CAM. If time permits, allow the equipment to weather to reduce chemical contamination to negligible risk levels.	Caustic decontaminants should be used only on those assemblies that have been removed from the aircraft. Assemblies must be rinsed thoroughly before replacing. This process has the same results as a deliberate decon.

To remove the BDU, follow the procedures for removing the protective overgarment as described in paragraphs 8-124 and 8-125. Do not remove the patient's identification tags.

Step 5 - Decon the Patient's Skin

8-130. **Spot decon.** Wash exposed skin surfaces with soap and warm water, to include areas around the neck, wrists, and lower parts of the face. Decon the patient's identification tags and chain, if necessary.

8-131. **Combat medic care.** The combat medic gently cuts away the bandage. He decontaminates the area around the wound and irrigates it with a 0.5 percent chlorine solution. If bleeding begins, he replaces the bandage with a clean one. He replaces the old tourniquet by placing a new one ½ to 1 inch above the old one. He then removes the old tourniquet and decontaminates the patient's skin with a 0.5 percent chlorine solution. He does not remove a splint. He decontaminates the splint by thoroughly rinsing it, to include the padding and cravats, with a 0.5 percent chlorine solution.

> **WARNING**
> **Do not use an SDK around any wounds. Do not remove splints. The splint will not be removed until the patient has been evacuated to a forward MTF or hospital.**

8-132. **Completeness of decon check.** Check the patient with M8 detector paper or the CAM for completeness of decon.

NOTE: Other monitoring devices may be used, if available.

8-133. **Contaminated-waste disposal.** Dispose of contaminated bandages and coverings by placing them in a contaminated-waste bag. Seal the bag and place it in the contaminated-waste dump.

Step 6 - Transfer the Patient to the Hot Line

8-134. Move the patient to the hot line. Two medics from the clean side of the hot line move the patient into the clean treatment area.

NOTE: Before decontaminating another patient, each decon team member drinks about half a quart of water. The exact amount of water consumed is increased or decreased according to the work level and temperature (see Table 8-2, page 8-11).

AMBULATORY PATIENT

8-135. The decon procedures for an ambulatory patient that is contaminated with radiation are discussed in the following paragraphs. Some procedures can be done with one soldier, while others require more than one.

8-136. All ambulatory patients requiring EMT or AMT in the clean area of the BAS will be decontaminated. Stable patients not requiring treatment at the BAS, but requiring evacuation to the medical company's clearing station or a corps hospital for treatment (for example, a patient with a broken arm), should be evacuated in their protective overgarments and masks by any

where it will pose a vapor hazard and be difficult, if not impossible, to decon later.

NOTE: Many (but not all) aircraft have been painted with CARC paint. Aircraft will be painted with CARC paint when sent to the appropriate maintenance shop. Contamination stays on top of CARC surfaces where it can be more easily neutralized or removed.

7-14. An aircraft washdown should be conducted within 6 hours and is most effective if conducted within 1 hour. Some amounts of chemical contamination may remain after the aircraft washdown. The aircrew should continue to wear, as a minimum, its protective masks and rubber gloves for protection until a deliberate decon can be conducted.

7-15. Presently, there is no timely detection method to check the effectiveness of decon for biological contamination. Medical personnel will monitor the aircrew's disposition if biological contamination is suspected.

Detailed Aircraft Decon

7-16. The goal of a DAD is to reduce contamination hazards to a negligible risk level so that air and ground crews can operate the aircraft without protective measures. The DTD can be conducted at the same time. The decon site should be set up in a similar fashion to the thorough-decon site (see Chapter 4). The decon site consists of the following areas: a predecon staging area, DAD and DTD areas, and a postdecon AA. For safety reasons, it is not possible to decon aircraft and ground vehicles at the same site. A separate DED site can be established in the vicinity to support the contaminated aviation unit's ground vehicles.

7-17. Aviation units must be supported by a chemical unit to set up and operate the DAD site. The chemical unit, assisted by the supported aviation unit, operates the DAD site while the aviation unit sets up and operates the DTD. See Chapter 4 for DTD operations.

7-18. Typically, a DAD is either performed as part of a reconstitution effort or before maintenance operations (after every 25 hours of flight time). When the aircraft is either overhauled or grounded for scheduled maintenance, contamination hazards should be reduced to negligible risk levels by going through a DAD. When components are removed from the aircraft for repair, care must be taken because of residual contamination. Decon these components either before cannibalization or overhaul.

7-19. Chemical units (decon and smoke/decon) are responsible for setting up, operating, and closing the DAD portion of the thorough-decon operation. The site setup is the same for all types of contamination (chemical, biological, and radiological). The DAD area is composed of the following five stations: primary wash, decon-solution application, contact time/interior decon, rinse, and check (see Figure 7-2).

DECON STATIONS

7-20. The actions that are taken at each of the stations are as follows:

Station 1 - Primary Wash

7-21. At this station, gross contamination is removed from the aircraft. Spray the aircraft for 2 to 3 minutes with hot, soapy water. Avoid hitting the

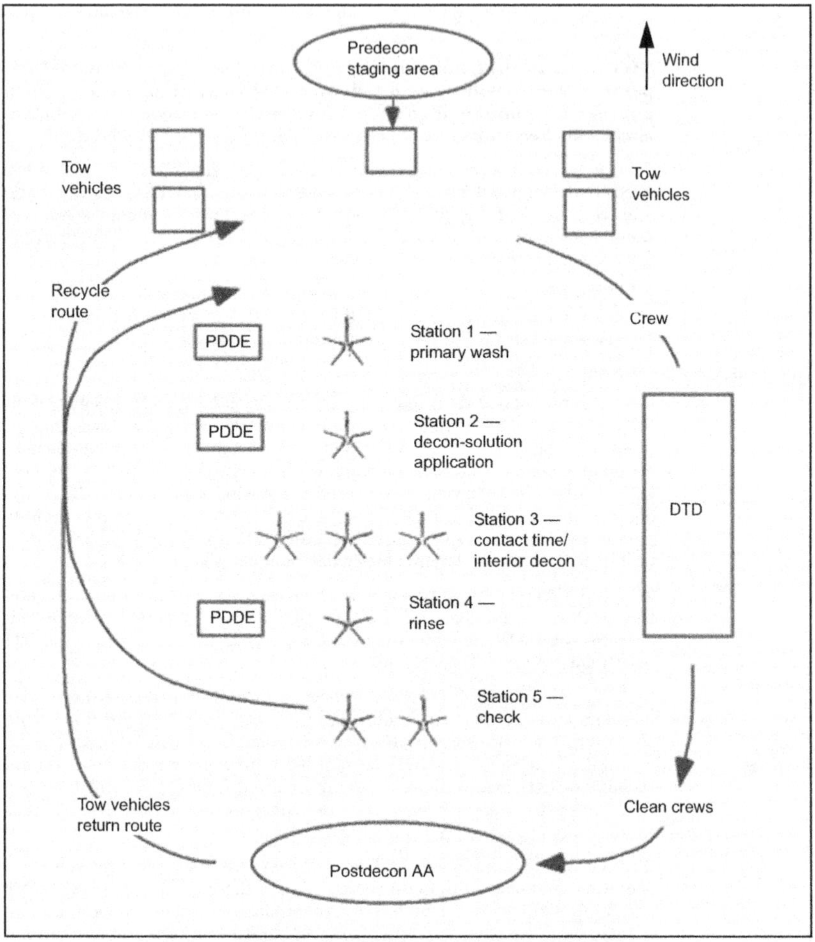

Figure 7-2. Optimum Layout for a DAD Station

aircraft's skin at a 90^0 angle to prevent damage. Also avoid the sensitive areas shown in Figures 7-3 through 7-6, pages 7-10 through 7-13.

Station 2 - Decon-Solution Application

7-22. At this station, a decon solution is applied using the M12A1 PDDE or the M17 LDS (in the siphinjector mode). After the solution is applied, a member of the scrubbing team scrubs the aircraft's skin. This allows the chemical agent

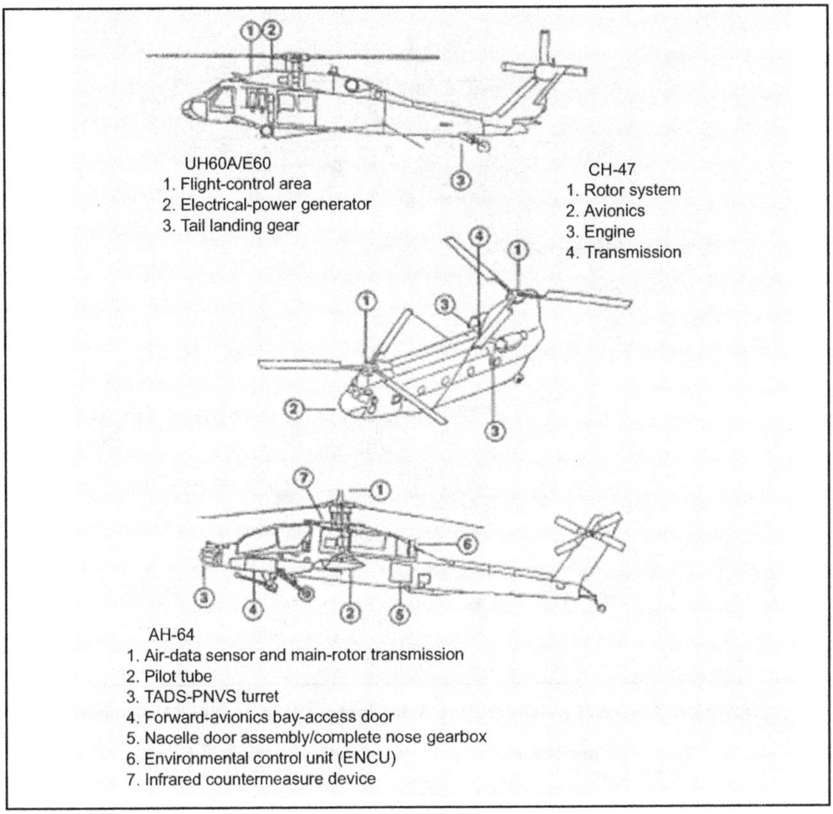

Figure 7-3. No Direct Water-Pressure Areas for the UH60A/E60, CH-47, and AH-64 Helicopters

to mix with the decon solution, especially when thickened chemical agents are present. For decontaminants, refer to Appendix B. STB and HTH should not be used because of their corrosive nature.

Station 3 - Contact Time/Interior Decon

7-23. At this station, the decon solution is allowed to completely neutralize the chemical agent; also, the interior of the aircraft is decontaminated, if necessary. The aircraft will remain in station 3 for no less than 30 minutes. The decon solution reacts with most chemical agents within 5 minutes. However, allowing the decon solution to remain on the contaminated surface for 30 minutes ensures a complete neutralization.

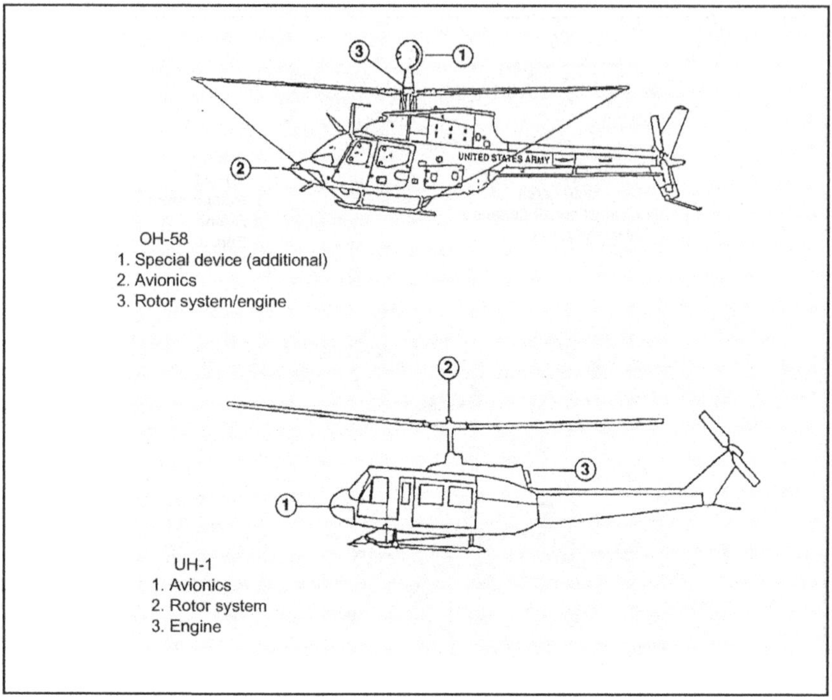

Figure 7-4. No Direct Water-Pressure Areas for the OH-58 and UH-1 Helicopters

7-24. While the aircraft is in this station, an attendant inspects the interior of the aircraft for liquid contamination using the CAM and M8 detector paper. If liquid-chemical contamination is identified, he will decon the interior of the aircraft.

WARNING
Personnel should not be in the aircraft during this operation.

7-25. Decon the aircraft's interior with hot, soapy water as long as maintenance personnel approve the use of it on certain pieces of equipment. Use IEDKs or damp, soapy washcloths to decon optical and electrical equipment and instruments.

NOTE: The IEDK leaves a charcoal residue on surfaces.

7-26. Field studies show the Herman-Nelson portable duct heaters to be effective in decontaminating chemically contaminated vehicle and aircraft

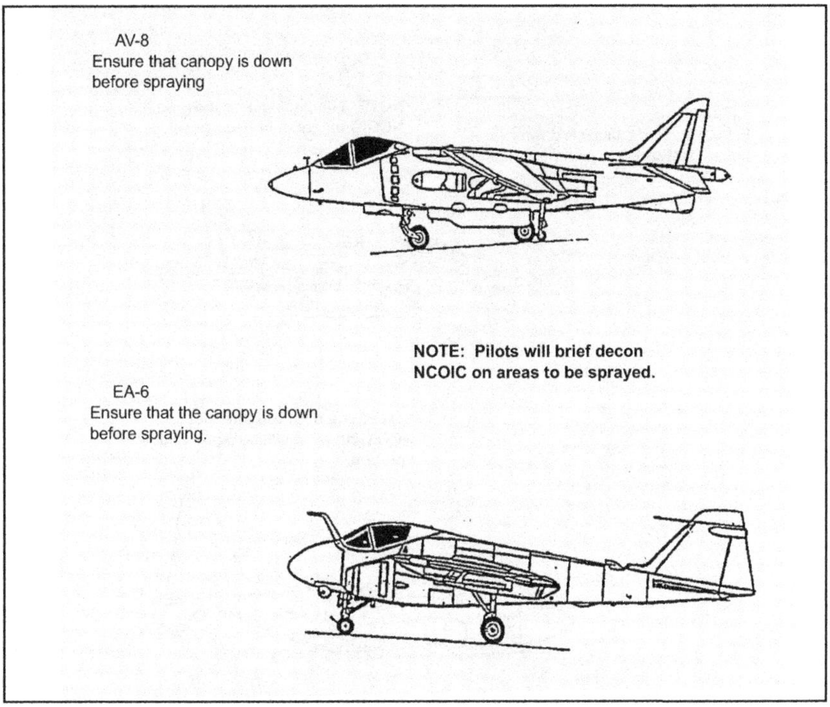

Figure 7-5. No Direct Water-Pressure Areas for the AV-8 and EA-6

interiors. However, using a hot-air decon has the following drawbacks: time required, heat damage, and downwind vapor hazards. A hot-air decon for a helicopter or light aircraft requires about 30 to 120 minutes to remove all of the contamination. Times are based on the amount of contamination and interior surface materials. Enough time must be provided to get the contaminated surface hot enough to vaporize the agent. The Herman-Nelson heaters discharge air at a temperature between 150^{0} and 280^{0}F. Pressurized containers, such as fire extinguishers, must be removed before starting a hot-air decon. These items will fail catastrophically and could cause severe damage. The heat will also damage sensitive electronic equipment. Additionally, chemical agents irreversibly bind to many materials, particularly gaskets, vinyl coatings, seat belts, and seat cushions. These items should be removed since they cannot be decontaminated. A hot-air decon does not destroy chemical agents. The air exiting the aircraft may contain hazardous levels of the agent. A downwind exclusion area of about 164 yards must be maintained to prevent unwanted casualties.

7-27. While a hot-air decon is an alternate to using hot, soapy water or an IEDK, the technique has many drawbacks. Maintenance personnel must be

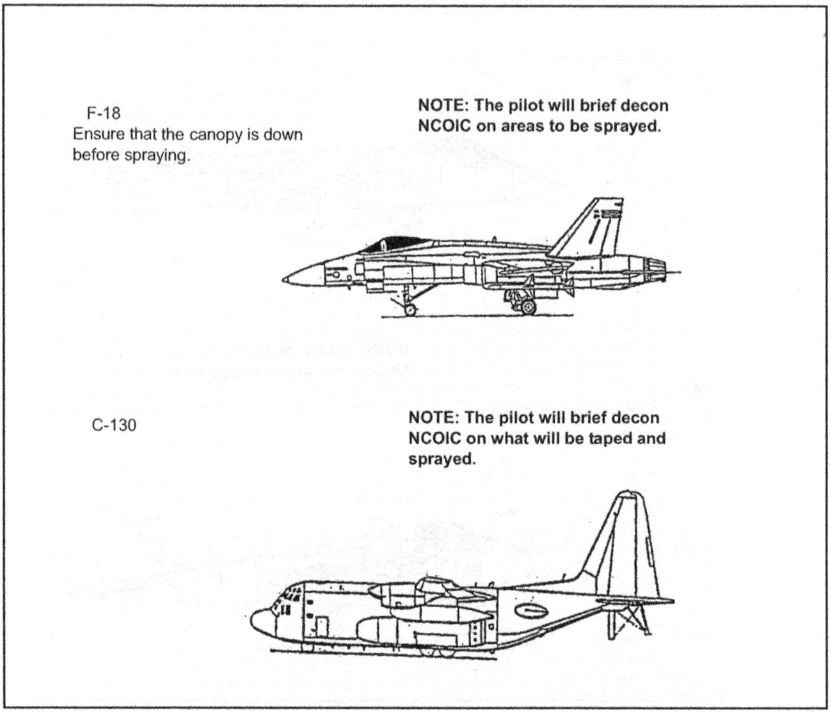

Figure 7-6. No Direct Water-Pressure Areas for the F-18 and C-130

consulted before using this technique to ensure that the aircraft is not damaged by the heat.

7-28. For radiological contamination, use a radiac meter to determine the extent and location of contamination inside the aircraft. If there is contamination, determine the intensity of the contamination. If the contamination has an intensity greater than 0.33 cGy, the interior of the aircraft must be decontaminated. Use hot, soapy water to wash the contaminated areas and a sponge to mop up the water and the contamination.

Station 4 - Rinse

7-29. At this station, the decon solution is removed from the aircraft. Spray the aircraft with water from top to bottom. Take care not to damage the aircraft's skin. This station will use about 250 gallons of water. Failure to remove all decon solution from the aircraft's skin may cause corrosion.

Station 5 - Check

7-30. At this station, the completeness of the decon is checked. Determine if the aircraft has a negligible risk or if it still has significant contamination remaining. Detection procedures vary depending on the type of contamination. If significant contamination is found on the aircraft, it will be recycled to station 2 for chemical contamination or station 1 for radiological contamination.

7-31. **Chemical.** Use the CAM to check for the presence of vapors. A one bar or lower reading on the CAM indicates a negligible risk. If the CAM indicates the presence of vapors, use M8 detector paper to check for liquid contamination. If it is suspected that these are producing a false positive, use an M256A1 detector kit to confirm or deny the presence of contamination. If the aircraft has significant contamination remaining, recycle it. Desorption of chemical agents may occur after the decon. On CARC-painted surfaces, the desorption of vapors will stop sooner than on alkyd-painted surfaces.

7-32. **Radiological.** Use the radiac meter to determine if any contamination remains. If so, determine the intensity of the contamination. If the contamination is greater than 0.33 cGy, recycle the aircraft to station 1.

RECYCLE CRITERIA

7-33. The commander, with the chemical unit leader's help, establishes the recycle criteria before starting a decon operation. If contamination is detected at station 5, the recycle criteria is used to determine which aircraft are returned to station 2 or, if radiologically contaminated, station 1. If the unit has sufficient time and resources available, any aircraft having more contamination than a negligible risk should be recycled. However, time and resources are usually limited and not all aircraft can be recycled.

AIRCRAFT DECON-SITE SELECTION

7-34. Operational- and thorough-decon sites for aircraft must be selected with care. The decon area must be big enough for the required number of aircraft, have a readily available water source, and allow for adequate drainage. The site should be relatively secure; however, it should be close enough to the AOs and FARP to allow for a reasonably quick turnaround of aircraft. The site must not have less than a 5 percent slope. Decon sites should be integrated into the tactical plan.

DECON-SITE LAYOUT

7-35. The equipment and personnel requirements for a DAD layout are identified in Table 7-3, page 7-16. Chemical units can establish other layouts based on METT-TC.

7-36. It may not be possible, for a variety of reasons, for a chemical platoon to use the optimum layout. Limited personnel or equipment will affect the layout of the thorough aircraft decon site. The platoon leader will use METT-TC and the concepts outlined in this chapter to establish a DAD station within his capabilities.

WARNING

Chemical unit leaders must consider the impact of the work/rest cycle on their operation to process aircraft. Failure to initiate a work/rest cycle could result in heat casualties and mission failure.

Table 7-3. Personnel and Equipment Requirements for an Optimum DAD Layout

Stations	Personnel		Equipment/Supplies
	Decon Platoon	Augmenter	
Station 1 - primary wash	1 squad leader 1 PDDE operator 2 sprayers	2 scrubbers	1 PDDE 1 3,000-gallon tank 2 65-GPM pumps 6 long-handled brushes 5 TAP aprons Liquid detergent
Station 2 - decon-solution application	1 squad leader 2 appliers 2 sprayers 1 PDDE operator	4 appliers	1 PDDE 18 long-handled brushes 9 mops with extra heads Liquid detergent
Station 3 - contact time/interior decon	1 NCO (CAM operator)	2 interior decon assis-tants	2 AN/VDR-2 or AN/PDR-77 3 TAP aprons 6 30-gallon containers 10 books of M8 detector paper 30 sponges 8 M256A1 detector kits 50 plastic trash bags 1 clipboard with a pen 1 stopwatch 1 CAM
Station 4 - rinse	1 squad leader 1 PDDE operator 2 sprayers		1 PDDE 1 3,000-gallon tank 3 65-GPM pumps 1 TPU 2 TAP aprons
Station 5 - check	1 NCO 2 CAM operators		1 CAM 10 M256A1 detector kits 20 books of M8 detector paper 2 AN/VDR-2 or AN/PDR-77 2 M8A1 or M22 ACAA
C^2	1 platoon leader 1 platoon sergeant		1 HMMWV 3 marking kits
Aircraft moving team		6 drivers 18 ground guides	
Total personnel	20	32	

Chapter 8

Patient Evacuation and Decon

Evacuating patients under NBC conditions forces the unit commander to consider to what extent he will commit evacuation assets to enter the contaminated area. Generally, if most or all of a supported force is operating in a contaminated area, most or all of the medical-evacuation assets will operate there also. If it is possible to keep some ambulances free from contamination, every effort should be made to do so.

EVACUATION CONSIDERATIONS

8-1. On the modern battle space, there are three basic modes of evacuating patients (personnel, ground vehicles, and aircraft). Personnel who physically carry the patients incur a great deal of inherent stress. Cumbersome MOPP gear, climate, increased workloads, and fatigue will greatly reduce the effectiveness of unit personnel.

8-2. If evacuation personnel are to be sent into an area that is contaminated with radiation, an OEG must be established. Radiation-exposure records must be maintained by the supported unit chemical NCO and made available to the commander, staff, and medical leader. Based on the OEG, the commander or medical leader decides which evacuation elements to send into the contaminated environment. Every effort is made to limit the number of evacuation assets that are contaminated; however, a number of these assets will become contaminated in the course of a battle. Therefore, optimize the use of resources, medical and nonmedical, which are already contaminated before employing uncontaminated resources.

8-3. Once a vehicle has entered a contaminated area, it is highly unlikely that it can be spared long enough to undergo a complete decon. This will depend on the contaminant, the tempo of the battle, and the resources available for casualty evacuation. Normally, contaminated vehicles (air and ground) will be confined to dirty environments.

8-4. Ground ambulances should be used instead of air ambulances in contaminated areas. They are more plentiful, are easier to decon, and can be replaced more easily. However, this does not preclude the use of aircraft, if required.

8-5. The patient's medical condition and the relative positions of the contaminated area, the forward line of troops (FLOT), and the threat's air-defense systems determine if and where air ambulances may be used in the evacuation process. One or more air ambulances may be restricted to the contaminated areas. To the greatest extent possible, use ground vehicles to cross the line separating the contaminated and clean areas.

8-6. After the patients are loaded, the medical-evacuation vehicles proceed to a medical-treatment facility (MTF) with a patient-decon station. (For the setup of a patient-decon station, see FM 8-10-7.) The patients are decontaminated and treated at this station. If further evacuation is required, transfer the patients to a clean ground or air ambulance. The routes that the ground vehicles use to cross between contaminated and clean areas should be considered dirty routes and should not be crossed by clean vehicles. The effects of the wind and time on the contaminants must be considered.

8-7. The rotor wash of the helicopters must always be kept in mind when evacuating patients. The intense winds will disturb the contaminants, increasing vapor hazards.

8-8. The helicopter should be allowed to land and reduce to a flat pitch before the patients are brought near. Additionally, the helicopter must not land too close to a decon station because any trace of contaminants in the rotor wash may be spread into the clean treatment area.

8-9. Evacuating patients must continue even under NBC conditions. The medical leader must recognize the constraints that NBC operations place on him, and he must plan and train to overcome these constraints.

PATIENT DECON

8-10. Patient decon presents special problems for units and combat-health-support (CHS) personnel. Under NBC conditions, contaminated wounded soldiers create increased hazards to rescuers and CHS personnel. In the following paragraph, decon procedures at unit level are discussed:

8-11. On the NBC battle space, two classifications of patients will be encountered—contaminated and uncontaminated. Those contaminated may suffer from the effects of an NBC agent, a conventional wound, or both. For more information on the treatment of NBC patients, see FM 8-9. Some may suffer battle fatigue or heat injuries induced by the stress of NBC conditions and the extended time spent in MOPP4. It is important to follow proper decon procedures to limit the spread of contamination. The most important decon is performed at the contamination site. Later decon may be too late to prevent injury. All agents should be promptly removed from the skin.

8-12. Patient decon must begin at the platoon and company level with the individual soldier. The soldier himself or members of his team must perform immediate decon and administer nerve-agent antidotes and convulsant antidotes, if required. Tag the patient with a DD Form 1380 or a field-expedient tag, noting the time and type of contamination (see Figure 8-1). If available, use the CAM or M8/M9 detector paper to determine the type and concentration of contamination. When the patient's condition and the battle permits, the patient may go through a MOPP-gear exchange (see Chapter 3 of this manual). The MOPP-gear exchange must not cause further injury to the patient.

PATIENT DECON AT AN MTF

8-13. In the following paragraphs, the types of MTFs are discussed:

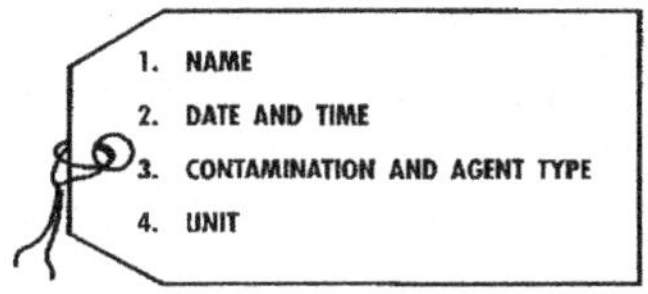

Figure 8-1. Field-Expedient NBC Patient Tag

BATTALION AID STATION (BAS)

8-14. When battle conditions prevent decon procedures forward, the patient may have to be evacuated to the BAS before decon. Contaminated patients arriving at the MTF must be decontaminated before admission into the clean treatment area.

8-15. Patient decon is the systematic removal of clothing and contaminants from a patient who is unable to decon himself. Patient decon is done by an 8-man, patient-decon team consisting of nonmedical personnel from the supported unit at the BAS. The team operates under the supervision of medical personnel to ensure that no further injury is caused to the patient by the decon process.

MEDICAL COMPANY'S CLEARING STATION

8-16. The medical company's clearing station may receive patients from the BAS or directly from other areas who may be contaminated. The clearing station must also have a patient-decon area. For more information on BAS operations under NBC conditions, see FMs 8-10-4 and 8-10-7. As with the BAS, the clearing station must have at least an 8-man, patient-decon team consisting of nonmedical personnel from the supported units to perform the decon. Procedures for patient decon at the clearing station are the same as for the BAS.

HOSPITAL

8-17. To the maximum extent possible, hospitals are located away from tactical or logistical targets. Patients evacuated from forward areas should already be decontaminated; however, contaminated patients may arrive from forward MTFs and units located within the geographical area of the hospital and require decon. Patient decon is done by 20 nonmedical personnel from units located in the geographical area/base cluster of the hospital.

8-18. If the hospital does not have CPSs and becomes contaminated with a persistent agent, patients are rerouted to other hospitals. All inpatients are evacuated, if possible, and the hospital is decontaminated. On completion of the decon, the hospital will return to normal operations.

8-19. A hospital with CPSs will decon the areas around the entry to these facilities, then continue receiving and caring for patients. Forward medical facilities and hospitals use the same patient-decon procedures. Several patient-decon stations can be operated at a hospital decon site. All patients arriving at the hospital will be considered contaminated. They must be decontaminated before being admitted into the clean areas. Perform decon as required.

CHLORINE-SOLUTION PREPARATION FOR PATIENT DECON

8-20. To decon a patient, use an SDK; however, if an SDK is not available, use a chlorine solution. If the chlorine solution is used, two concentrations of it are required. A 5 percent chlorine solution is required to decon gloves, aprons, litters, scissors, patient's hood, and other nonskin contact areas. A 0.5 (½) percent chlorine solution is required to decon the patient's mask, skin, and splints and to irrigate his wounds. To prepare the solutions, use calcium hypochlorite granules or sodium hypochlorite (household bleach) (see Table 8-1).

Table 8-1. Preparation of Chlorine Solution for Patient Decon

HTH Ounces	HTH Spoonfuls*	Household Bleach	Solution Percentage in 5 Gallons of Water
6	5	2 quarts	0.5
48	40	**	5

*These measurements are used when bulk HTH is used. To measure this preparation, use the plastic spoon supplied with your meal, ready-to-eat (MRE). Use a heaping spoonful of chlorine (all that the spoon will hold). Do not shake any granules off the spoon before adding to the water.

**Do not dilute in water because household bleach is about a 5 percent solution.

NOTE: HTH is supplied in a 6-ounce jar in the chemical-agent patient's treatment and decon medical-equipment set.

CAUTION

Only use a 0.5 percent chlorine solution on the patient's skin. Avoid vigorous scrubbing because it can force the agent into the skin.

CHEMICAL-AGENT PATIENT DECON PROCEDURES

8-21. Decon procedures for chemical-agent patients are discussed in the following paragraphs:

LITTER PATIENT

8-22. Before patients receive medical treatment in the clean treatment area, the 8-man patient-decon team decontaminates them. Figure 8-2 shows one way to establish the patient-decon station. Place the bandage scissors in a container of 5 percent chlorine solution between each use. The team members decon their gloves and aprons with a 5 percent chlorine solution.

NOTE: Litter patients requiring emergency medical treatment (EMT) or ambulatory medical treatment (AMT) in the clean area of the MTF will be completely decontaminated. However, a patient requiring immediate evacuation should have only his wound area and MOPP gear spot decontaminated to remove any gross contamination (for example, a stable patient with a partial amputation of a lower extremity). The patient should be evacuated in his MOPP gear.

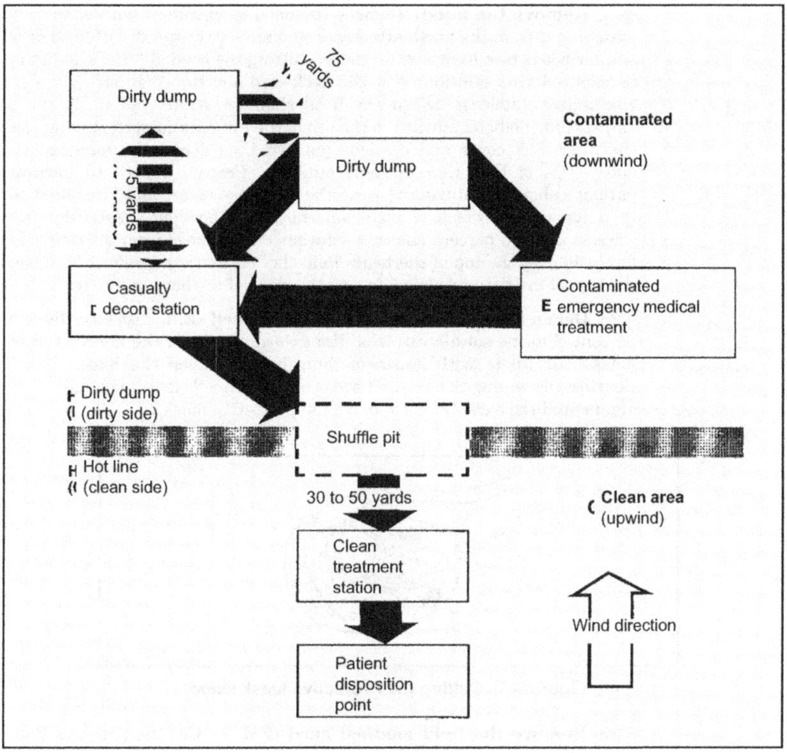

Figure 8-2. Layout for a Patient-Decon Station and a Clean Treatment Area Without a CPS

8-23. Decon the patient's skin, bandages, wounds, mask, identification tags with chain, and splints by wiping them with a 0.5 percent chlorine solution. For treatment procedures, refer to FMs 8-9, 8-33, and 8-285.

8-24. Some procedures in the following steps can be done with one soldier, while others require more than one soldier.

Step 1 - Decon the Patient's Mask and Hood

8-25. **Move the patient to the clothing removal station.** After the patient has been triaged and stabilized (if necessary) by the senior medic in the patient-decon area, move him to the litter stands at the clothing removal station.

8-26. **Decon the hood.** Use either an IEDK or a 5 percent chlorine solution (or household bleach) to wipe down the front, sides, and top of the hood.

8-27. **Remove the hood.** Remove the hood by cutting it with scissors or by loosening it from the mask attachment points for the quick-doff hood or other similar hoods (see Figure 8-3). Before cutting the hood, dip the scissors in a 5 percent chlorine solution. Cut the neck cord and the small string under the voicemitter. Release or cut the hood shoulder straps and unzip the hood zipper. Cut the hood upward to the top of the eye-lens outsert, staying close to the filter-inlet cover and eye-lens outsert, then across the forehead to the outer edge of the other eye-lens outsert. Proceed downward toward the patient's shoulder, staying close to the eye lens outsert and filter-inlet cover, then across the lower part of the voicemitter to the zipper. After dipping the scissors in the 5 percent chlorine solution, cut the hood from the center of the forehead over the top of the head. Fold the left and right sides of the hood to the side of the patient's head, laying the sides of the hood on the litter.

8-28. **Decon the protective mask and exposed skin.** Use an SDK or a 0.5 percent chlorine solution to wipe the external parts of the mask. Cover the mask's air inlets with gauze or your hand to keep the mask filter dry. Continue by wiping the exposed areas of the patient's face, to include the neck and behind the ears. Do not remove the protective mask.

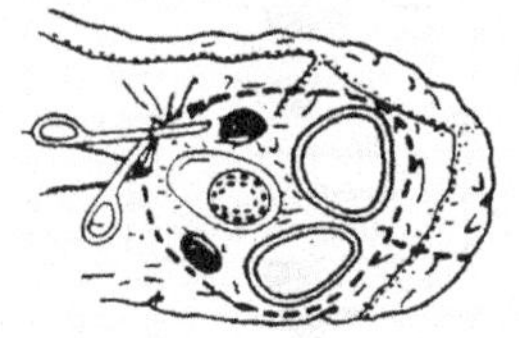

Figure 8-3. Cutting the Protective Mask Hood

8-29. **Remove the field medical card (FMC).** Cut the patient's FMC tie wire, allowing the FMC to fall into a plastic bag. Seal the plastic bag and rinse the outside of the bag with a 5 percent chlorine solution. Place the plastic bag under the back of the protective mask's head straps. The FMC will remain with the patient in the contaminated area and a clean copy will be made before the patient is moved to the clean area.

Step 2 - Remove Gross Contamination From the Patient's Protective Overgarment

8-30. Remove all visible contamination spots from the overgarment with an SDK or a 5 percent chlorine solution.

Step 3 - Remove the Patient's Personal Effects and Protective Overgarment

8-31. **Remove the patient's personal effects.** Remove the patient's personal effects from his protective overgarment and BDU pockets. Place the articles in a plastic bag, label the bag with the patient's identification, and seal the bag. If the articles are not contaminated, they are returned to the patient. If the articles are contaminated, place them in the contaminated holding area until they can be decontaminated, then return them to the patient.

CAUTION

Bandages may have been applied to control severe bleeding; therefore, they are treated like tourniquets. Only medical personnel will remove bandages, tourniquets, and splints.

NOTE: The overgarment jacket and trousers will be cut simultaneously. One soldier cuts the jacket, while another soldier cuts the trousers.

8-32. **Cut and remove the overgarment jacket.** Before cutting the overgarment jacket and trousers, dip the scissors in a 5 percent chlorine solution to prevent contamination of the patient's BDU or undergarments. Make two cuts, one up each sleeve from the wrist up to the shoulder and then across the shoulder through the collar (Figure 8-4). Cut around bandages, tourniquets, and splints, leaving them in place. Do not allow your gloves to touch the patient along the cut line. Keep the cuts close to the inside of the arms so that most of the sleeve material can be folded outward. Unzip the jacket and roll the chest sections to the respective sides, with the inner surface outward. Continue by tucking the clothing between the arm and chest. Roll the cut sleeves away from the arms, exposing the black liner.

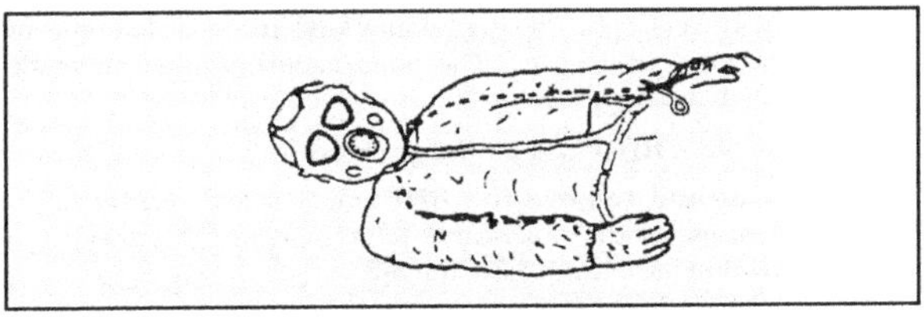

Figure 8-4. Cutting the Overgarment Jacket

8-33. **Cut and remove the overgarment trousers.** Cut both trousers legs starting at the ankle as shown in Figure 8-5, page 8-8. Keep the cuts near the inside of the legs, along the inseam, to the crotch. With the left leg, cut all the way to the waist, avoiding the pockets. With the right leg, cut across at the crotch to the left leg cut. Cut around bandages, tourniquets, and splints, leaving them in place. Place the scissors in a 5 percent chlorine solution. Fold the cut trouser halves away from the patient and allow the halves to drop to the litter with the contaminated (green) side down. Roll the inner leg portion under and between the legs.

8-34. **Remove the outer gloves.** Before touching the patient, the patient-decon team decontaminates its gloves with a 5 percent chlorine solution. Lift the patient's arms up and out of the cutaway sleeves unless detrimental to his condition. Grasp the fingers of the gloves, roll the cuffs over the fingers, and turn the gloves inside out. Do not remove the inner cotton gloves at this time. Carefully lower the patient's arms across the chest after the outer gloves have been removed (Figure 8-6, page 8-8). Do not allow the patient's arms to come into contact with the exterior of his overgarment. Drop his gloves into the

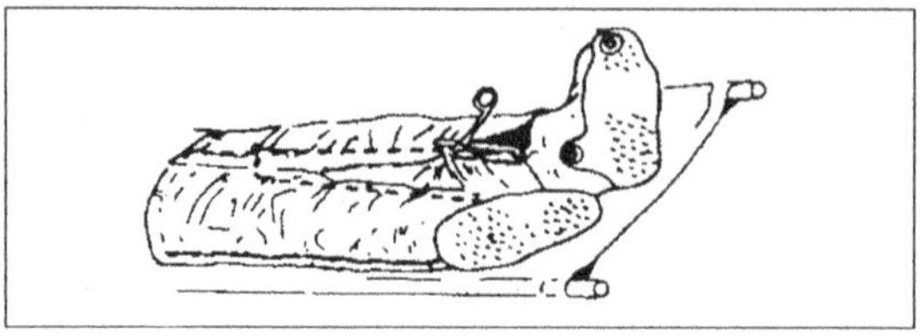

Figure 8-5. Remove the Overgarment Trousers

contaminated-waste bag. The team members decon their gloves with the 5 percent chlorine solution.

8-35. **Remove the overboots.** Cut the overboot laces and fold the lacing eyelets flat outward. While standing at the foot of the litter, hold the patient's heel with one hand. Pull the overboot downward, then toward you to remove it. Remove the two overboots simultaneously. This reduces the likelihood of contaminating one of the combat boots.

8-36. While holding the patient's heels off the litter, have a team member wipe the end of the litter with a 5 percent chlorine solution to neutralize any contamination that was transferred to the litter from the overboots. Lower the patient's heels onto the decontaminated litter. Place the overboots in the contaminated-waste bag. The team members decon their gloves with the 5 percent chlorine solution.

Step 4 - Remove the Patient's BDU and Undergarments

8-37. **Cut and remove the BDU.** To cut and remove the BDU jacket and trousers, follow the procedures for removing the protective overgarment as described in paragraph 8-32 and 8-33.

8-38. **Remove the combat boots.** Cut the bootlaces along the tongue. Remove the boots by pulling them toward you. Place the boots in the contaminated-waste bag. Do not touch the patient's skin with your contaminated gloves when removing his boots.

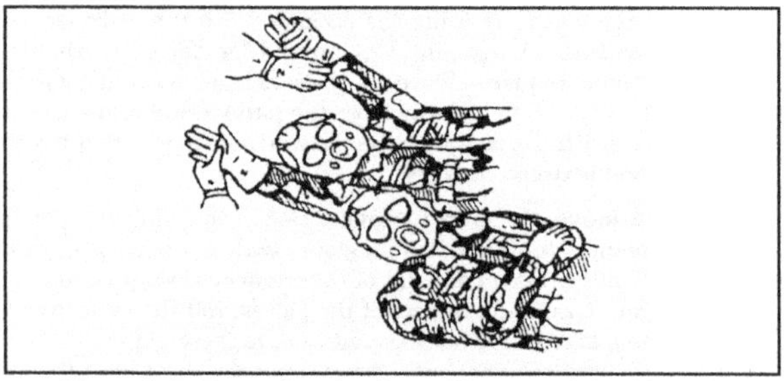

Figure 8-6. Removing the Outer Gloves and Positioning the Arms After the Glove Removal

8-39. **Cut and remove the undergarments.** Follow the procedures for cutting away the protective overgarment and rolling it away from the patient (see paragraph 8-32 and 8-33). If the patient is wearing a brassiere, cut it between the cups. Cut both shoulder straps where they attach to the cups and lay them back off of the shoulders. Remove the socks and cotton gloves. Do not remove the identification tags.

Step 5 - Transfer the Patient to a Decon Litter

8-40. After the patient's clothing has been cut away, transfer him to a decon litter or a canvas litter with a plastic-sheeting cover. Three decon team members decon their gloves and aprons with a 5 percent chlorine solution. One member places his hands under the patient's legs at the thighs and Achilles tendons, a second member places his arms under the patient's back and buttocks, and a third member places his arms under the patient's shoulders and supports the head and neck. They carefully lift the patient using their knees (not their backs) to minimize back strain. While the patient is elevated, another decon team member removes the litter from the litter stands and replaces it with a decontaminated (clean) litter. The team members carefully lower the patient onto the clean litter. The clothing and overgarments are placed in a contaminated-waste bag and moved to the contaminated-waste dump. The dirty litter is rinsed with the 5 percent chlorine solution and placed in the litter storage area.

Step 6 - Decon the Patient's Skin

8-41. **Spot decon.** With the patient in a supine position, spot decon the skin by using an SDK or a 0.5 percent chlorine solution. Decon areas of potential contamination, to include areas around the neck, wrists, and lower parts of the face. Decon the patient's identification tags and chain, if necessary.

NOTE: A complete body wash is not appropriate and may be harmful to the patient. During a complete body wash, the patient would have to be rolled over to reach all areas of the skin. This is not necessary for an adequate decon.

8-42. **Combat medic care.** The combat medic gently cuts away the bandage. He decontaminates the area around the wound and irrigates it with a 0.5 percent chlorine solution. If bleeding begins, he replaces the bandage with a clean one. He replaces the old tourniquet by placing a new one ½ to 1 inch above the old one. He then removes the old tourniquet and decontaminates the patient's skin with an SDK or a 0.5 percent chlorine solution. He does not remove a splint. He decontaminates the splint by thoroughly rinsing it, to include the padding and cravats, with a 0.5 percent chlorine solution.

WARNING

Do not use an SKD around any wounds. Do not remove splints. Splints will not be removed until the patient has been evacuated to a forward MTF or hospital.

8-43. **Completeness of decon check.** Check the patient with M8 detector paper or the CAM for completeness of decon.

NOTE: Other monitoring devices may be used, if available.

8-44. **Contaminated-waste disposal.** Dispose of contaminated bandages and coverings by placing them in a contaminated-waste bag. Seal the bag and place it in the contaminated-waste dump.

Step 7 - Transfer the Patient Across the Shuffle Pit

8-45. The patient's clothing has been cut away and his skin, bandages, and splints have been decontaminated. Transfer the patient to the shuffle pit and place the litter on the litter stands. The shuffle pit is wide enough to prevent the decon team members from straddling it while carrying the litter. A third team member will assist with transferring the patient to a clean treatment litter in the shuffle pit. Decon personnel rinse or wipe down their aprons and gloves with a 5 percent chlorine solution.

8-46. The three team members lift the patient off the decon litter (see step 5 for lifting procedures). While the patient is elevated, another team member removes the litter from the stands and returns it to the decon area. A medic from the clean side of the shuffle pit replaces the litter with a clean one. The patient is lowered onto the clean litter. Two medics from the clean side of the shuffle pit move the patient to the clean treatment area. The patient is treated in this area or awaits processing into the CPS. The litter is wiped down with a 5 percent chlorine solution in preparation for reuse. Once the patient is in the air lock of the CPS and the air lock has been purged, his protective mask is removed. Place the mask in a plastic bag and seal it.

NOTE: Before decontaminating another patient, each decon team member drinks about half a quart of water. The exact amount of water consumed is increased or decreased according to the work level and temperature (see Table 8-2).

AMBULATORY PATIENT

8-47. All ambulatory patients requiring EMT or AMT in the clean treatment area of the BAS will be decontaminated. Stable patients not requiring treatment at the BAS, but requiring evacuation to the medical company's clearing station or a corps hospital for treatment (for example, a patient with a broken arm), should be evacuated in their protective overgarments and masks by any available transportation. However, before evacuation, spot remove all thickened agents from their protective clothing.

NOTE: Place the bandage scissors that are used in this procedure in a container of 5 percent chlorine solution when not in use. Most ambulatory patients will be treated in the contaminated treatment area and returned to duty. Upon removal of an ambulatory patient's clothing, he becomes a litter patient. The BAS and clearing station do not have clothing to replace those that are cut off during the decon process. The patient must be placed in a patient protective wrap (PPW) for protection during evacuation (Figure 8-7, page 8-12).

8-48. The ambulatory patient is decontaminated and undressed as follows. Some procedures in the following steps can be done with one soldier, while others require more than one.

Table 8-2. Heat-Injury Prevention and Water Consumption

Criteria			Controls	
Heat		Water Intake Quart/Hour	Physical Activity for Soldiers/Trainees	
Condition	WBGT Index °F*		Acclimatized to Work/Rest**	Not Acclimatized to Work/Rest
White	78-81.9	At least ½	Continuous	
Green	82-84.9	At least ½	50/10 minutes	Use discretion in planning heavy exercise.
Yellow	85-87.9	At least 1	45/15 minutes	Suspend strenuous exercise during the first 3 weeks of training. Training activities may be continued on a reduced scale after the second week of training. Avoid activity in the direct sun.
Red	88-89.9	At least 1½	30/30 minutes	Curtail strenuous exercise for all personnel with less than 12 weeks of hot-weather training.
White	78-81.9	At least ½	Continuous	
Green	82-84.9	At least ½	50/10 minutes	Use discretion in planning heavy exercise.
Black	90 and up	More than 2	20/40 minutes	Suspend physical training and strenuous exercise. Essential operational commitments (not for training), where the risk of heat casualties may be warranted, are excluded from suspension. Enforce water intake to minimize expected heat injuries.

*MOPP gear or body armor adds 10°F to the wet-bulb gradient-temperature (WBGT) index.
**An acclimatized soldier is one who has worked in the given heat condition for 10 to 14 days.

NOTE: "Rest" means minimal physical activity. Rest should be accomplished in the shade, if possible. Any activity requiring only minimal physical activity can be performed during rest periods (for examples, training by lecture or demonstration, minor maintenance procedures on vehicles or weapons, and personal-hygiene activities such as skin and foot care).

Step 1 - Remove the LCE

8-49. Remove the LCE by unfastening/unbuttoning all connectors or tie straps and then place the equipment in a plastic bag. Place the plastic bag in the designated storage area for later decon.

Step 2 - Decon the Patient's Mask and Hood

8-50. **Begin the clothing removal process.** After the patient has been triaged and treated (if necessary) by the senior medic in the patient-decon station, the clothing removal process begins.

8-51. **Decon the hood.** Use either an IEDK or a 5 percent chlorine solution (or household bleach) to wipe down the front, sides, and top of the hood.

8-52. **Remove the hood.** Remove the hood by cutting it with scissors or by loosening it from the mask attachment points for the quick-doff hood or other similar hoods (see Figure 8-3, page 8-6). Before cutting the hood, dip the scissors in a 5 percent chlorine solution. Cut the neck cord and the small

To remove the BDU, follow the procedures for removing the protective overgarment as described in paragraphs 8-124 and 8-125. Do not remove the patient's identification tags.

Step 5 - Decon the Patient's Skin

8-130. **Spot decon.** Wash exposed skin surfaces with soap and warm water, to include areas around the neck, wrists, and lower parts of the face. Decon the patient's identification tags and chain, if necessary.

8-131. **Combat medic care.** The combat medic gently cuts away the bandage. He decontaminates the area around the wound and irrigates it with a 0.5 percent chlorine solution. If bleeding begins, he replaces the bandage with a clean one. He replaces the old tourniquet by placing a new one ½ to 1 inch above the old one. He then removes the old tourniquet and decontaminates the patient's skin with a 0.5 percent chlorine solution. He does not remove a splint. He decontaminates the splint by thoroughly rinsing it, to include the padding and cravats, with a 0.5 percent chlorine solution.

WARNING
Do not use an SDK around any wounds. Do not remove splints. The splint will not be removed until the patient has been evacuated to a forward MTF or hospital.

8-132. **Completeness of decon check.** Check the patient with M8 detector paper or the CAM for completeness of decon.

NOTE: Other monitoring devices may be used, if available.

8-133. **Contaminated-waste disposal.** Dispose of contaminated bandages and coverings by placing them in a contaminated-waste bag. Seal the bag and place it in the contaminated-waste dump.

Step 6 - Transfer the Patient to the Hot Line

8-134. Move the patient to the hot line. Two medics from the clean side of the hot line move the patient into the clean treatment area.

NOTE: Before decontaminating another patient, each decon team member drinks about half a quart of water. The exact amount of water consumed is increased or decreased according to the work level and temperature (see Table 8-2, page 8-11).

AMBULATORY PATIENT

8-135. The decon procedures for an ambulatory patient that is contaminated with radiation are discussed in the following paragraphs. Some procedures can be done with one soldier, while others require more than one.

8-136. All ambulatory patients requiring EMT or AMT in the clean area of the BAS will be decontaminated. Stable patients not requiring treatment at the BAS, but requiring evacuation to the medical company's clearing station or a corps hospital for treatment (for example, a patient with a broken arm), should be evacuated in their protective overgarments and masks by any

articles in a plastic bag, label the bag with the patient's identification, and seal the bag. If the articles are not contaminated, they are returned to the patient. If the articles are contaminated, place them in the contaminated holding area until they can be decontaminated, then return them to the patient.

8-57. **Remove the overgarment jacket.** Have the patient stand with his feet spread apart at shoulder width. Unsnap the front flap of the jacket and unzip the jacket. If the patient can extend his arms, have him clinch his fist and extend his arms backward at about a 30° angle. Move behind the patient, grasp his jacket collar at the sides of the neck, and peel the jacket off the shoulders at a 30° angle down and away from the patient. Avoid any rapid or sharp jerks, which spread contamination. Gently pull the inside sleeves over the patient's wrists and hands.

8-58. If the patient cannot extend his arms, you must cut the jacket to aid in its removal. Before cutting the overgarment jacket, dip the scissors in a 5 percent chlorine solution to prevent contamination of the patient's BDU or underclothing. As with the litter patient, make two cuts, one up each sleeve from the wrist up to the shoulder and then across the shoulder through the collar. Cut around bandages, tourniquets, and splints, leaving them in place. Do not allow your gloves to touch the patient along the cut line. Peel the jacket back and downward to avoid spreading contamination. Ensure that the outside of the jacket does not touch the patient or his inner clothing.

8-59. **Cut and remove the overgarment trousers.** Unfasten or cut all ties, buttons, or zippers before grasping the trousers at the waist and peeling them down over the patient's combat boots. Again, the trousers are cut to aid in removal. If necessary, cut both trouser legs starting at the ankle. Keep the cuts near the inside of the legs, along the inseam, to the crotch. Cut around all bandages, tourniquets, and splints. Continue to cut up both sides of the zipper to the waist and allow the narrow strip with the zipper to drop between the legs. Peel or allow the trouser halves to drop to the ground. Have the patient step out of the trouser legs one at a time. Place the trousers in the contaminated-waste bag. Place the scissors in a 5 percent chlorine solution.

8-60. **Remove the outer gloves.** Grasp the fingers of the gloves, roll the cuffs over the fingers, and turn the gloves inside out. Do not remove the inner cotton gloves at this time. Drop the gloves into the contaminated-waste bag. Do not allow the patient to touch his clothing or other contaminated objects with his hands.

8-61. **Remove the overboots.** Cut the overboot laces and fold the lacing eyelets flat on the ground. Step on the toe and heel eyelets to hold the overboot on the ground and have the patient step out of it. Repeat this procedure for the other overboot. If the overboots are in good condition, they can be decontaminated and reissued.

8-62. **Remove the patient's cotton glove liners.** Instruct the patient to remove his cotton glove liners to reduce the possibility of spreading contamination. Have the patient grasp the heel of one glove liner with the other gloved hand, peeling it off of his hand. Hold the removed glove by the inside and grasp the heel of the other glove, peeling it off of his hand. Place both gloves in the contaminated-waste bag.

Step 5 - Check the Patient for Contamination

8-63. After the patient's overgarment has been removed, check his BDU by using M8 detector paper or the CAM. Carefully survey the patient, paying particular attention to discolored areas, damp spots, and tears on the uniform; areas around the neck, wrists, and ears; and bandages, tourniquets, and splints. Remove contaminated spots by using an SDK or a 0.5 percent chlorine solution or, if possible, by cutting away the contaminated area. Always dip the scissors in a 5 percent chlorine solution after each cut. Recheck the area with the detection equipment. If significant contamination is found on the BDU, then remove it and spot decon the skin. To remove the BDU, follow the procedures for removing the overgarment as described in paragraphs 8-57 through 8-59. Do not remove the patient's identification tags.

Step 6 - Decon the Patient's Skin

8-64. **Spot decon.** Use an SDK or a 0.5 percent chlorine solution to spot decon the skin and areas of potential contamination, to include areas around the neck, wrists, and lower parts of the face. Decon the patient's identification tags and chain, if necessary.

8-65. Have the patient hold his breath, close his eyes, and lift, or assist him with lifting, his mask at the chin. Wipe his face and exposed areas of the skin with an SDK or a 0.5 percent chlorine solution. Starting at the top of the ear and quickly wiping downward, wipe all folds in the skin, ear lobes, upper lip, chin, dimples, and nose. Continue up the other side of the face to the top of the other ear. Wipe the inside of the mask where it touches the face. Have the patient reseal and check his mask.

CAUTION

Keep the decon solution out of the patient's eyes.

8-66. **Combat medic care.** The combat medic gently cuts away the bandage. He decontaminates the area around the wound and irrigates it with a 0.5 percent chlorine solution. If bleeding begins, he replaces the bandage with a clean one. He replaces the old tourniquet by placing a new one ½ to 1 inch above the old one. He then removes the old tourniquet and decontaminates the patient's skin with an SDK or a 0.5 percent chlorine solution. He does not remove a splint. He decontaminates the splint by thoroughly rinsing it, to include the padding and cravats, with a 0.5 percent chlorine solution.

8-67. **Contaminated-waste disposal.** Dispose of contaminated bandages and coverings by placing them in a contaminated-waste bag. Seal the bag and place it in the contaminated-waste dump.

Step 7 - Proceed Through the Shuffle Pit to the Clean Treatment Area

8-68. Have the decontaminated patient proceed through the shuffle pit to the clean treatment area. To ensure that the patient's boots are well decontaminated, have him stir the contents of the shuffle pit with his boots as he crosses it. The patient's combat boots and protective mask will be removed at the entrance of the CPS or clean treatment area.

BIOLOGICAL-AGENT PATIENT DECON PROCEDURES

8-69. The decon station as established for chemical-agent patients is also used for biologically contaminated patients. The 8-man, patient-decon team is required for biologically contaminated patient-decon procedures.

LITTER PATIENT

8-70. The decon procedures for a litter patient that is contaminated with a biological agent are described in the following paragraphs. Some procedures can be done with one soldier, while others require more than one.

Step 1 - Decon the Patient's Mask and Hood

8-71. **Move the patient to the clothing removal station.** After the patient has been triaged and stabilized (if necessary) by the senior medic in the patient-decon area, move him to the litter stands at the clothing removal station.

8-72. **Decon the hood.** Use a 0.5 percent chlorine solution to wipe down the front, sides, and top of the hood.

8-73. **Remove the hood.** Remove the hood by cutting it with scissors or by loosening it from the mask attachment points for the quick-doff hood or other similar hoods (see Figure 8-3, page 8-6). Before cutting the hood, dip the scissors in a 5 percent chlorine solution. Cut the neck cord and the small string under the voicemitter. Release or cut the hood shoulder straps and unzip the hood zipper. Cut the hood upward to the top of the eye-lens outsert, staying close to the filter-inlet cover and eye-lens outsert, then across the forehead to the outer edge of the other eye-lens outsert. Proceed downward toward the patient's shoulder, staying close to the eye-lens outsert and filter-inlet cover, then across the lower part of the voicemitter to the zipper. After dipping the scissors in the 5 percent chlorine solution, cut the hood from the center of the forehead over the top of the head. Fold the left and right sides of the hood to the side of the patient's head, laying the sides of the hood on the litter.

8-74. **Decon the protective mask and exposed skin**. Use a 0.5 percent chlorine solution to wipe the external parts of the mask. Cover the mask's air inlets with gauze or your hand to keep the mask filter dry. Wash the exposed areas of the patient's face, to include the neck and behind the ears, with soap and warm water or wipe them with a 0.5 percent chlorine solution. Do not remove the protective mask.

8-75. **Remove the FMC.** Cut the patient's FMC tie wire, allowing the FMC to fall into a plastic bag. Seal the plastic bag and rinse the outside of the bag with a 0.5 percent chlorine solution. Place the plastic bag under the back of the protective mask's head straps. The FMC will remain with the patient in the contaminated area and a clean copy will be made before the patient is moved to the clean area.

Step 2 - Remove Gross Contamination From the Patient's Protective Overgarment

8-76. Remove all visible contamination spots from the overgarment by using a 0.5 percent chlorine solution.

Step 3 - Remove the Patient's Personal Effects and Protective Overgarment

8-77. **Remove the patient's personal effects.** Remove the patient's personal effects from his protective overgarment and BDU pockets. Place the articles in a plastic bag, label the bag with the patient's identification, and seal the bag. If the articles are not contaminated, they are returned to the patient. If the articles are contaminated, place them in the contaminated holding area until they can be decontaminated, then return them to the patient.

CAUTION

Bandages may have been applied to control severe bleeding; therefore, they are treated like tourniquets. Only medical personnel will remove bandages, tourniquets, and splints.

NOTE: The overgarment jacket and trousers will be cut simultaneously. One soldier cuts the jacket, while another soldier cuts the trousers.

8-78. **Cut and remove the overgarment jacket.** Before cutting the overgarment jacket and trouser, dip the scissors in a 5 percent chlorine solution to prevent contamination of the patient's BDU or undergarments. Make two cuts, one up each sleeve from the wrist up to the shoulder and then across the shoulder through the collar (Figure 8-4, page 8-7). Cut around bandages, tourniquets, and splints, leaving them in place. Do not allow your gloves to touch the patient along the cut line. Keep the cuts close to the inside of the arms so that most of the sleeve material can be folded outward. Unzip the jacket and roll the chest sections to the respective sides, with the inner surface outward. Continue by tucking the clothing between the arm and chest. Roll the cut sleeves away from the arms, exposing the black liner.

8-79. **Cut and remove the overgarment trousers.** Cut both trousers legs starting at the ankle as shown in Figure 8-5, page 8-8. Keep the cuts near the inside of the legs, along the inseam, to the crotch. Cut the left leg all the way to the waist, avoiding the pockets. Cut the right leg across at the crotch to the left leg cut. Cut around bandages, tourniquets, and splints, leaving them in place. Place the scissors in a 5 percent chlorine solution. Fold the cut trouser halves away from the patient and allow the halves to drop to the litter with the contaminated (green) side down. Roll the inner leg portion under and between the legs.

8-80. **Remove the outer gloves.** Before touching the patient, the patient-decon team decontaminates its gloves with a 5 percent chlorine solution. Lift the patient's arms up and out of the cutaway sleeves unless detrimental to his condition. Grasp the fingers of the gloves, roll the cuffs over the fingers, and turn the gloves inside out. Do not remove the inner cotton gloves at this time. Carefully lower the patient's arms across the chest after the outer gloves have been removed (Figure 8-6, page 8-8). Do not allow the patient's arms to come into contact with the exterior of his overgarment. Drop his gloves into the contaminated-waste bag. The team members decon their gloves with the 5 percent chlorine solution.

8-81. **Remove the overboots.** Cut the overboot laces and fold the lacing eyelets flat outward. While standing at the foot of the litter, hold the patient's heel with one hand. Pull the overboot downward, then toward you to remove it. Remove the two overboots simultaneously. This reduces the likelihood of contaminating one of the combat boots.

8-82. While holding the patient's heels off the litter, have a team member wipe the end of the litter with a 0.5 percent chlorine solution to neutralize any contamination that was transferred to the litter from the overboots. Lower the patient's heels onto the decontaminated litter. Place the overboots in the contaminated-waste bag. The team members decontaminate their gloves with the 5 percent chlorine solution.

Step 4 - Remove the Patient's BDU and Undergarments

8-83. **Cut and remove the BDU.** To cut and remove the BDU jacket and trousers, follow the procedures (described in paragraphs 8-78 and 8-79) for removing the protective overgarment.

8-84. **Remove the combat boots.** Cut the bootlaces along the tongue. Remove the boots by pulling them toward you. Place the boots in the contaminated-waste bag. Do not touch the patient's skin with your contaminated gloves when removing his boots.

8-85. **Cut and remove the undergarments.** Follow the procedures for cutting away the protective overgarment and rolling it away from the patient (see paragraphs 8-78 and 8-79). If the patient is wearing a brassiere, cut it between the cups. Cut both shoulder straps where they attach to the cups and lay them back off of the shoulders. Remove the socks and cotton gloves. Do not remove the identification tags.

Step 5 - Transfer the Patient to a Decon Litter

8-86. After the patient's clothing has been cut away, transfer him to a decon litter or a canvas litter with a plastic-sheeting cover. Three decon team members decon their gloves and aprons with a 5 percent chlorine solution. One member places his hands under the patient's legs at the thighs and Achilles tendons, a second member places his arms under the patient's back and buttocks, and a third member places his arms under the patient's shoulders and supports the head and neck. They carefully lift the patient using their knees (not their backs) to minimize back strain. While the patient is elevated, another decon team member removes the litter from the litter stands and replaces it with a decontaminated (clean) litter. The patient is carefully lowered onto the clean litter. The clothing and overgarments are placed in a contaminated-waste bag and moved to the contaminated-waste dump. The dirty litter is rinsed with the 0.5 percent chlorine solution and placed in the litter storage area.

Step 6 - Decon the Patient's Skin

8-87. **Spot decon.** With the patient in a supine position, spot decon the skin by washing it with soap and warm water or by wiping it with a 0.5 percent chlorine solution. Decon areas of potential contamination, to include areas around the neck, wrists, and lower parts of the face. Decon the patient's identification tags and chain, if necessary.

NOTES:

1. Use a 0.5 percent chlorine solution to decon patients suspected of being contaminated with mycotoxins.

2. A complete body wash is not appropriate and may be harmful to the patient. During a complete body wash, the patient would have to be rolled over to reach all areas of the skin. This is not necessary for an adequate decon.

8-88. **Combat medic care.** The combat medic gently cuts away the bandage. He decontaminates the area around the wound and irrigates it with a 0.5 percent chlorine solution. If bleeding begins, he replaces the bandage with a clean one. He replaces the old tourniquet by placing a new one ½ to 1 inch above the old one. He then removes the old tourniquet and decontaminates the patient's skin with a 0.5 percent chlorine solution. He does not remove a splint. He decontaminates the splint by thoroughly rinsing it, to include the padding and cravats, with a 0.5 percent chlorine solution.

WARNING

Do not use an SDK around any wounds. Do not remove splints. The splint will not be removed until the patient has been evacuated to a forward MTF or hospital.

8-89. **Completeness of decon check.** Check the patient with M8 detector paper or the CAM for completeness of decon.

NOTE: Other monitoring devices may be used, if available.

8-90. **Contaminated-waste disposal.** Dispose of contaminated bandages and coverings by placing them in a contaminated-waste bag. Seal the bag and place it in the contaminated-waste dump.

Step 7 - Transfer the Patient to the Hot Line

8-91. The patient's clothing has been cut away and his skin, bandages, and splints have been decontaminated. Transfer the patient to the hot line and place the litter on the litter stands. A third team member will assist with transferring the patient to a clean treatment litter at the hot line. Decon personnel rinse or wipe down their aprons and gloves with a 5 percent chlorine solution.

8-92. Three team members lift the patient off the decon litter (see step 5 for lifting procedures). While the patient is elevated, another team member removes the litter from the stands and returns it to the decon area. A medic from the clean side of the hot line replaces the litter with a clean one. The patient is lowered onto the clean litter. Two medics from the clean side of the hot line move the patient to the clean treatment area. The patient is treated in this area or awaits processing into the CPS. The litter is wiped down with a 0.5 percent chlorine solution in preparation for reuse. Once the patient is in the air lock of the CPS and the air lock has been purged, his protective mask is removed. Place the mask in a plastic bag and seal it.

NOTE: Before decontaminating another patient, each decon team member drinks about half a quart of water. The exact amount of water consumed is increased or decreased according to the work level and temperature (see Table 8-2, page 8-11).

AMBULATORY PATIENT

8-93. The decon procedures for an ambulatory patient that is contaminated with a biological agent are described in the following paragraphs. Some procedures can be done with one soldier, while others require more than one.

8-94. All ambulatory patients requiring EMT or AMT in the clean treatment area of the BAS will be decontaminated. Stable patients not requiring treatment at the BAS, but requiring evacuation to the medical company's clearing station or a corps hospital for treatment (for example, a patient with a broken arm), should be evacuated in their protective overgarments and masks by any available transportation. However, before evacuation, spot remove all thickened agents from their protective overgarment.

NOTE: Place the bandage scissors that are used in this procedure in a container of 5 percent chlorine solution when not in use. Most ambulatory patients will be treated in the contaminated treatment area and returned to duty. Upon removal of an ambulatory patient's clothing, he becomes a litter patient. The BAS and clearing station do not have clothing to replace those that are cut off during the decon process. The patient must be placed in a PPW for protection during evacuation (Figure 8-7, page 8-12).

Step 1 - Remove the LCE

8-95. Remove the LCE by unfastening/unbuttoning all connectors or tie straps and then place the equipment in a plastic bag. Place the plastic bag in the designated storage area for later decon.

Step 2 - Decon the Patient's Mask and Hood

8-96. **Begin the clothing removal process.** After the patient has been triaged and treated (if necessary) by the senior medic in the patient-decon station, the clothing removal process begins.

8-97. **Decon the hood.** Use a 0.5 percent chlorine solution to wipe down the front, sides, and top of the hood.

8-98. **Remove the hood.** Remove the hood by cutting it with scissors or by loosening it from the mask attachment points for the quick-doff hood or other similar hoods (Figure 8-3, page 8-6). Before cutting the hood, dip the scissors in a 5 percent chlorine solution. Cut the neck cord and the small string under the voicemitter. Release or cut the hood shoulder straps and unzip the hood zipper. Cut the hood upward to the top of the eye-lens outsert, staying close to the filter-inlet cover and eye-lens outsert, then across the forehead to the outer edge of the other eye-lens outsert. Proceed downward toward the patient's shoulder, staying close to the eye-lens outsert and filter-inlet cover, then across the lower part of the voicemitter to the zipper. After dipping the scissors in the 5 percent chlorine solution, cut the hood from the center of the forehead over the top of the head. Fold the left and right sides of the hood away from the patient's head and remove the hood.

8-99. **Decon the protective mask and exposed skin.** Use a 0.5 percent chlorine solution to decon the external parts of the mask. Cover the mask's air inlets with gauze or your hands to keep the mask filters dry. Wash the exposed areas of the patient's face, to include the neck and behind the ears, with soap and warm water or wipe them with a 0.5 percent chlorine solution. Do not remove the protective mask.

8-100. **Remove the FMC.** Cut the patient's FMC tie wire, allowing the FMC to fall into a plastic bag. Seal the plastic bag and rinse the outside of the bag with a 5 percent chlorine solution. Place the plastic bag under the back of the protective mask's head straps. The FMC will remain with the patient in the contaminated area and a clean copy will be made before the patient is moved to the clean area.

Step 3 - Remove Gross Contamination From the Patient's Protective Overgarment

8-101. Remove all visible contamination spots from the overgarment by using a 0.5 percent chlorine solution.

Step 4 - Remove the Patient's Personal Effects and Protective Overgarment

8-102. **Remove the patient's personal effects.** Remove the patient's personal effects from his protective overgarment and BDU pockets. Place the articles in a plastic bag, label the bag with the patient's identification, and seal the bag. If the articles are not contaminated, they are returned to the patient. If the articles are contaminated, place them in the contaminated holding area until they can be decontaminated, then return them to the patient.

8-103. **Remove the overgarment jacket.** Have the patient stand with his feet spread apart at shoulder width. Unsnap the front flap of the jacket and unzip the jacket. If the patient can extend his arms, have him clinch his fist and extend his arms backward at about a 30° angle. Move behind the patient, grasp his jacket collar at the sides of the neck, and peel the jacket off the shoulders at a 30° angle down and away from the patient. Avoid any rapid or sharp jerks, which spread contamination. Gently pull the inside sleeves over the patient's wrists and hands.

8-104. If the patient cannot extend his arms, you must cut the jacket to aid in its removal. Before cutting the overgarment jacket, dip the scissors in a 5 percent chlorine solution to prevent contamination of the patient's BDU or undergarment. As with the litter patient, make two cuts, one up each sleeve from the wrist up to the shoulder and then across the shoulder through the collar. Cut around bandages, tourniquets, and splints, leaving them in place. Do not allow your gloves to touch the patient along the cut line. Peel the jacket back and downward to avoid spreading contamination. Ensure that the outside of the jacket does not touch the patient or his inner clothing.

8-105. **Cut and remove the overgarment trousers.** Unfasten or cut all ties, buttons, or zippers before grasping the trousers at the waist and peeling them down over the patient's combat boots. Again, the trousers are cut to aid in removal. If necessary, cut both trouser legs starting at the ankle. Keep the cuts near the inside of the legs, along the inseam, to the crotch. Cut around all bandages, tourniquets, and splints. Continue to cut up both sides of the zipper to the waist and allow the narrow strip with the zipper to drop between the legs.

Peel or allow the trouser halves to drop to the ground. Have the patient step out of the trouser legs one at a time. Place the trousers in the contaminated-waste bag. Place the scissors in a 5 percent chlorine solution.

8-106. **Remove the outer gloves.** Grasp the fingers of the gloves, roll the cuffs over the fingers, and turn the gloves inside out. Do not remove the inner cotton gloves at this time. Drop the gloves into the contaminated-waste bag. Do not allow the patient to touch his clothing or other contaminated objects with his hands.

8-107. **Remove the overboots.** Cut the overboot laces and fold the lacing eyelets flat on the ground. Step on the toe and heel eyelets to hold the overboot on the ground and have the patient step out of it. Repeat this procedure for the other overboot. If the overboots are in good condition, they can be decontaminated and reissued.

8-108. **Remove the patient's cotton glove liners.** Instruct the patient to remove his cotton glove liners to reduce the possibility of spreading contamination. Have the patient grasp the heel of one glove liner with the other gloved hand, peeling it off of his hand. Hold the removed glove by the inside and grasp the heel of the other glove, peeling it off of his hand. Place both gloves in the contaminated-waste bag.

Step 5 - Check the Patient for Contamination

8-109. After the patient's overgarment has been removed, check his BDU by using M8 detector paper or the CAM. Carefully survey the patient, paying particular attention to discolored areas, damp spots, and tears on the uniform; areas around the neck, wrists, and ears; and bandages, tourniquets, and splints. Remove contaminated spots by washing with soap and water, by wiping with a 0.5 percent chlorine solution or, if possible, by cutting away the contaminated area. Always dip the scissors in a 5 percent chlorine solution after each cut. Recheck the area with the detection equipment. If significant contamination is found on the BDU, then remove it and spot decon the skin. To remove the BDU, follow the procedures for removing the protective overgarment as described in paragraphs 8-103 through 8-105. Do not remove the patient's identification tags.

Step 6 - Decon the Patient's Skin

8-110. **Spot decon.** Use soap and water or a 0.5 percent chlorine solution to spot decon the skin and areas of potential contamination, to include areas around the neck, wrists, and lower parts of the face. Decon the patient's identification tags and chain, if necessary.

8-111. Have the patient hold his breath, close his eyes, and lift, or assist him with lifting, his mask at the chin. Wash his face and exposed areas of the skin with soap and water or wipe them with a 0.5 percent chlorine solution. Starting at the top of the ear and quickly wiping downward, wipe all folds in the skin, ear lobes, upper lip, chin, dimples, and nose. Continue up the other side of the face to the top of the other ear. Wipe the inside of the mask where it touches the face. Have the patient reseal and check his mask.

> **CAUTION**
> Keep the decon solution out of the patient's eyes.

8-112. **Combat medic care.** The combat medic gently cuts away the bandage. He decontaminates the area around the wound and irrigates it with a 0.5 percent chlorine solution. If bleeding begins, he replaces the bandage with a clean one. He replaces the old tourniquet by placing a new one ½ to 1 inch above the old one. He then removes the old tourniquet and decontaminates the patient's skin with an SDK or a 0.5 percent chlorine solution. He does not remove a splint. He decontaminates the splint by thoroughly rinsing it, to include the padding and cravats, with a 0.5 percent chlorine solution.

NOTE: Use a 0.5 percent chlorine solution to decon ambulatory patients suspected of being contaminated with mycotoxins.

8-113. **Contaminated-waste disposal.** Dispose of contaminated bandages and coverings by placing them in a contaminated-waste bag. Seal the bag and place it in the contaminated-waste dump.

Step 7 - Cross the Hot Line to the Clean Treatment Area

8-114. Have the decontaminated patient cross the hot line to the clean treatment area. The patient's boots must be decontaminated at the hot line before enters the clean area. The patient's combat boots and protective mask will be removed at the entrance of the CPS or clean treatment area.

RADIOLOGICAL-AGENT PATIENT DECON PROCEDURES

8-115. The decon of patients that are contaminated with radiation is easily accomplished without interfering with the required medical care. The 8-man, patient-decon team is required to perform the decon procedures on patients contaminated with radiation.

NOTE: Patients must be monitored by using a radiac meter before, during, and after each step of the decon procedure.

LITTER PATIENT

8-116. The decon procedures for a litter patient that is contaminated with radiation are discussed in the following paragraphs. Some procedures can be done with one soldier, while others require more than one soldier.

Step 1 - Decon the Patient's Mask and Hood

8-117. **Move the patient to the clothing removal station.** After the patient has been triaged and stabilized (if necessary) by the senior medic in the patient-decon area, move him to the litter stands at the clothing removal station.

8-118. **Decon the hood.** Brush contamination off or use soap and water to wipe down the front, sides, and top of the hood.

8-119. **Remove the hood.** Remove the hood by cutting it with scissors or by loosening it from the mask attachment points for the quick-doff hood or other similar hoods (see Figure 8-3, page 8-6). Before cutting the hood, dip the scissors in a 5 percent chlorine solution. Cut the neck cord and the small string under the voicemitter. Release or cut the hood shoulder straps and unzip the hood zipper. Cut the hood upward to the top of the eye-lens outsert, staying close to the filter-inlet cover and eye-lens outsert, then across the forehead to the outer edge of the other eye-lens outsert. Proceed downward toward the patient's shoulder, staying close to the eye-lens outsert and filter-inlet cover, then across the lower part of the voicemitter to the zipper. After dipping the scissors in the 5 percent chlorine solution, cut the hood from the center of the forehead over the top of the head. Fold the left and right sides of the hood to the side of the patient's head, laying the sides of the hood on the litter.

8-120. **Decon the protective mask and exposed skin.** Use soap and water to wipe the external parts of the mask. Cover the mask's air inlets with gauze or your hand to keep the mask filter dry. Wash the exposed areas of the patient's face, to include the neck and behind the ears, with soap and warm water. Do not remove the protective mask.

8-121. **Remove the FMC.** Cut the patient's FMC tie wire, allowing the FMC to fall into a plastic bag. Seal the plastic bag and rinse the outside of the bag with soap and water. Place the plastic bag under the back of the protective mask's head straps. The FMC will remain with the patient in the contaminated area and a clean copy will be made before the patient is moved to the clean area.

Step 2 - Remove Gross Contamination From the Patient's Protective Overgarment

8-122. Remove all visible contamination spots from the overgarment by brushing off or by washing with soap and water.

Step 3 - Remove the Patient's Personal Effects and Protective Overgarment

8-123. **Remove the patient's personal effects.** Remove the patient's personal effects from his protective overgarment and BDU pockets. Place the articles in a plastic bag, label the bag with the patient's identification, and seal the bag. If the articles are not contaminated, they are returned to the patient. If the articles are contaminated, place them in the contaminated holding area until they can be decontaminated, then return them to the patient.

CAUTION

Bandages may be applied to control severe bleeding; therefore, they are treated like tourniquets. Only medical personnel will remove bandages.

NOTE: The overgarment jacket and trousers will be cut simultaneously. One soldier cuts the jacket, while another soldier cuts the trousers.

8-124. **Cut and remove the overgarment jacket.** Before cutting the overgarment jacket and trousers, dip the scissors in a 5 percent chlorine solution to prevent contamination of the patient's BDU or undergarment. Make two cuts, one up each sleeve from the wrist up to the shoulder and then across the shoulder through the collar (Figure 8-4, page 8-7). Cut around bandages, tourniquets, and splints, leaving them in place. Do not allow your gloves to touch the patient along the cut line. Keep the cuts close to the inside of the arms so that most of the sleeve material can be folded outward. Unzip the jacket and roll the chest sections to the respective sides, with the inner surface outward. Continue by tucking the clothing between the arm and chest. Roll the cut sleeves away from the arms, exposing the black liner.

8-125. **Cut and remove the overgarment trousers.** Cut both trousers legs starting at the ankle as shown in Figure 8-5, page 8-8. Keep the cuts near the inside of the legs, along the inseam, to the crotch. Cut the left leg all the way to the waist, avoiding the pockets. Cut the right leg across at the crotch to the left leg cut. Cut around bandages, tourniquets, and splints, leaving them in place. Place the scissors in a 5 percent chlorine solution. Fold the cut trouser halves away from the patient and allow the halves to drop to the litter with the contaminated- (green-) side down. Roll the inner leg portion under and between the legs.

8-126. **Remove the outer gloves.** Before touching the patient, the patient-decon team decontaminates its gloves with a 5 percent chlorine solution. Lift the patient's arms up and out of the cutaway sleeves unless detrimental to his condition. Grasp the fingers of the gloves, roll the cuffs over the fingers, and turn the gloves inside out. Do not remove the inner cotton gloves at this time. Carefully lower the patient's arms across the chest after the outer gloves have been removed (Figure 8-6, page 8-8). Do not allow the patient's arms to come into contact with the exterior of his overgarment. Drop his gloves into the contaminated-waste bag. The team members decon their gloves with the 5 percent chlorine solution.

8-127. **Remove the overboots.** Cut the overboot laces and fold the lacing eyelets flat outward. While standing at the foot of the litter, hold the patient's heel with one hand. Pull the overboot downward, then toward you to remove it. Remove the two overboots simultaneously. This reduces the likelihood of contaminating one of the combat boots.

8-128. While holding the patient's heels off the litter, have a team member wipe the end of the litter with a 0.5 percent chlorine solution to neutralize any contamination that was transferred to the litter from the overboots. Lower the patient's heels onto the decontaminated litter. Place the overboots in the contaminated-waste bag. The team members decon their gloves with the 5 percent chlorine solution.

NOTE: Patients arriving at the MTF in MOPP gear will only have their MOPP gear removed. They will remain in their BDUs unless contamination is found on them. If contamination is found, follow the procedures for removing the protective overgarment as described in paragraphs 8-124 and 8-125.

Step 4 - Check the Patient for Contamination

8-129. After the patient's overgarment has been removed, check his BDU by using M8 detector paper or the CAM. Carefully survey the patient, paying particular attention to discolored areas, damp spots, and tears on the uniform; areas around the neck, wrists, and ears; and bandages, tourniquets, and splints. Remove contaminated spots by washing with soap and water, by wiping with a 0.5 percent chlorine solution or, if possible, by cutting away the contaminated area. Always dip the scissors in a 5 percent chlorine solution after each cut. Recheck the area with the detection equipment. If significant contamination is found on the BDU, then remove it and spot decon the skin.

To remove the BDU, follow the procedures for removing the protective overgarment as described in paragraphs 8-124 and 8-125. Do not remove the patient's identification tags.

Step 5 - Decon the Patient's Skin

8-130. **Spot decon.** Wash exposed skin surfaces with soap and warm water, to include areas around the neck, wrists, and lower parts of the face. Decon the patient's identification tags and chain, if necessary.

8-131. **Combat medic care.** The combat medic gently cuts away the bandage. He decontaminates the area around the wound and irrigates it with a 0.5 percent chlorine solution. If bleeding begins, he replaces the bandage with a clean one. He replaces the old tourniquet by placing a new one ½ to 1 inch above the old one. He then removes the old tourniquet and decontaminates the patient's skin with a 0.5 percent chlorine solution. He does not remove a splint. He decontaminates the splint by thoroughly rinsing it, to include the padding and cravats, with a 0.5 percent chlorine solution.

WARNING
Do not use an SDK around any wounds. Do not remove splints. The splint will not be removed until the patient has been evacuated to a forward MTF or hospital.

8-132. **Completeness of decon check.** Check the patient with M8 detector paper or the CAM for completeness of decon.

NOTE: Other monitoring devices may be used, if available.

8-133. **Contaminated-waste disposal.** Dispose of contaminated bandages and coverings by placing them in a contaminated-waste bag. Seal the bag and place it in the contaminated-waste dump.

Step 6 - Transfer the Patient to the Hot Line

8-134. Move the patient to the hot line. Two medics from the clean side of the hot line move the patient into the clean treatment area.

NOTE: Before decontaminating another patient, each decon team member drinks about half a quart of water. The exact amount of water consumed is increased or decreased according to the work level and temperature (see Table 8-2, page 8-11).

AMBULATORY PATIENT

8-135. The decon procedures for an ambulatory patient that is contaminated with radiation are discussed in the following paragraphs. Some procedures can be done with one soldier, while others require more than one.

8-136. All ambulatory patients requiring EMT or AMT in the clean area of the BAS will be decontaminated. Stable patients not requiring treatment at the BAS, but requiring evacuation to the medical company's clearing station or a corps hospital for treatment (for example, a patient with a broken arm), should be evacuated in their protective overgarments and masks by any

available transportation. However, before evacuation, spot remove all thickened agents from their protective overgarment.

NOTE: Place the bandage scissors that are used in this procedure in a container of 5 percent chlorine solution when not in use. Most ambulatory patients will be treated in the contaminated treatment area and returned to duty. Upon removal of an ambulatory patient's clothing, he becomes a litter patient. The BAS and clearing station do not have clothing to replace those that are cut off during the decon process. The patient must be placed in a PPW for protection during evacuation (Figure 8-7, page 8-12).

Step 1 - Remove the LCE

8-137. Remove the LCE by unfastening/unbuttoning all connectors or tie straps and then place the equipment in a plastic bag. Place the plastic bag in the designated storage area for later decon.

Step 2 - Decon the Patient's Mask and Hood

8-138. **Begin the clothing removal process.** After the patient has been triaged and treated (if necessary) by the senior medic in the patient-decon station, the clothing removal process begins.

8-139. **Decon the hood.** Brush off contamination or use soap and water to wipe down the front, sides, and top of the hood.

8-140. **Remove the hood.** Remove the hood by cutting it with scissors or by loosening it from the mask attachment points for the quick-doff hood or other similar hoods (see Figure 8-3, page 8-6). Before cutting the hood, dip the scissors in a 5 percent chlorine solution. Cut the neck cord and the small string under the voicemitter. Release or cut the hood shoulder straps and unzip the hood zipper. Cut the hood upward to the top of the eye-lens outsert, staying close to the filter-inlet cover and eye-lens outsert, then across the forehead to the outer edge of the other eye-lens outsert. Proceed downward toward the patient's shoulder, staying close to the eye-lens outsert and filter-inlet cover, then across the lower part of the voicemitter to the zipper. After dipping the scissors in the 5 percent chlorine solution, cut the hood from the center of the forehead over the top of the head. Fold the left and right sides of the hood away from the patient's head and remove the hood.

8-141. **Decon the protective mask and exposed skin.** Use soap and water to wipe the external parts of the mask. Cover the mask's air inlets with gauze or your hands to keep the mask filters dry. Wash the exposed areas of the patient's face, to include the neck and behind the ears, with soap and warm water. Do not remove the protective mask.

8-142. **Remove the FMC.** Cut the patient's FMC tie wire, allowing the FMC to fall into a plastic bag. Seal the plastic bag and rinse the outside of the bag with a 5 percent chlorine solution. Place the plastic bag under the back of the protective mask's head straps. The FMC will remain with the patient in the contaminated area and a clean copy will be made before the patient is moved to the clean area.

Step 3 - Remove Gross Contamination From the Patient's Protective Overgarment

8-143. Remove all visible contamination spots by brushing off or by washing with soap and water.

Step 4 - Remove the Patient's Personal Effects and Protective Overgarment

8-144. **Remove the patient's personal effects.** Remove the patient's personal effects from his protective overgarment and BDU pockets. Place the articles in a plastic bag, label the bag with the patient's identification, and seal the bag. If the articles are not contaminated, they are returned to the patient. If the articles are contaminated, place them in the contaminated holding area until they can be decontaminated, then return them to the patient.

8-145. **Remove the overgarment jacket.** Have the patient stand with his feet spread apart at shoulder width. Unsnap the front flap of the jacket and unzip the jacket. If the patient can extend his arms, have him clinch his fist and extend his arms backward at about a 30° angle. Move behind the patient, grasp his jacket collar at the sides of the neck, and peel the jacket off the shoulders at a 30^0 angle down and away from the patient. Avoid any rapid or sharp jerks, which spread contamination. Gently pull the inside sleeves over the patient's wrists and hands.

8-146. If the patient cannot extend his arms, you must cut the jacket to aid in its removal. Before cutting the overgarment jacket, dip the scissors in a 5 percent chlorine solution to prevent contamination of the patient's BDU or undergarment. As with the litter patient, make two cuts, one up each sleeve from the wrist up to the shoulder and then across the shoulder through the collar. Cut around bandages, tourniquets, and splints, leaving them in place. Do not allow your gloves to touch the patient along the cut line. Peel the jacket back and downward to avoid spreading contamination. Ensure that the outside of the jacket does not touch the patient or his inner clothing.

8-147. **Cut and remove the overgarment trousers.** Unfasten or cut all ties, buttons, or zippers before grasping the trousers at the waist and peeling them down over the patient's combat boots. Again, the trousers are cut to aid in removal. If necessary, cut both trouser legs starting at the ankle. Keep the cuts near the inside of the legs, along the inseam, to the crotch. Cut around all bandages, tourniquets, and splints. Continue to cut up both sides of the zipper to the waist and allow the narrow strip with the zipper to drop between the legs. Peel or allow the trouser halves to drop to the ground. Have the patient step out of the trouser legs one at a time. Place the trousers in the contaminated-waste bag. Place the scissors in a 5 percent chlorine solution.

8-148. **Remove the outer gloves.** Grasp the fingers of the gloves, roll the cuffs over the fingers, and turn the gloves inside out. Do not remove the inner cotton gloves at this time. Drop the gloves into the contaminated-waste bag. Do not allow the patient to touch his clothing or other contaminated objects with his hands.

8-149. **Remove the overboots.** Cut the overboot laces and fold the lacing eyelets flat on the ground. Step on the toe and heel eyelets to hold the overboot on the ground and have the patient step out of it. Repeat this procedure for the other overboot. If the overboots are in good condition, they can be decontaminated and reissued.

8-150. **Remove the patient's cotton glove liners.** Instruct the patient to remove his cotton glove liners to reduce the possibility of spreading contamination. Have the patient grasp the heel of one glove liner with the other gloved hand, peeling it off of his hand. Hold the removed glove by the inside and grasp the heel of the other glove, peeling it off of his hand. Place both gloves in the contaminated-waste bag.

Step 5 - Check the Patient for Contamination

8-151. After the patient's overgarment has been removed, check his BDU by using M8 detector paper or the CAM. Carefully survey the patient, paying particular attention to discolored areas, damp spots, and tears on the uniform; areas around the neck, wrists, and ears; and bandages, tourniquets, and splints. Remove contaminated spots by washing with soap and water. Always dip the scissors in a 5 percent chlorine solution after each cut. Recheck the area with the detection equipment. If significant contamination is found on the BDU, then remove it and spot decon the skin. To remove the BDU, follow the procedures for removing the protective overgarment as described in paragraphs 8-145 through 8-147. Do not remove the patient's identification tags.

Step 6 - Decon the Patient's Skin

8-152. **Spot decon.** Wash exposed skin surfaces with soap and warm water, to include areas around the neck, wrists, and lower parts of the face. Decon the patient's identification tags and chain, if necessary.

8-153. Have the patient hold his breath, close his eyes, and lift, or assist him with lifting, his mask at the chin. Wipe his face and exposed areas of the skin with soap and water. Starting at the top of the ear and quickly wiping downward, wipe all folds in the skin, ear lobes, upper lip, chin, dimples, and nose. Continue up the other side of the face to the top of the other ear. Wipe the inside of the mask where it touches the face. Have the patient reseal and check his mask.

8-154. **Combat medic care.** The combat medic gently cuts away the bandage. He decontaminates the area around the wound and irrigates it with a 0.5 percent chlorine solution. If bleeding begins, he replaces the bandage with a clean one. He replaces the old tourniquet by placing a new one ½ to 1 inch above the old one. He then removes the old tourniquet and decontaminates the patient's skin with a 0.5 percent chlorine solution. He does not remove a splint. He decontaminates the splint by thoroughly rinsing it, to include the padding and cravats, with a 0.5 percent chlorine solution.

8-155. **Contaminated-waste disposal.** Dispose of contaminated bandages and coverings by placing them in a contaminated-waste bag. Seal the bag and place it in the contaminated-waste dump.

Step 7 - Transfer the Patient to the Hot Line

8-156. Have the decontaminated patient cross the hot line to the clean treatment area. The patient's boots must be decontaminated at the hot line before he enters the clean area. The patient's combat boots and protective mask will be removed at the entrance of the CPS or clean treatment area.

NOTE: Before decontaminating another patient, each decon team member drinks about half a quart of water. The exact amount of water consumed is increased or decreased according to the work level and temperature (see Table 8-2, page 8-11).

Chapter 9

Logistics

The materials that are needed to conduct decon operations are identified in this chapter.

SKIN DECON AND PERSONAL WIPE DOWN

9-1. Each soldier carries his own skin-decon supplies. The SDK is stored in the right cargo pocket of the overgarment trousers or BDU trousers. The basis of issue per soldier is two SDKs (12 packets). Each kit contains six individual decon packets. The unit maintains at least one SDK per assigned person for resupply. Personal wipe down is also done with these kits.

INDIVIDUAL-GEAR DECON

9-2. The standard decon kit for individual gear is the IEDK. Each soldier carries an IEDK in the right cargo pocket of his overgarment trousers or BDU trousers. The basis of issue per soldier is one IEDK. Each kit contains four individual decon packets, enough to decon two complete sets of individual gear. The unit maintains at least one IEDK per assigned soldier for resupply.

OPERATOR'S SPRAY DOWN

9-3. The M11 or M13 is the decon apparatus used for the operator's spray down. Fill the M11 with a 1 $^1/_3$-quart can of DS2 before use. The IEDK can be used on equipment that the DS2 may cause damage to by corrosive action.

9-4. Four nitrogen cylinders and two 1 $^1/_3$-quart cans of DS2 should be carried in the vehicle with each M11 (two nitrogen cylinders are authorized per can of DS2). Use the nitrogen cylinders to pressurize the M11 during use. Operating the M11 in below-freezing weather (below 32°F) requires two nitrogen cylinders to expend its contents.

9-5. Units maintain limited spare parts and accessories. They are not required to maintain additional stocks of 1 $^1/_3$-quart cans of DS2. The 5-gallon pails that are authorized to each unit can be used to refill empty M11s until a resupply of 1 $^1/_3$-quart cans of DS2 is received from the battalion.

9-6. The M13 comes with a 3.7 gallon DS2 container.

MOPP-GEAR EXCHANGE

9-7. Each soldier wears or carries one complete set of MOPP gear. The unit stocks a second complete set, which is sized and identified for each soldier. When a squad or platoon undergoes a MOPP-gear exchange, the unit sends a supply vehicle with replacement MOPP gear and any decontaminants (STB) to rendezvous with the contaminated element at the operational-decon site. (The procedure on how to prepare the STB dry mix is discussed in Appendix B.) Use about 40 pounds of the STB dry mix per squad-sized element. In wartime, 5-pound quantities of the STB dry mix can be put in 1-gallon airtight containers and stored in the company's supply section.

NOTE: Do not leave a 50-pound drum of STB dry mix open because it loses its effectiveness if left open to the air for extended periods. The unit is resupplied with MOPP gear and decontaminants from its battalion.

9-8. Units maintain a 5 percent overage of MOPP gear based on their personnel strength or authorization (whichever is greater) to ensure a complete range of sizes and replacement gear. All MOPP gear, serviceable and unserviceable, is handled as organizational clothing and equipment (Class II supplies). The higher HQ is responsible for resupply of these items.

VEHICLE WASHDOWN

9-9. Like the MOPP-gear exchange, a vehicle washdown is done in the unit's AO. The battalion's PDDE crew, stationed in the battalion trains, conducts the washdown. The crew moves to the operational-decon site, rendezvous with the contaminated element, and conducts the washdown. The crew uses its PDDE to spray about 100 to 150 gallons of hot, soapy water on each vehicle to wash off the gross contamination. For vehicles such as the M1-series armored fighting vehicles, about 200 gallons of water may be required per vehicle. One hundred gallons of water per vehicle provides a 2- to 3-minute wash. To speed up the operation, the crew should heat the water before arriving at the operational-decon site. See Chapters 2 and 3 and Table 9-1 for the planning and coordination of and the requirements for a vehicle washdown.

9-10. The M12A1 PDDA injects detergent into the water as it operates. It uses 2.5 quarts of detergent for every 1,200 gallons of water (about 1 quart of soap per 450 gallons of water). The battalion decon crew should maintain a basic load of liquid all-purpose detergent, which is sufficient to decon 30 percent of the battalion's organic vehicles.

DETAILED TROOP DECON

9-11. Generally, units conduct DTD in the brigade support area. Battalions conduct it in the division/corps support area. Materials for this technique usually are stocked in the battalion or brigade trains. Reconstitution operations should be closely associated with decon operations. The battalion's assessment and recovery team (ART) ensures that the material and equipment are available for the decon operation as part of the reconstitution effort.

DETAILED EQUIPMENT DECON

9-12. A chemical platoon must have access to large water sources—rivers, ponds, and public water systems—to conduct the DED. The chemical unit leader estimates the amounts of decontaminants that are needed. The platoon sets up the DED site, supplies the decontaminants, and conducts the DED. The platoon should carry enough decontaminants to service one company/team/battery. A decon platoon that is assigned to support a maneuver company (about 16 vehicles per tank company) should carry a minimum of 4 gallons of liquid detergent, 48 5-gallon cans of DS2, and 10 50-pound drums of STB. The chemical platoon is usually resupplied through its parent unit. Command-assignment relationships can change the resupply channels.

Table 9-1. Estimated Water Consumption for Decon

Required Equipment	Time	Water Consumption
Operational Decon[1]		
M12A1 PDDA or M17 LDS	1-3 minutes	100 to 150 gallons per regular vehicle 150 to 200 gallons per armored or larger vehicle Example: 15 (contaminated vehicles) x 150 (gallons of water) = 2,250 gallons[2]
Detailed Equipment Decon[3]		
M12A1 PDDA[4] or M17 LDS[4]	See Chapter 4 for the time allowed during a primary wash and rinse. Use the proper decon method.	Vehicles: Station 1 - primary wash, 250 gallons Station 4 - rinse, 200 gallons Armored or larger vehicle: Station 1 - primary wash, 300 gallons Station 4 - rinse, 200 gallons Example: 6 vehicles x 450 gallons of water = 2,700 gallons 4 tanks x 500 gallons of water = 2,000 gallons Total: 4,700 gallons[2]
Detailed Troop Decon		
30-gallon container		The initial setup requires 258 gallons of water. The water must be exchanged after 10 troops have been decontaminated through the DTD to avoid the transfer of contamination. Station 1 (120 gallons, 4 30-gallon containers).
3-gallon container		Station 2 (6 gallons, 2 3-gallon containers).
30-gallon container		Station 4 (180 gallons, 6 30-gallon containers).
3-gallon container		Station 7 (12 gallons, 4 3-gallon containers). Example: About 150 troops are to be decontaminated through the DTD. You will need 258 gallons of water per every 10 troops. Example: 150 (number of troops) ÷ 10 (required water exchange) = 15 (the amount of times the water will need to be exchanged) 15 x 258 (gallons of water per every 10 troops) = 3,870 gallons of water for 150 troops[2]

[1]To reduce contamination, conduct the operator's spray down before the operational decon. This process requires less water consumption during a thorough decon.

[2]Always include a 10 percent planning factor to the total estimate of the water consumption for the DED and the DTD. Example: 2,250 (gallons of water for the DTD) x 10 percent = 225 (additional gallons of water). A total of 2,250 + 225 = 2,475 (gallons of water required for the DTD).

[3]The planner should consider vehicle predecon action to estimate water consumption. He should consider vehicles that were processed through operational decon because they will normally increase the weatheriza-tion process and may reduce water usage.

[4]The M17 LDS with two wands uses 14 gallons of water per minute. The M17 LDS with only one wand, which is fitted with an injector, uses 25 gallons per minute. The M12A1 PDDA pumps 25 gallons of water per minute.

9-13. Chemical platoons attached to division engineers may be resupplied through the division engineer battalion instead of their parent company. See Table 9-2 for the decon resources that are available at each organizational level. For the equipment and supplies that are needed for decon operations, see Table 9-3, pages 9-5 through 9-7.

Table 9-2. Decon Resources Available at Each Organizational Level

Organizational Level	Decon Resources
Individual soldiers	2 SDKs 1 canteen of water 2 IEDKs
Operators and crews	2 SDKs 2 IEDKs 1 on-board decon apparatus (M11 or M13) Soap and water 2 50-pound drums STB
Companies	2 5-gallon pails of DS2 2 immersion heaters 2 to 3 30-gallon containers 6 3-gallon containers 6 long-handled brushes 6 sponges 300 plastic trash bags
Battalion PDDE crews	PDDE (M17 LDS) Basic load, liquid detergent
Chemical company decon squad	PDDE (M12A1 PDDA/M17 LDS) Basic load, liquid detergent
Chemical company decon platoon	PDDE (M12A1 PDDA/M17 LDS) Interior decon equipment Sufficient materials to set up a DTD

Table 9-3. Equipment and Supplies Needed for Decon Operations

Minimum Amounts of Equipment and Supplies Needed for Decon Techniques							Nomenclature	NSN	Class of Supply	Unit of Issue	Basis of Issue (See TOE for Actual Authorization)***
SD*	PW*	OS*	MGX**	VW**	DTD**	DED**					
		1				***	Decon apparatus, DS2, ABC-M11 or M13 DAP	4230-00-720-1618 4230-01-113-4124	II II	Each Each	1 per every major equipment
		2					Cylinder, nitrogen filled	4230-00-775-7541	II	Box	2 per can, DS2 5 per box
1	1	1	2		35	2	Decon kit, M291 SDK M295 IEDK	4230-01-0276-1905 6850-01-3577-8456	II II	Each Each	1 per mask 2 per soldier
		1					Fluid-filled container, DS2	4230-01-136-8888	II	Each	1 per M13 DAP
					1		Filter canister, C2, or filter canister, C2A1	4240-01-119-2315 4240-01-361-1319	II	Each	1 per M40-series mask
					1		Hood, M40 mask	4240-01-376-3152	II	Each	1 per mask
			1		2		Shears	5111-00-223-6371	II	Each	As needed
							Knife	5110-00-240-5943	II	Each	As needed
			***			***	Axe, single bit	5110-00-293-2336	II	Each	1 per most vehicles
			***		***	***	Shovel, hand, RD, PT, D handle	5120-00-293-3336	VII	Each	1 each for most vehicles
						***	NAAK, MK1	6505-01-140-6455	VII	Each	3 per individual
						***	CANA	6505-01-274-0951	VII	Each	1 per individual 5 per combat life saver 10 per combat medic
					***	3	Paper, chemical agent, detector, M9	6665-00-049-8982	II	Roll	1 per squad 3 per platoon

Table 9-3. Equipment and Supplies Needed for Decon Operations (Continued)

Minimum Amounts of Equipment and Supplies Needed for Decon Techniques							Nomenclature	NSN	Class of Supply	Unit of Issue	Basis of Issue (See TOE for Actual Authorization)***
SD*	PW*	OS*	MGX**	VW**	DTD**	DED**					
					***	***	Paper, chemical agent, detector, M8	6665-00-050-8529	II	Book	6 books per company
					***	***	Radiac meter, AN/PDR-2	6665-01-222-1425	VII	Each	Per MTOE
					2	***	Radiac meter, IM-93/UD	6665-00-752-7759	VII	Each	Per MTOE
					1	2	Alarm, chemical agent, M8A1 or M22	6665-00-935-6955 6665-01-438-6963	VII VII	Each Each	Per MTOE
					4	2	CAM	6665-01-199-4153	VII	Each	Per MTOE
					4		Detector kit, chemical agent, M256A1	6665-01-016-8399	II	Kit	1 per squad
					1		Mask sanitizing solution	6810-00-266-6979	III	Tube	4 per 10 masks
					**	1	Decontaminating agent, STB	6850-00-297-6653	III	Drum	2 50-pound drums per company
						***	Decontaminating agent, DS2	6850-00-753-4870	II	Can	5 gallons, 25 cans per decon
		1			2	24	Brush, scrub, long-handled	7920-00-141-5452	II	Each	As required
					7		Pail, metal, 14-quart	7240-00-160-0455	II	Each	As required
			1		9	4	Garbage can, galvanized, 30-gallon	7240-00-160-0440	II	Each	2 per company
						6	Mop	7920-00-224-8756		Each	As required
					5		Sponge, cellulose	7920-00-240-2559	II	Each	As required
			2		4	6	Brush, scrub	7920-00-240-7171	II	Each	As required
			2		2		Towels, paper	7920-00-823-6931	II	Box	As required
			1	1	1	1	Detergent, GP, liquid	7930-00-282-9699	II	Gallon	As required

Table 9-3. Equipment and Supplies Needed for Decon Operations (Continued)

Minimum Amounts of Equipment and Supplies Needed for the Decon Techniques							Nomenclature	NSN	Class of Supply	Unit of Issue	Basis of Issue (See TOE for Actual Authorization)***
SD*	PW*	OS*	MGX**	VW**	DTD**	DED**					
			1		1	1	Bag, plastic	8105-00-655-8286	II	Box	125 count
				**		10	TAP apron	8415-00-281-7813 through 8415-00-281-7816	II	Each	10 per platoon
			1		1		Suit, clothing, protective, See CTA 50-909 for clothing tariff	8415-01-137-1700 through 8415-01-137-1707	II	Each	2 per soldier
			1		1		Glove set, CP, See SB 10-523 for clothing tariff	8415-01-033-3517 through 8415-01-033-3520	II	Pair	1 per soldier
							Cover, helmet, chemical protective	8415-01-111-9028	II	Each	1 per soldier
			1		1		Overboots	8430-01-048-6305 through 8430-01-049-0887	II	Pair	1 per soldier
			2	2	2	2	NBC marking kit	9905-12-124-5955	II	Kit	1 per squad
					2		Immersion heater		II	Each	2 per company
*Techniques executed by individual soldiers							SD – Skin decon PW – Personal wipe down OS – Operator's spray down		Amount is consumption rate for 1 soldier		
**Techniques executed by units							MGX – MOPP-gear exchange VW – Vehicle washdown DTD – Detailed troop decon DED – Detailed equipment decon		Amount is consumption rate for 1 platoon (4 vehicles, 40 personnel) Amount is consumption rate for 1 company (20 vehicles, 150 personnel)		
***Amounts vary, depending on situation											

Chapter 10

Decon Procedures for Individual and Crew-Served Weapons

Decon procedures for light infantry crew-served weapons and some individual major weapons that are critical to the unit's mission are discussed in this chapter. For those units that are not authorized decon devices such as the M11 or M13 DAP, alternative measures are described. Decon procedures are specified for the following weapons: M60 machine gun; M249 squad automatic weapon (SAW); M47 Dragon; tube-launched, optically tracked, wire-guided (TOW) missile (M220 series); 81-millimeter mortar; 60-millimeter mortar; M203 grenade launcher; AT4 light antitank weapon (LAW); LAW (M72 series); and 66-millimeter rocket launcher.

OPERATIONAL-DECON PROCEDURES USING THE IEDK

10-1. Operational decon is conducted to sustain combat operations, remove gross contamination, and reduce the transfer/spread of contamination. Before decon, soldiers should visually inspect the weapon system for contamination. If liquid contamination is present, follow the procedures prescribed for each weapon system. The protection level required for operational-decon operations is MOPP4.

NOTE: The United States Marine Corp (USMC) does not possess the IEDK; therefore, it will have to use additional SDKs to decon individual and crew-served weapons.

10-2. The following procedures reduce the spread/transfer of liquid contamination. These procedures are unique to the IEDK. Using other decon equipment with these procedures could result in a malfunction of the weapons/weapon systems. Powder from the IEDK could affect the mechanical parts of the weapons/weapon systems if not used properly. See the maintenance manual for proper lubrication after decon.

WARNING
Do not apply an IEDK to any optic lens. The abrasive effect of the charcoal inside the IEDK will damage the lens.

M60 MACHINE GUN

NOTE: Ensure that the feed tray of the weapon is closed. This prevents the powder from the IEDK from falling inside the feed tray and jamming the weapon. Do not touch the barrel of the weapon with the chemical protective gloves when it is hot. The heat of the barrel could melt the gloves.

10-3. To decon the M60 machine gun, follow the steps below:

- Step 1. Decon your gloves using the IEDK.
- Step 2. Use the IEDK to remove all liquid contamination from the bipod, barrel assembly, forearm assembly, and carrying handle. Use a second IEDK if necessary.
- Step 3. Use the IEDK to decon the feed-tray cover, shoulder stock, and trigger assembly.
- Step 4. Repeat step 1.

NOTE: Do not decon the flash suppressor with the IEDK. Powder debris may fall into the barrel and cause the weapon to malfunction.

M249 SQUAD AUTOMATIC WEAPON

NOTE: Ensure that the feed tray of the weapon is closed. This prevents the powder from the IEDK from falling inside the feed tray and jamming the weapon. Do not touch the barrel of the weapon with chemical protective gloves when it is hot. The heat of the barrel could melt the gloves.

10-4. To decon the M249 SAW, follow the steps below:

- Step 1. Decon your gloves using the IEDK.
- Step 2. Use the IEDK to remove all liquid contamination from the bipod, barrel assembly, gas regulator, and carrying handle.
- Step 3. Use the IEDK to decon the feed-tray cover, magazine, trigger assembly, and shoulder stock.
- Step 4. Repeat step 1.

NOTE: Ensure that the weapon is upright when decontaminating the feed-tray cover with the IEDK. This prevents powder from the IEDK from falling into the ammunition feeder. Place the magazine in the well before the decon.

M203 GRENADE LAUNCHER

NOTE: The M203 is not a crew-served weapon; however, this weapon is considered critical for accomplishing the infantry squad's mission. Decon the M16A2 according to STP 21-24-SMCT. Before decontaminating the M203 grenade launcher, ensure that the dust cover is closed and the magazine is stored in the well. Ensure that the barrel assembly is closed.

10-5. To decon the M203 grenade launcher, follow the steps below:

- Step 1. Decon your gloves using the IEDK.
- Step 2. Use the IEDK to decon the barrel, hand guard, and receiver.
- Step 3. Use the IEDK to decon the lower receiver group and butt stock.
- Step 4. Repeat step 1.

M47 DRAGON

10-6. Use this procedure only when liquid contamination is present on encased missile or tracker components. If the launcher has been fired and chemical contamination is present, follow the procedures shown in steps 1, 4, and 5 of paragraph 10-7 to decon the tracker. Gather all the contaminated launching tubes in a single area before departing the contaminated area and mark them properly.

NOTE: Do not use the IEDK on any optic lens (exclude rubber eyecups), battery connector, electrical connectors, or lubricated components. Using the IEDK could cause corrosion if the powder from it is left on the equipment for an extended period of time. However, if these items are stored in a protective container/box, decon the storage container with the IEDK. Rinse or flush the lens with water only if it becomes contaminated. The rubber eyecups can be rinsed with water or decontaminated with the IEDK.

10-7. To decon the M47 Dragon, follow the steps below:

- Step 1. Decon your gloves using the IEDK.
- Step 2. Use the IEDK to decon the bipod.
- Step 3. Use the IEDK to decon the exterior of the launcher.
- Step 4. Rinse or flush any liquid contamination off the exterior of the day/night sight with water. Use a rag or any available towelettes to dry the sight but do not touch the lens.
- Step 5. Repeat step 1.

TOW MISSILE (M220 SERIES)

10-8. Gather all the contaminated launching tubes in a single area before departing the contaminated area and mark them properly.

NOTE: Do not use the IEDK on any optic lens (exclude rubber eyecups), battery connector, electrical connectors, or lubricated components. However, if these items are stored in a protective container/box, decon the storage container with the IEDK or spray it with the M11 DAP. Do not touch or decon the front lens of the night sight. Rinse or flush the lens with water if it becomes contaminated with liquid chemical agent. The rubber eyecups can be rinsed with water.

10-9. To decon the TOW missile (M220 series), follow the steps below:

- Step 1. Decon your gloves using the IEDK.
- Step 2. Use an M11 DAP filled with bleach or soapy water to spray down the launch tube, traversing unit, and tripod respectively. Protect or cover the day/night sights when spraying.
- Step 3. Use the IEDK to decon the exterior of the day-sight tracker and night-sight tracker. Avoid touching the lens. Once the sights are decontaminated, brush off any powder residue.
- Step 4. Use the IEDK to decon the collimator.
- Step 5. Use the IEDK to decon the exterior of the missile-guidance set and cable surface. Ensure that the electrical connectors are covered before the decon.
- Step 6. Use the IEDK to decon the power conditioner and its cable for the battery component.
- Step 7. Repeat step 1.

AT4 LIGHT ANTITANK WEAPON

NOTE: Ensure that the transport safety pin, the cocked-level safety, and the forward safety on the launcher are engaged.

10-10. To decon the AT4 LAW, follow the steps below:

- Step 1. Decon your gloves using the IEDK.
- Step 2. Use the IEDK to wipe the exterior of the launcher and the front and rear sights.
- Step 3. Repeat step 1.

66-MILLIMETER ROCKET LAUNCHER

NOTE: Ensure that the launcher is in the safe position and is not loaded before decontaminating. Omit step 2 if the launcher has not been fired in a contaminated environment.

10-11. To decon the 66-millimeter rocket launcher, follow the steps below:

- Step 1. Decon your gloves using the IEDK.
- Step 2. Extend the launcher into the firing position.
- Step 3. Use the IEDK to decon the exterior of the launcher and around the reflecting sight.
- Step 4. Repeat step 3 using the IEDK.
- Step 5. Repeat step 1.

81-MILLIMETER MORTAR

NOTE: Following the decon of the mortar with the IEDK, ensure that the mechanical parts of the mortar are dried. Lubricate the mortar when the mission permits. Do not touch the mortar tube with chemical protective gloves when it is hot. The heat of the tube could melt the gloves. Do not apply the IEDK to any optic lens due to the abrasive effect of the charcoal in the IEDK. Rinse or flush the lens with water, when required.

10-12. To decon the 81-millimeter mortar, follow the steps below:

- Step 1. Decon your gloves using the IEDK.
- Step 2. Use the IEDK to decon from the blast attenuator device to the lower half of the mortar tube.
- Step 3. Use the IEDK to decon the base plate.
- Step 4. Use the IEDK to decon the bipod.
- Step 5. Use the IEDK to decon the elbow telescope and telescope mount. Do not use the IEDK on the optic lens. Flush the lens with water if contamination is present.
- Step 6. Repeat step 1.

60-MILLIMETER MORTAR

NOTE: Following the decon of the mortar with the IEDK, ensure that the mechanical parts of the mortar are dried. Lubricate the mortar when the mission permits. Do not touch the mortar tube with chemical protective gloves when it is hot. The heat of the tube could

melt the gloves. Do not apply the IEDK to any optic lens due to the abrasive effect of the charcoal in the IEDK. Rinse or flush the lens with water, when required.

10-13. To decon the 60-millimeter mortar, follow the steps below:

- Step 1. Decon your gloves using the IEDK.
- Step 2. Use the IEDK to decon the mortar tube and handgrip. If the mortar is heavily contaminated, you may need to use another IEDK.
- Step 3. Use the IEDK to decon the bipod and base plate.
- Step 4. Use the IEDK to decon the exterior of the telescope. Do not apply the IEDK to the lens. Flush the lens with water if contamination is present.
- Step 5. Repeat step 1.

JAVELIN

10-14. The decon procedures for the Javelin and its components are listed in Table 10-1, page 10-6.

> **WARNING**
> **The IEDK only removes a liquid hazard. Decontaminated items may still present a vapor hazard. Do not unmask until it has been determined safe to do so.**

> **CAUTION**
> Close the lens covers on the Javelin before decontaminating because decon powders on the lens can obscure your vision. Do not use the IEDK on the eyecup because the decon powder entering the eyecup can obscure your vision.

THOROUGH-DECON PROCEDURES USING THE HTH SOLUTION

10-15. When conducting the DED, the equipment is decontaminated with a 10 percent HTH solution (see Table 10-2, page 10-6) or with pure bleach that is diluted. Use the M256A1 detector kit, CAM, or M8 detector paper to determine the effectiveness of the decon. The crew-served weapons should be decontaminated before the DTD by the individual weapons' crew (the preferred method). Alternatively, the commander could task his own decon team to decon the unit's light weapons. The contaminated unit is responsible for establishing an area about 109 to 164 yards upwind from the DTD and DED sites for decon of the crew-served weapons. Unit leaders supervise this procedure. The unit chemical NCO plans and coordinates the preparation of this area.

10-16. The preferred decontaminant for a thorough decon is a 10 percent HTH solution (see Table 10-2). Use a contact time for the HTH solution of 15 minutes for chemical and biological agents. Prepare the amount of solution required according to the number and size of the weapon systems that need to be decontaminated.

Table 10-1. Decon Procedures for the Javelin and Its Components

Immediate Decon	
Round	**Command Launch Unit (CLU)**
Step 1. Decon your gloves using the IEDK. Step 2. Decon the round by patting it with the IEDK, working from the front to the rear until the entire round is covered. **NOTE: When the shoulder strap is grossly contaminated, discard it.** Step 3. Repeat step 1. Step 4. Discard the IEDK.	Step 1. Decon your gloves using the IEDK. Step 2. Decon the CLU by patting it with the IEDK. Pay special attention to the handgrips, CLU carrying handle, battery cover, round interface-connector protective cover, and switch boots. Step 3. Use a knife to cut the protective-cover lanyard from the protective cover and remove it. Step 4. Repeat step 1. Step 5. Discard the IEDK.
Thorough Decon	
Predecon Round	**Predecon CLU**
Step 1. Check for contamination. Step 2. Remove the shoulder strap. Step 3. Use a knife to cut the connector boot. Step 4. Use a knife to cut the protective-cover lanyard from the protective cover and remove it. Step 5. Use a knife to scrape about 1/4 inch of the shoulder pad away from the launcher-tube assembly (LTA). Step 6. Ensure that the protective cover is installed on the round/CLU interface connector.	**NOTE: Ensure that the on/off switch is in the off position. If the connector was not covered before contamination, place the protective cover on the connector before decon. After the decon, remove the protective cover and allow the connector to weather. The foam around the connector may be cut away as required.** Step 1. Check for contamination. Step 2. Remove the rubber from the handles and remove the lanyards from the CLU. Step 3. Ensure that the protective covers are installed on the connectors.
Decon LTA	**Decon CLU**
Step 1. Submerge one half of the LTA in the decon solution. Step 2. Submerge the other half of the LTA in the decon solution. Step 3. Rinse one half of the LTA by submerging it in the rinse solution. Step 4. Rinse the other half of the LTA by submerging it in the rinse solution. Step 5. Use the CAM to check the LTA, ensuring that all contaminants have been removed. Step 6. Repeat steps 1 through 5 above if the LTA is still contaminated.	Step 1. Open the lens covers. Step 2. Submerge the CLU in the decon solution. Step 3. Rinse the CLU by submerging it in the rinse solution. Step 4. Use the CAM to check the CLU, ensuring that all contaminants have been removed. Step 5. Repeat steps 1 through 4 above if the CLU is still contaminated. If only the switch boots remain contaminated, remove them by using a knife.

Table 10-2. HTH Mixture

HTH Solution	Amount of HTH per 1 Gallon of Water	Amount of HTH per 5 Gallons of Water
5 percent	9.6 ounces	48 ounces
10 percent	12 ounces	60 ounces

NOTE: Refer to Appendix B for instruction on how to use sodium hypochlorite (household bleach) and HTH, and see Table 10-2 for instructions on how to prepare the 5 and 10 percent HTH solution.

10-17. Follow the procedures below during the DED and DTD. Always wear chemical gloves when performing these procedures.

- Use the HTH solution to decon up to 10 weapon systems. Use the required number of rags on a weapon system and then dispose of them properly to avoid transferring the contamination.
- Do not apply the HTH solution or bleach on electrical connectors and lens.
- Use bleach or a 10 percent HTH solution on the exterior of the optic systems that are not sensitive to bleach and then rinse them with water.

10-18. Once the HTH solution has been used on the weapon systems, thoroughly dry and lubricant them. Avoid spilling the HTH solution or bleach on the chemical overgarment.

MATERIALS

10-19. Listed in the following paragraphs are the different types of materials that will be needed to clean the weapon systems:

Cleaning Rags

10-20. The recommended number of clean rags to use with the HTH solution on each weapon system follows:

- M60 machine gun/M249 SAW (three rags)
  - 1: Bipod and barrel.
  - 1: Forearm, carrying handle, and rear sight.
  - 1: Feed-tray cover, shoulder stock, pistol grip, and trigger assembly.
- M203 grenade launcher (two rags)
  - 1: M16A1/2 barrel, hand guard, receiver group, and butt stock.
  - 1: M203 receiver group, sight, and exterior of the barrel.
- M47 Dragon (three rags)
  - 1: Exterior of the sight.
  - 1: Bipod (if weapon is not fired).
  - 1: Round (canister) (if weapon is not fired).

NOTE: Keep the electrical connector covered before the decon. The Dragon's rubber eyepiece can be removed and submerged in the decontaminant.

- TOW missile (M220 series) (seven rags)
  - 1: Exterior of the night sight.
  - 1: Exterior of the day sight.
  - 1: Launch tube.
  - 1: Traversing unit.
  - 1: Tripod assembly.
  - 1: Battery-power conditioner and cable. Ensure that the battery is in a waterproof container or case.
  - 1: Missile-guidance set and its cable.

NOTE: Lock the bridge clamp after removing the round. Do not get the HTH solution on the electrical connector under the bridge clamp or on the post-amplifier cable.

- 66-millimeter rocket launcher (one rag)
- AT4 LAW (one rag)
- 81-millimeter mortar (four rags)
  - 1: Blast denature device and tube.
  - 1: Bipod.
  - 1: Base plate.
  - 1: Exterior of the telescope.
- 60-millimeter mortar (three rags)
  - 1: Tube (barrel) and handgrip.
  - 1: Base plate and bipod.
  - 1: Exterior of the telescope.

Water

3-21. Use five gallons of water (as a minimum) for the HTH mix. Change the solution after 10 weapons are decontaminated.

Buckets

3-22. Use a minimum of two buckets for the procedures below.

Trash Bags

3-23. Dispose of the contaminated rags in plastic trash bags.

NBC Marking Kit

3-24. Mark the contaminated area once the decon procedures are completed.

Lubricants

3-25. Use the required lubricants for the weapon systems. Follow the procedures specified in the applicable TMs.

PROCEDURES

3-26. The procedures on how to decon weapons and weapon systems with the HTH solution are described below:

- Step 1. Submerge the rag or cloth in the HTH solution and ensure that the material is completely dampened.
- Step 2. Wipe down the weapons and weapon systems with the rag until the contamination is no longer suspected.
- Step 3. Submerge the rag or cloth into a container of clean water (use hot water when available) and wipe down the weapons and weapon systems once more to remove the HTH or bleach solution.
- Step 4. Completely dry the weapons and weapon systems once step 3 is completed.
- Step 5. Lubricate the weapons and weapon systems.

Appendix A

Conversions and Measurements

This appendix contains an English/Metric conversion chart (see Table A-1). It also contains the measurements and weights of decontaminant containers that are replacements for used containers (see Table A-2, page A-2).

Table A-1. English/Metric Conversion Chart

To Convert	To	Multiply By	To Convert	To	Multiply By
Feet	Centimeters Miles Kilometers Meters	30.48 0.00019 0.00003 0.3048	Gallons	Milliliters Cubic meters	3,785 0.00379
Cubic yards	Cubic meters	0.76456	Quarts (fluid)	Millimeters Liters	946.358 0.946333
Cubic feet	Cubic meters Liters	0.02831702 28.32	Ounces	Grains Drams Pounds Grams Kilograms	437.6 16 0.0625 28.34952 0.02835
Yards	Meters	0.9144	Pounds	Grams Kilograms Metric tons	453.6 0.453592 0.0004536
Inches	Centimeter Kilometers Millimeters	2.54000508 $2.54000508 \times 10^{-5}$ 25.4000508	Ounces (fluid)	Milliliters Liters	29.57353 0.02957
			Square yards	Square meters	0.836
NOTE: For temperatures, Celsius = 5/9 (F° - 32°) and Fahrenheit = 9/5 (C° + 32°)					

Table A-2. Measurements and Weights of Decontaminant Containers

Containers	Measurements	Weights
Drum, 55 gallons, 16 gauge NSN 8110-00-597-2353	Volume—12 cubic feet Length—35 inches Width—27.5 inches	Empty—70 pounds Filled with water—526 pounds Filled with STB (slurry)—620 pounds Filled with DS2—507 pounds
Drum, 55 gallons, 18 gauge NSN 8110-00-292-9783	Volume—12 cubic feet Length—35 inches Width—27.5 inches	Empty—50 pounds Filled with water—506 pounds Filled with STB (slurry)—600 pounds Filled with DS2—487 pounds
Ash and garbage can, 32 gallons (GI can) NSN 7240-00-160-0440	Volume—7 cubic feet Length—26.5 inches Width—20 inches	Empty—33 pounds Filled with water—300 pounds Filled with STB (slurry)—353 pounds Filled with DS2—288 pounds
Gasoline can, 5 gallons NSN 7240-00-178-8286	Volume—1 cubic foot Length—18.50 inches Width—6.75 inches	Empty—10.5 pounds Filled with water—52 pounds Filled with STB (slurry)—66 pounds Filled with DS2—50 pounds
Pail, 5 gallons, DS2 NSN 7240-00-575-2243	Volume—1cubic foot Length—13.75 inches Width—11.12 inches	Empty—6 pounds Filled with water—48 pounds Filled with STB (slurry)—61 pounds Filled with DS2—46 pounds
Drum, 8 gallons, STB, 16 gauge NSN 6850-00-297-6693	Volume—1.4 cubic feet Length—14 inches Width—14 inches	Empty—11 pounds Filled with water—78 pounds Filled with STB (slurry)—91 pounds Filled with DS2—71 pounds

Appendix B

Decontaminants

The decontaminants listed in Tables B-1 through B-3, pages B-1 through B-14, are grouped according to their classification: standard, nonstandard, or natural. Each is identified by name and national stock number (NSN) (where applicable). Also listed are brief directions on preparing the decontaminants and any appropriate cautions and information on application, coverage, substitute solutions, and related effectiveness of the decontaminants.

Table B-1. Standard Decontaminants Available in the Supply System

Decontaminants	Use	Remarks	Cautions	Preparation
Decontaminating solution number 2; DS2 in a 1 1/3-quart can (NSN 6850-00-753-4827); DS2 in a 14-liter container for the M13 DAP; (NSN 4230-01-136-8888); DS2 in a 5-gallon pail (NSN 6850-00-753-4870)	Bio Chem	Is effective against all known toxic chemical and biological agents (except bacterial spores) if sufficient contact time is allowed. Must remain in contact with contaminated surface for about 30 minutes. Must be rinsed off with water. Recheck the item for contamination. Can be used at temperatures above -15°F. Is used with the M11 or M13 DAP or can be applied with brooms and swabs. Is most effective when the application is accompanied by scrubbing action.	Is extremely irritating to the eyes and skin. Must wear protective mask and rubber gloves. If DS2 contacts skin, wash area with water. Do not inhale vapors. Will cause a green or black color change upon contact with M8 detector paper and cause a false/positive reading with M9 detector paper. Will ignite spontaneously on contact with STB or HTH. Should not be spilled on the chemical protective overgarment. Is a combustible liquid with a flash point of 160°F. Do not confuse it with a fire extinguisher. Spraying DS2 on surfaces above 168°F will ignite it. Must not be used on individual mask (damages Mylar diaphragm in voicemitter assembly).	Mixing is not required. Issued in ready-to-use solutions.

Table B-1. Standard Decontaminants Available in the Supply System (Continued)

Decontaminants	Use	Remarks	Cautions	Preparation
Decontaminating solution number 2 (continued)			Will corrode aluminum, cadmium, tin, and zinc; will soften leather. May soften, remove, or discolor paint. Rinse well after use and oil metal surfaces. Is ineffective against bacterial spores.	
Supertropical bleach (NSN 6850-00-297-6653)	Bio Chem	Is effective against lewisite, V and G agents, and biological agents. Must remain in contact with contaminated surface for at least 30 minutes; then, wash off with clear water. Does not effectively decon mustard if it has solidified at low temperatures. Should be applied several times to porous surfaces.	Will ignite spontaneously on contact with liquid blister agent or DS2. Will give off toxic vapors on contact with G agent. Is not recommended for ship use. Store on top deck only. Is corrosive to most metals and damaging to most fabrics (rinse thoroughly and oil metal surfaces). Should not be inhaled or allowed to touch the skin. When preparing a slurry, wear a protective mask or respiratory protective device. Should be stored in an unheated warehouse away from combustibles and metals subject to corrosion.	Slurry paste: mix one 50-pound drum of STB with 6 gallons of water. Slurry paste consists of about equal parts (by weight) of STB and water. Dry mix: mix two shovels of STB to three shovels of earth or inert material (ashes). Slurry mix, chemical: mix will consist of 40 parts of STB to 60 parts of water (by weight). To mix in the M12A1 PDDA, use 1,300 pounds of STB, 225 gallons of water, 12½ pounds of antiset, 24 ounces of antifoam. Slurry mix, biological: mix will consist of 7 parts of STB to 93 parts of water (by weight). To mix in the M12A1 PDDA, use 150 pounds of STB, 225 gallons of water, 12½ pounds of antiset, and 24 ounces of antifoam. Camouflage: lampblack or dye mixes may be added for camouflage.

Table B-1. Standard Decontaminants Available in the Supply System (Continued)

Decontaminants	Use	Remarks	Cautions	Preparation
Mask sanitizing solution	Bio Chem	Use on a previously cleaned mask with filter elements/canisters removed. Place the mask face up; attach the canteen to the mask at the drinking tube. Drain one canteen full of sanitizing solution through the mask. Rinse the mask with two canteens of clear water. Immerse the mask and outserts in the sanitizing solution. Agitate the mask for 5 minutes. Rinse it twice in clear water, agitating 2 to 3 minutes each time. Dry all parts of the mask and reassemble. Use one gallon of solution for every 10 masks.		Fill a standard plastic canteen to the shoulder with water. Add a 0.5-gram tube of calcium hypochlorite from the water-purification kit (NSN 6810-00-266-6976). Cover the canteen and shake vigorously for 30 seconds. Mix bulk quantities as follows: add 2.0 grams of calcium hypochlorite from a 6-ounce jar (NSN 6810-00-255-0471) to 1 gallon of water. Use a ratio of about 1 pound of soap per gallon of water for smaller amounts of solution. Mix 2 pints of detergent to 450 gallons of water in the M12A1 PDDA.
Soaps and detergents: detergent, GP, liquid (NSN 7930-00-282-9699)	Rad Bio Chem	Scrub or wipe the contaminated surface with a hot, soapy water solution or immerse the item in the solution.	Are effective in physically removing contamination. However, casualty-producing levels of contamination may remain in the runoff water and must be considered contaminated.	Mix 75 pounds of powdered soap in 350 gallons of water. If powdered soap is not available, use bar laundry soap (75 pounds of soap cut into 1-inch pieces and dissolved in 350 gallons of hot water). Use a ratio of about 1 pound of soap per gallon of water for smaller amounts of soap solution. Mix 2 pints of detergent to 450 gallons of water in an M12A1 PDDA.

Table B-1. Standard Decontaminants Available in the Supply System (Continued)

Decontaminants	Use	Remarks	Cautions	Preparation
Calcium hypochlorite (high-test hypochlorite [HTH] or high-test bleach [HTB]) (NSN 6810-01-225-2682) (25 pounds), (NSN 6810-00-225-0472) (100 pounds)	Bio Chem	Is effective against mustards, lewisite, V agents, and all biological agents, including bacterial spores. Will react rapidly (within 5 minutes) with mustards and lewisite. Should be allowed a 15-minute contact time for biological agents. Will act faster than STB. Can be used as a dry mix or slurry.	Precautions are the same as for STB. Pure calcium hypochlorite will burn on contact with VX, HD, or DS2. Agent is more corrosive than STB. Will destroy clothing, has a toxic vapor, and will burn the skin. Protective mask and rubber gloves are the minimum protective equipment needed when handling.	Chemical: mix 5 pounds of decontaminant to 6 gallons of water (10 percent solution). Biological: mix 1 pound of decontaminant to 6 gallons of water (2 percent solution). PDDE: mix a slurry of 1 part decontaminant to 2 parts water (any heavier slurry will clog the decon apparatus). HTH should only be used if STB is not available.

Table B-2. Nonstandard Decontaminants Available in the Supply System

Decontaminants	Use	Remarks	Cautions	Preparation
Oxidizing agents (nitric acid, aqua regia, sodium dichromate, and potassium permanganate)	Rad	Are effective in dissolving surfaces containing absorbed radioactive contamination. Are applied to a surface or the item is dipped. Exposure must be limited due to the corrosive nature of the solution. Rinse thoroughly with water and detergent and then with clear water.	Are extremely corrosive. Use only under the supervision of a trained individual. Will require the use of a neoprene or rubber protective apron, gloves, boots, and safety glasses when handling. (Rubber offers only limited protection.)	Aqua regia is prepared by mixing 3 parts of concentrated hydrochloric acid and 1 part of concentrated nitric acid. Other oxidizing agents do not require mixing.
Iodine water purification tablets	Bio	Are used when it is impractical to boil drinking water. Two iodine tablets per canteen are effective against most biological agents.	None	None

Table B-2. Nonstandard Decontaminants Available in the Supply System (Continued)

Decontaminants	Use	Remarks	Cautions	Preparation
Complexing (chelating) agents (versene, citric acid, sequesterene, sodium citrate, tartanic acid, sodium oxalate, sodium tartrategoxalic acid, orthophosphoric acid, and similar agents)	Rad	Will aid in the physical removal of absorbed contamination on surfaces. Should be applied as a film over the surface using the PDDE, firefighting apparatus, or tree or garden sprayer. Should be allowed a 30-minute contact time and then flushed with water.	Do not neutralize contamination. Runoff will be contaminated.	Mix 3 to 5 percent of the agent (by weight) in water.
Disinfectant chlorine (NSN 6840-00-270-8172)	Bio	Use to decon utensils, mess gear, exteriors of sealed containers, and food products that can withstand soaking. Dispose of any food or vegetables that are damaged and any outer leaves that are bruised or torn. Do not cut or peel fruits and vegetables before disinfecting them. Leave items in the solution for 30 minutes and stir occasionally. Rinse thoroughly in potable water.	None	Dissolve one package of disinfectant in 20 gallons of warm potable water (100°F).

Table B-2. Nonstandard Decontaminants Available in the Supply System (Continued)

Decontaminants	Use	Remarks	Cautions	Preparation
Disinfectant chlorine (continued)	Bio	Use the solution only once. Prepare an emergency solution by mixing one level MRE spoonful of calcium hypochlorite (water disinfecting powder) to each 10 gallons of water. If liquid chlorine bleach is available, use about $^{1}/_{3}$-canteen cup of 5 percent chlorine bleach to each 10 gallons of water. Make fresh solutions for rinsing and disinfecting utensils for each 100 persons.	None	
Ethylene oxide	Bio	Is effective against all microorganisms, to include bacterial spores. Should be applied in the strength of 30 pounds for every 1,000 cubic feet. Should be allowed a 6-hour contact time (contact time must be doubled for each 20°F drop in temperature below 75°F). Should be used in an airtight enclosure.	Is flammable. Not recommended for interior use.	None
Formalin (formaldehyde)	Bio	Agent is effective against all microorganisms, to include bacterial spores. Agent is used for interior decon of relatively close areas. Vapors should remain 16 hours in a closed structure; then, aerate until the odor is no longer objectionable.	Vapors are very toxic. Agent will curl and discolor paper. Leaves a white residue. A self-containing breathing apparatus is required in closed spaces. Personnel handling or spraying the agent should wear impermeable protective clothing.	No mixing is required. However, less residue remains and less aeration is required if the mixture of 5 parts formalin and 3 parts methanol is used. (Use this mixture at a rate of 4/5 quart per 1,000 cubic feet of space).

Table B-2. Nonstandard Decontaminants Available in the Supply System (Continued)

Decontaminants	Use	Remarks	Cautions	Preparation
Formalin (formaldehyde) (continued)	Bio	Optimum conditions for spraying formalin are 70° to 80°F with an 85 percent relative humidity. The minimum effective relative humidity is 70 percent. At 60°F, the exposure should be increased to 24 hours. Agent is applied as a vapor from standard insecticide sprayers or is vaporized by heat or a bubbling steam from a pan.	Personnel entering an area containing formalin vapors should— • Wear a protective mask. • Wear washable outer clothing, fastened to prevent vapors from entering at wrists, ankles, or neck. • Remove outer clothing after emerging from vapors. • Shower and put on clean clothing as soon as possible. Vapors are not flammable; open flame should not be used for vaporizing when methanol has been added to the agent. When steam is used, the source of the steam should be outside the area being decontaminated.	
Detrochlorite	Bio	Agent is a thickened bleach that is very useful on vertical surfaces. Coverage is 1 gallon per 8 square yards. Apply by means of the PDDA. Allow 30 minutes contact time, then rinse with water.	Agent is very corrosive. See *Preparation* column.	Mix by weight 19.3 percent diatomaceous earth, 0.5 percent anionic wetting agent, 2.9 percent calcium hypochlorite (70 percent available chlorine), and 77.3 percent water. Mix wetting agent and diatomaceous earth with water before adding the calcium hypochlorite. Mixing the wetting agent and calcium hypochlorite in a dry and undiluted state may cause an explosion.

Table B-2. Nonstandard Decontaminants Available in the Supply System (Continued)

Decontaminants	Use	Remarks	Cautions	Preparation
Peracetic acid (PAA)	Bio	Agent is effective against all microorganisms, to include bacterial spores. A contact time of 10 minutes must be allowed. Item should be wiped with a rag or swab. Remove excess acid and aerate for 10 to 15 minutes or until no objectionable odor remains. Small items should be immersed for 10 minutes. Remove excess acid and aerate until no objectionable odor remains.	Fumes are highly irritating. Prolonged exposure will damage most material. A 40 percent solution has a low flash point (105°F); a 3 percent solution is nonflammable. Burns and blistered on the skin will occur. Agent must be stored in original containers under refrigeration to prevent decomposition. Protective mask and clothing are required. A violent explosion may result if heavy metal ions come in contact with the agent. Prolonged exposure will corrode iron and deteriorate rubber, plastic, and leather.	Available as a 40 percent solution. Mix 1 quart of PAA to 3½ gallons of water (add PAA to the water).
Hyamine (benzethonium chloride)	Bio	Agent is effective against all bacteria. A contact time of 5 to 30 minutes should be allowed.	Agent is very toxic. Estimated fatal dose to man is 1 to 3 grams. Care should be taken when mixing to avoid inhalation of powder. Agent is not to be used on aircraft or ships.	Use a 0.1 to 1 percent solution (1 pound of hyamine for every 12 gallons of water yields a 1 percent solution).
Carbon dioxide and ethylene oxide mixture (with more than 87 percent ethylene oxide)	Bio	Should be applied in the strength of 30 pounds for every 1,000 cubic feet. Should be allowed a 12-hour contact time (doubled for each 20°F drop in temperature below 75°F). Is nonflammable and is recommended for interior use. Should be used in an airtight enclosure.	Will blister the skin. Aerate items next to the skin 18 to 24 hours.	None

Table B-2. Nonstandard Decontaminants Available in the Supply System (Continued)

Decontaminants	Use	Remarks	Cautions	Preparation
Sodium hypochlorite solution (household bleach)	Bio Chem	Is effective against blister and V agents and all biological agents. Will react rapidly (within 5 minutes) with blister and V agents. Should be allowed a 10- to 15-minute contact time for biological agents. Should be applied undiluted with brooms, brushes, or swabs. Is the preferred decon for ship use. A 5:1 concentration is recommended. Has a limited storage problem.	Is harmful to the skin and clothing if undiluted. Remove from the skin and clothing by flushing with water. Is corrosive to metals unless rinsed, dried, and lubricated after decon. Should be stored in a cool place.	For chemical decon, no mixing is required. For biological decon, dilute by adding 2 parts bleach to 10 parts water. For decon of cotton clothing and utensils, dilute 2 cups of bleach to 1 gallon of water. For application, dilute half and half with water and spray from the PDDE.
2-Propanone (acetone)	Chem	Freezing point is -203°F; boiling point is 133°F (evaporates rapidly). Good decontaminant for use in arctic regions. Commonly obtained as fingernail polish remover or paint thinners. Scrubbing increases its effectiveness.	Is extremely flammable. Does not neutralize agents. Is effective for dissolving and flushing agent by physically removing it.	None
Dietyl ether	Chem	Freezing point is -241°F; boiling point is 93°F. Good decontaminant for use in arctic regions. Available through medical supply facilities. Scrubbing increases its effectiveness.	Is the same as for 2-propanone.	None

Table B-2. Nonstandard Decontaminants Available in the Supply System (Continued)

Decontaminants	Use	Remarks	Cautions	Preparation
Ethylene glycol	Chem	Scrub on contaminated surfaces and rinse thoroughly.	Removes contamination, but does not neutralize it. Therefore, runoff residue must be considered contaminated.	Mix equal amounts of solution and water.
Solvents (gasoline, JP-4, diesel fuel, kerosene, and similar solvents)	Chem	Scrub on contaminated surfaces and rinse thoroughly.	Is the same as for ethylene glycol. May damage materials such as rubber and plastic.	None
Sodium hydroxide (caustic soda or lye) (NSN 6810-00-174-6581) (100 pounds)	Bio Chem	Is effective against G agents, lewisite, and all biological agents, including bacterial spores. Will neutralize G agents on contact. Should be allowed to remain in contact with the chemically contaminated surface for about 15 minutes.	Agent will damage the skin, eyes, and clothes and can cause upper respiratory or lung damage if inhaled. Full rubber protective clothing, gloves, boots, and mask are required when using. Affected area must be washed immediately with large amounts of water and flushed with diluted acetic acid or vinegar. Remove affected clothing. If eyes are involved, flush them at once with large amounts of warm water and seek medical attention. Runoff from decon operations is highly corrosive and toxic. Drain runoff into a sump and bury. Equipment must be flushed with large amounts of clear water.	Small amount—10 pounds of lye to 12 gallons of water (10 percent solution). Mix in an iron or steel container (never aluminum, zinc, or tin). Add lye to the water to prevent boiling and splattering due to heat being emitted. Do not handle mixing container with bare hands. Large amount (PDDE use)—Prepare a solution of 227 grams (½ pound) of lye for each gallon of water. Pump 350 gallons of water into the tank unit. Connect the tank unit, pump unit, and heater together. Heat the water to 122°F. Disconnect the heater unit and add 175 pounds of lye to the heated water. Circulate the solution with the pump unit until all the lye is dissolved. The temperature will increase noticeably. Use while hot.

Table B-2. Nonstandard Decontaminants Available in the Supply System (Continued)

Decontaminants	Use	Remarks	Cautions	Preparation
Sodium hydroxide (caustic soda or lye) (NSN 6810-00-174-6581) (100 pounds) (continued)			Sodium hydroxide— • Is not recommended for ship use. Store on top deck only. • Is corrosive to most metals. • Is not recommended if less toxic caustic decontaminants are available. Will cause a red color change upon contact with M8 detector paper. Agent's effectiveness is directly proportional to the strength of the solution.	Simultaneous mixing and applying—Sprinkle dry lye on the contaminated area and then dissolve it with a spray of steam or hot water. Do not wash the lye off the surface while applying the steam or hot water. Paint removal—1 pound of lye per 2½ gallons of water is capable of removing an average coat of paint from about 11 square yards of surface. This solution is effective in removing paint on which chemical contamination has absorbed. (Can substitute calcium hydroxide, potassium hydroxide, or trisodium phosphate for sodium hydroxide.)
Sodium carbonate (washing soda, soda ash, sal soda, or laundry soda)	Chem	Is effective against G agents and CN. Will react rapidly with G agents, normally within 5 minutes. Is the preferred decontaminant for ship use. The recommended concentration is 5 percent by weight. There is no storage problem. Should be used with a hot solution to decon CN effectively.	Should not be used for VX. It cannot detoxify VX and creates extremely toxic by-products. Does not dissolve mustard agents or detoxify them.	Mix 10 pounds of washing soda to 12 gallons of water (10 percent solution).
Potassium hydroxide (caustic potash)	Bio Chem	Is the same as for sodium hydroxide.	Is the same as for sodium hydroxide.	Is the same as for sodium hydroxide.
Hexachloramelamine	Chem	Is effective against mustard agents.	May require the use of a protective mask and rubber gloves when used. Is corrosive to metal.	Is not soluble in water but is soluble in organic solvents such as gasoline, kerosene, and paint thinner.

Table B-2. Nonstandard Decontaminants Available in the Supply System (Continued)

Decontaminants	Use	Remarks	Cautions	Preparation
Ammonia or ammonium hydroxide (household ammonia)	Chem	Is effective against G agents. Is slower acting than sodium hydroxide or potassium hydroxide.	May require the use of a self-contained breathing apparatus or special purpose mask.	Ammonium hydroxide is a water solution of ammonia. No further mixing is required.
Dichloramine-B and Dichloramine-T	Chem	Is effective against mustard agents.	May require the use of a protective mask and rubber gloves when used. Is corrosive to metal.	Is not soluble in water, but is soluble in certain organic solvents. Is normally mixed as a 10 percent solution in dichloroethane.
Perchloroethylene (tetrachloroethylene)	Chem	Freezing point is -8°F; boiling point is 250°F. Agent— • Is good for use in arctic climates. • Is a nonflammable, synthetic solvent widely used in dry cleaning plants. • Dissolves H and V agents but not G. • Has a low toxicity. Scrubbing increases its effectiveness.	Physically dissolves and removes contamination, but does not neutralize it.	Requires no mixing (practically insoluble in water).
Acids (sulfuric acid, hydrochloric acid, oxalic acid, and similar acids)	Rad	Are effective solvents for rust and mineral deposits holding radioactive material on metal surfaces. Should be allowed a 1-hour contact time. Must be flushed with water, scrubbed with a water-detergent solution, and flushed again with water.	Are difficult to handle and are harmful to the body, especially the eyes. Can produce boiling and splattering of the solution when mixed. May require the use of rubber boots, gloves, aprons, and goggles when used. Will require the use of respiratory protection when used in closed areas. Can cause serious injury when they come in contact with the body. Flush the area immediately with water. Use a 5 percent solution of water and baking soda (sodium bicarbonate).	None

Table B-3. Natural Decontaminants

Decontaminants	Use	Remarks	Cautions	Preparation
Water	Rad Bio Chem	Can be used to flush contamination from surfaces.	Is effective in physically removing contamination, but does not neutralize it. Do not use water on lewisite.	Hot water with soap makes water more effective in removing agents. Boiling for 15 minutes (30 minutes at high altitude) destroys biological agents.
Steam	Rad Bio Chem	Is more effective when accompanied by scrubbing.	Is effective in physically removing contamination, but does not neutralize it.	None
Absorbents (earth, sawdust, ashes, rags, and similar materials)	Chem	Are used to remove gross contamination, physically, from surfaces.	The contamination is transferred from the surface to the absorbent. The absorbent becomes contaminated and must be disposed of accordingly. Sufficient contamination to produce casualties may well remain on surfaces.	
Sealants (concrete, asphalt, earth, paint, and similar materials)	Rad Bio Chem	Sealants are used to physically seal in or shield contamination. They are effective as follows: • 12 inches of earth provides a good protection from fallout (3 inches will reduce the dose rate about one half). • 1 inch of asphalt or concrete completely absorbs alpha and beta radiation.	A break in the surface of the sealant will expose the contamination. Contaminated areas covered with sealants must be marked with appropriate NBC warning signs.	None

Figure B-3. Natural Decontaminants (Continued)

Decontaminants	Use	Remarks	Cautions	Preparation
Sealants (concrete, asphalt, earth, paint, and similar materials) (continued)	Rad Bio Chem	• 1/4 inch of grout shields alpha and bata radiation. • 4 inches of earth provides good protection from chemical contamination. Burying items contaminated with biological agents is an effective means of sealing off contamination.	A break in the surface of the sealant will expose the contamination. Contaminated areas covered with sealants must be marked with appropriate NBC warning signs.	None
Weather/time	Rad Bio Chem	UV light kills most bio-organism agents, organisms, and radiation decay overtime.	Should be used when time and the mission permits.	None
Burning	Bio Chem	Must be approved by the first general officer in the chain of command.	Creates downwind hazards. Requires that sentries be posted to keep people out of the danger area.	None

Appendix C

Decon of Specific Surfaces and Materials

Table C-1, pages C-1 through C-7, lists specific surfaces or materials and explains briefly how to best decon each for chemical, biological, or radiological contamination.

Table C-1. Decon Procedures for Specific Surfaces and Materials

Surface or Material	Types of Contamination and How to Decon		
	Chemical	**Biological**	**Radiological**
Asphalt roads (applicable to small vital areas only)	Flush with water. Spray with an STB slurry from the PDDE. Cover with STB (pure form). When liquid contamination is visible and personnel are nearby, use the STB dry mix. Weather. Cover small areas or paths across roads with 4 inches of earth.	Weather (remain masked). Wet with water (will help prevent secondary aerosols but does not decon). Apply 2 percent household-bleach solution. Spray with an STB slurry from the PDDE. Pour, spray, or spread oil on the surface (will help prevent secondary aerosols, but does not decon).	Brush or sweep. Flush with water (this may drive some of the contamination into the surface; waste must be controlled). Clean with a vacuum.
Roofs	Follow the same directions as for asphalt roads.	Follow the same directions as for asphalt roads. Apply detrochlorite; leave on at least 30 minutes, then flush with water.	Follow the same directions as for asphalt roads.
Brick and stone roads (applicable to small vital areas only)	Weather. Spray with an STB slurry from the PDDE or apply with brushes and brooms. Leave on for 24 hours, then flush with water. Wash with soapy water, preferably hot. Cover small areas or paths across roads with 4 inches of earth.	Follow the same directions as for asphalt roads.	Follow the same directions as for asphalt roads. Use abrasion (sand blasting). This provides direct and complete removal of contaminated dust; however, sand and equipment being used becomes contaminated.

Table C-1. Decon Procedures for Specific Surfaces and Materials (Continued)

Surface or Material	Types of Contamination and How to Decon		
	Chemical	Biological	Radiological
Brick and stone buildings, bunkers, gun emplacements, and tank obstacles	Spray with an STB slurry from the PDDE or apply with brushes and brooms. Leave on for 24 hours, then flush with water. Use STB (pure form) or an STB dry mix around buildings where waste water runs. Wash with soapy water, preferably hot. Weather.	Follow the same directions as for asphalt roads. Apply an STB slurry to vertical surfaces by manual means or with the PDDE. The slurry may be left on exteriors.	Follow the same directions as for brick and stone roads.
Concrete roads (applicable to small vital areas only)	Spray with an STB slurry from the PDDE. Cover with an STB slurry or STB dry mix. Weather. Cover small areas or paths across roads with 4 inches of earth. Scrape the layer of contaminated earth to the side of the road.	Follow the same directions as for asphalt roads.	Follow the same directions as for brick and stone roads.
Earth (applicable to small vital areas only): roads, gun emplacements, bivouac areas, pathways, and bomb craters	Spray with an STB slurry from the PDDE. Cover with STB (pure form). When liquid contamination is visible and personnel are nearby, use an STB dry mix. Weather. Cover small areas or paths across roads with 4 inches of earth. Scrape the layer of contaminated earth to the side of the road.	Follow the same directions as for asphalt roads. Burn.	Earth moving (removal): Try to control contaminated dust, as equipment may become contaminated. Consider waste disposal. Sealing (with earth): Does not create a waste-disposal problem; however, equipment may be contaminated.
Grass and low vegetation: fields and open terrain	Burn. Spray with an STB slurry from the PDDE. Cover with STB (pure form) or an STB dry mix. Explode drums of STB. Clear paths through area using detonating cord or other detonating devices.	Burn. Follow the same directions as for asphalt roads.	Follow the same directions as for earth.

Table C-1. Decon Procedures for Specific Surfaces and Materials (Continued)

Surface or Material	Types of Contamination and How to Decon		
	Chemical	**Biological**	**Radiological**
Undergrowth, tall grass (applicable to small vital areas only): meadows, jungles, and forests	Burn (may cause a downwind vapor hazard). Spray slurry with PDDE. Explode drums of STB. Clear a path with detonating cord, bangalore torpedoes, or demolition snakes.	Burn. Follow the same directions as for sand.	Follow the same directions as for earth.
Sand (applicable to small vital areas only): beaches and deserts	Flush with water. Use DS2. Spread STB (pure form) or spray an STB slurry over the surface. Weather. Cover paths with roofing paper. Scrape off 2 to 4 inches of contaminated top layer.	Burn.	Follow the same directions as for earth.
Fabrics: canvas, covers, tarpaulins, tentage, mask carriers, web gear, and clothing	Cotton: **NOTE: Not for MOPP gear.** Immerse in boiling, soapy water for 1 hour (1 pound of soap to 10 gallons of water) and stir. Use a 5 percent sodium-carbonate solution for G agents. Immerse in boiling water for 1 hour. Launder by standard methods. Use an STB slurry. Weather except for V agents. Woolen: Do not use DS2, as it is not recommended. Immerse in warm, soapy water (100°F) for 1 hour or longer with light agitation. Dry items slowly because the fabric may shrink.	Cotton: Boil in water for 15 minutes. Autoclave for 45 minutes at 253°F. Immerse in a 2 percent household-bleach solution for 30 minutes and rinse immediately. Launder (destroys or inactivates all but highly resistant spores). Woolen: Do not use DS2, as it is not recommended. Launder (fabric may shrink).	Cotton and woolen: Using DS2 is not recommended. Brushing removes contamination dust, but it presents a dust hazard to personnel. Laundering is the most practical procedure; however, the fabric may shrink. Try to control waste.

Table C-1. Decon Procedures for Specific Surfaces and Materials (Continued)

Surface or Material	Types of Contamination and How to Decon		
	Chemical	**Biological**	**Radiological**
Leather: boots, gloves, and other items	Scrub with hot, soapy water and rinse. Immerse in warm, soapy water at 120°F for 4 hours and rinse. Use a 5 percent sodium-carbonate solution for G agents. Air.	Immerse in a 2 percent household-bleach solution and rinse. Immerse in a 2 percent PAA for 10 minutes, rinse, and air 10 to 15 minutes. Wipe with a 2 percent PAA, remove excess agent, and air 10 to 15 minutes.	Brush. Flush with water or soapy water.
Glass, windows	Use an IEDK or DS2. Wash with hot, soapy water. Rinse with clear water or an organic solvent. Blot off surface. Air. Weather.	Use an IEDK. Wash with soap and water. Wipe with a disinfectant solution or a 2 percent PAA (see similar procedures on page C-7 for mess gear).	Use an IEDK. Wash with a detergent. Flush with water. Wipe with solvents.
Glass, lens	Wash with hot, soapy water. Rinse with clear water or an organic solvent. Blot off surface. Air. Weather.	Wash with hot, soapy water. Wipe with alcohol or household bleach.	Brush or wipe (be careful so as not to scratch the lens). Use compressed air to blow contamination from surface.
Metal (unpainted), ammunition	Wipe with soapy water. Wipe with organic solvent and dry. Air.	Wipe with soapy water. Wipe with a 2 percent household-bleach solution. Air.	Brush or wipe.
Metal (unpainted), machinery	Use DS2. Follow the same directions as for ammunition.	Use DS2. Wipe with a 2 percent PAA, rinse, and air 10 to 15 minutes.	Brush or wipe. Wash with a detergent. Flush with water.
Metals (painted): vehicles, weapons, and equipment	Use DS2 (may soften paint). Wash with hot, soapy water and rinse. Spray with an STB slurry from the PDDE then remove it in 1 hour and oil the surface. Weather. Air. Use an IEDK to decon individual gear.	Wash with a detergent and a high-pressure water system. Apply detrochlorite. Leave it on for 30 minutes then remove by washing with a stream of water. Steam clean using a detergent. Use a household-bleach solution. Use a 2 percent PAA.	Brush or wipe. Wash. Use organic solvents, caustic agents (not on aluminum or magnesium surfaces), complexing agents (of small value on weathered surfaces), or abrasives.

Table C-1. Decon Procedures for Specific Surfaces and Materials (Continued)

Surface or Material	Types of Contamination and How to Decon		
	Chemical	Biological	Radiological
Wood (unpainted surfaces): buildings, vehicle bodies, boxes, crates, and similar items	Apply an STB slurry with the PDDE, brooms, or swabs. Let the slurry remain 12 to 24 hours then flush with water. Repeat the application and flush again. Scrub with hot, soapy water and rinse. Weather.	Apply detrochlorite. Leave it on for at least 30 minutes then flush with water. Apply an STB slurry to vertical surfaces. The slurry may be left on interiors. Weather (sun and rain eliminate most microorganisms within 1 day). Burn.	Wash the interior with large amounts of water (some contamination may soak into surfaces).
Wood (painted surfaces): buildings and boxes	Apply an STB slurry with the PDDE, brooms, or swabs. Let the slurry remain 12 to 24 hours then flush with water. Scrub with hot, soapy water and rinse. Use DS2 and rinse. DS2 may soften paint. Weather.	Follow the same directions as for wood (unpainted surfaces).	Wash the interior with large amounts of water. Wipe the contamination from the surface.
Plastics (opaque): insulation, telephones, and panel boards	Use DS2 (may soften or damage some plastics). Wash with hot, soapy water and rinse. Weather. Air.	Follow the same directions as for glass lens.	Wash with a detergent. Flush with water. Wipe or brush.
Plastics (transparent): eyepieces and airplane canopies	Use DS2 (may soften or damage some plastics). Wash with hot, soapy water and rinse. Weather. Air.	Follow the same directions as for glass lens.	Wash with a detergent. Flush with water. Wipe or brush.
Nonsealed electronic equipment	Use a mild, evaporative solvent such as alcohol. Use DS2. Use a mild detergent and water. Rinse with distilled water. Keep the solution out of the adjustable switches, connectors, and relays. Use hot-air blowers.	Use a mild, evaporative solvent such as alcohol. Use DS2. Use a mild detergent and water. Rinse with distilled water. Keep the solution out of the adjustable switches, connectors, and relays. Use hot-air blowers.	Use hot-air blowers. Wipe or brush.

Table C-1. Decon Procedures for Specific Surfaces and Materials (Continued)

Surface or Material	Types of Contamination and How to Decon		
	Chemical	Biological	Radiological
Rubber (impermeable): aprons, suits, and other items	Spray with DS2 and rinse after 30 minutes. Immerse in soapy water (just below the boiling point) for 1 hour. Do not agitate. Rinse with clear water and hang to dry. Use a 10 percent sodium-carbonate solution for G agents, rinse, and air. Apply hot, soapy water with brushes and rinse. Spray with an STB slurry from the PDDE. Wash off, after a few minutes, with clear water.	Follow the same directions as for leather.	Brush. Scrub or flush with water or soapy water.
Rubber (natural and synthetic): gloves and boots	Spray with a 10 percent mixture of HTH and rinse. Immerse in STB slurry for 4 hours, rinse, and dry. Use an IEDK in emergencies. Air.	Follow the same directions as for leather.	Brush. Scrub or flush with water or soapy water.
Rubber: mask facepieces and other rubber articles coming in direct contact with the skin	Use IEDK in emergencies. Wash with warm, soapy water.	Follow the same directions as for leather.	Brush. Scrub or flush with water or soapy water.
Rubber: tires, hoses, mats, and insulation	Spray with a 10 percent mixture of HTH and rinse. Apply an STB slurry. Allow the slurry to remain at least 30 minutes then flush with clear water (may be left on tires). Apply hot, soapy water. Air. Weather.	Follow the same directions as for leather.	Brush. Scrub or flush with water or soapy water.
Water	Use trained water-purification personnel to decon.	Boil small amounts 15 minutes. Chlorinate using a chlorination kit. Add iodine water-purification tablets to small amounts.	Flocculation (requires special chemicals to remove suspended matter). Ion exchange (removes radians from solution).

Table C-1. Decon Procedures for Specific Surfaces and Materials (Continued)

Surface or Material	Types of Contamination and How to Decon		
	Chemical	Biological	Radiological
Mess gear and canned rations	Immerse in boiling, soapy water for 30 minutes and rinse. Immerse in boiling water for 30 minutes. Spray with DS2. Wash in hot, soapy water; rinse; and air.	Wash with soap and water, then immerse in disinfectant solution (disinfectant, chlorine or 1/3 cup of household bleach per 10 gallons of water). Boil in water for 15 minutes (not effective on toxins and bacterial spores.) Immerse in a 5 percent sodium-carbonate solution (4 pounds of washing soda to 10 gallons of water) and rinse with potable water. Immerse in a household-bleach solution (2 gallons of bleach to 25 gallons of water) for 30 minutes, rinse, and air for 10 to 15 minutes. Immerse in an HTH solution (½ pound to 25 gallons of water) for 30 minutes and rinse. Immerse in a 2 percent PAA for 10 minutes, rinse, and air 10 to 15 minutes.	Wash with soap and water and rinse. Brush and then wipe contamination from surfaces and containers.
Food: not canned or protected by impermeable container	Do not consume food that is suspected to be contaminated with chemical agents until veterinary personnel approve it.	Boil water 15 minutes. Cook food thoroughly. Immerse in or spray with a 2 percent household-bleach solution. (Packaged food or food that is peeled or pared can be immersed or sprayed with the solution.)	Wash or trim contamination from unpackaged food.
Food: canned, bottled, or protected by impermeable container	Follow the same directions as for mess gear and canned rations.	Follow the same directions as for mess gear and canned rations.	Follow the same directions as for mess gear and canned rations.
Paper currency	Destroy by burning. Do not decon.	Destroy by burning. Do not decon.	Destroy by burning. Do not decon.
Coins	Wipe with soapy water. Wipe with an organic solvent.	Use UV rays. Wash with soapy water. Wipe with a 2 percent household-bleach solution. Air.	Brush or wipe. Wash with a detergent. Flush with water.

Appendix D

Decon Kits, Apparatuses, and Equipment

Various materials and equipment are used in decon operations. Some are simple to use and are readily available to individual soldiers. Others are very complex to use and are available only to specially trained teams. Table D-1, pages D-1 and D-2, lists the decon equipment and materials for the different use levels; Table D-2, pages D-3 through D-5, lists the detection equipment and materials for the different use levels.

Table D-1. Decon Equipment and Materials

Item and Description	Use	Limitations	Reference
	Individual		
Decontaminating kit, skin, M291 SDK, (NSN 4230-01-276-1905) (20 kits per box)	To decon your skin, completely, through physical removal, absorption, and neutralization of toxic agent with no long-term effects **NOTE: Use this kit for both actual combat and training purposes.**	Is for external use only. **WARNING** **Keep decon powder out of eyes, cuts, or wounds. It could slightly irritate the skin or eyes.**	TM 3-4230-229-10
Decon kit, individual equipment, M295, IEDK, (NSN 6850-01-357-8456)	To decon your chemical protective gloves, mask, hood, overboots, LCE, and weapon.	**WARNING** **Do not use for skin decon. Keep off the skin and out of wounds, eyes, and mouth.**	TM 3-4230-235-10
Decontaminating apparatus, portable, M13 DAP, (3.7 gal), (NSN 4230-01-133-4124). It has a 3.7-gallon disposable DS2 container. The apparatus can be mounted to a standard 5-gallon fuel can mount on vehicles and equipment.	To spray DS2 on surfaces of vehicles and equipment.	Cannot refill DS2 container. See Table A-2, page A-2, for a list of replacement containers.	TM 3-4230-214-12&P TM 43-0001-26-1

Table D-1. Decon Equipment and Materials (Continued)

Item and Description	Use	Limitations	Reference
Battalion Decon Crew and Chemical Company			
Decontaminating apparatus, power driven, skid mounted, multipurpose, integral, 500 gallon, M12A1 PDDA, (NSN 4230-00-926-9488), LIN F81880. Apparatus includes pump unit, tank unit, and M2 water heater (all mounted on skids).	To spray decon agent, STB slurries, and solutions as well as hot, soapy water rinses during field decon operations. To pump water or foam to fight fires, de-ice items, wash vehicles, and pump various fluids.	Do not use with defoliants, herbicides, or insecticides.	LO 3-4230-209-12 LO 5-2805-259-12 TM 3-4230-209-20&P TM 43-0001-26-1
Decontaminating system, lightweight, M17 LDS, (NSN 4230-01-251-8702). The M17 is a portable pump and water-heating unit for producing hot water and steam. The system incorporates a 1,580- to 3,000-gallon collapsible water tank, two wand assemblies, and connecting hoses.	To perform operational and thorough decon of vehicles and equipment. Can also be used for troop showers, as necessary.		TM 3-4230-228-10

Table D-2. Detection Equipment and Materials

Item and Description	Use	Limitations	Reference
		Individual	
Paper, chemical agent, detector, VGH, M8, (NSN 6665-00-050-8529). Paper is issued in a book of 25 sheets, perforated for easy removal. A color-comparison bar chart is printed on the inside front cover.	To detect the presence of liquid V, G, and H chemical agents	Cannot be used to detect vapors or chemical agents in water, DS2, or petroleum products. May cause false readings.	TM 3-6665-254-12 TM 3-6665-268-10 TM 3-6665-307-10 TM 43-0001-26-1
Paper, chemical agent, detector, M9, (NSN 6665-01-226-5589). Paper is issued in a 7-ounce dispenser box that contains one 30-foot roll of 2-inch-wide detector paper and plastic storage bags. The paper has an adhesive back for attaching to equipment and clothing.	To detect the presence of liquid V, G, and H chemical agents	Cannot be used to detect vapors or chemical agents in water. Will not stick to dirty, oily, or greasy surfaces. Contamination indications cannot be read under red light or by color-blind soldier. The following can cause false readings: • Temperatures above 125°F. • Brake fluid. • Aircraft cleaning compound. • DS2. • Petroleum products. • Insect repellent.	TM 3-6665-311-10
		Company	
Chemical-agent alarm, automatic, M22 ACAA, (NSN 6665-01-438-6983)	To detect chemical nerve agents in the air.	**WARNING** **Radiation hazard. Contains beta emitters.**	TM 3-6665-321-12&P
Chemical-agent alarm, automatic, M8A1 ACAA, (NSN 6665-01-105-5623). It can be vehicle-mounted, back-packed, or ground-emplaced.	To detect chemical nerve agents in the air.	**WARNING** **Radiation hazard. Contains Americium (AM241).**	TM 3-6665-312-12&P

Table D-2. Detection Equipment and Materials (Continued)

Item and Description	Use	Limitations	Reference
Chemical-agent monitor system (CAM) (NSN 6665-01-199-4153)	To search out areas; to search and locate contamination on personnel, equipment, ship's structure, aircraft, land vehicles, buildings, and terrain; and to monitor for effectiveness of decon. Can also be used for monitoring collective protection. The CAM responds to nerve- and blister-agent vapors down to the lowest concentration that could affect personnel over a short period.	Cannot be used as a detector. The CAM is a monitor and not a detector. Since it is a monitor, it can become contaminated and overloaded (saturated). The CAM can only report conditions at the front of the inlet probe. It is, therefore, a point monitor only and cannot give a realistic assessment of the vapor hazard over an area from one position. **WARNING** **Beta radiation hazard.**	TM 3-6665-331-10
Detector kit, chemical agent, M256A1 (NSN 6665-01-438-6983). It contains M8 detector paper for liquid agents and samplers/detectors for vapors.	To detect liquid G, V, and H chemical agents using M8 detector paper and to detect and determine the type of vapor (G, V, H, and AC) using samplers/detectors.		TM 3-6665-307-10

Table D-2. Detection Equipment and Materials (Continued)

Item and Description	Use	Limitations	Reference
Radiac set, AN/VDR-2 (NSN 6665-01-222-1425)	To measure gamma radiation dose rates from 0.01 uGy/hr to 100 Gy/hr; to detect and display the level of beta particle dose rates from 0.1 uGy/hr to 5 cGy/hr; and to measure, store, and display accumulated dose rates from 0.01 uGy to 9.99 Gy. The instrument consists of a radiac meter with an internal sensor for obtaining dose rates during both mounted and dismounted operations. It has a second sensor housed in a probe and attached to the radiac meter with a cable and input connector. It is used for monitoring personnel, supplies, and equipment. The radiac set uses a presettable, an audible, and a visual warning device integral to the radiac meter. The system is air-transportable and organic to all units.		TM 11-6665-251-10
Computer indicator, radiac, CP696/PDR-75; detector, radiac/ DT236/PDR-75; radiac set, AN/PDR-75, (NSN 6665-01-211-4217)	To measure the accumulated neutron and gamma radiation dose recorded by the DT236. A person who may be exposed to radiation from tactical nuclear weapons wears the DT236 on his wrist.		

Appendix E

DTD Station Charts for Chemical Unit Leaders and NBC NCOs

This appendix provides station charts for the DTD. They can be laminated and posted at each station. They provide the station number and on the reverse side, instructions for the station supervisor/attendant. They are not meant to be all-inclusive and do not include setup procedures or equipment requirements. They are an easy guide for soldiers going through a decon and for station supervisors/attendants.

STATION 1

INDIVIDUAL-GEAR DECON

TASK: Remove contamination to a negligible risk from individual gear (LCE, mask carrier, helmet, and weapon)

- **The soldier—**
  - **Decontaminates his own gloves with hot, soapy water or an STB slurry.**
  - **Decontaminates his hose and canister (if wearing the M42 mask) with either hot, soapy water and a sponge or an STB slurry mix.**
  - **Decontaminates his own gear by washing and scrubbing it for 4 minutes in a decontaminant container with hot, soapy water or an STB slurry mix.**
- **The attendant—**
  - **Takes the gear to the equipment checkpoint and places it on the "dirty" side of the contamination control line (engineer tape).**
  - **Waits as the monitor holds the detection device 1 inch from the gear and checks it for residual contamination.**
  - **Recycles any gear that is above one bar.**
  - **Places the gear on the clean side of the contamination control line if it passes the check.**
  - **Carries the gear to the reissue point.**

NOTE: Change the slurry mix after 20 soldiers have decontaminated their gear and the rinse water after 10 soldiers or when it appears dirty (place the waste in the sump).

STATION 2

OVERBOOT AND HOOD DECON

TASK: Neutralize gross contamination on overboots, lower trouser legs, protective mask, and hood

- The soldier walks into the shuffle pit and spreads his legs apart (double-shoulder width), bends at the waist, and uses his hands to thoroughly rub the STB dry mix or the STB slurry on his overboots and lower trouser legs.
- Buddy #1 cuts the shoulder straps and drawcord on Buddy #2's hood. He pulls Buddy #2's hood inside out over the front of the mask, being careful not to touch the exposed neck or head. Buddy #1 gathers Buddy #2's hood in one hand and using a cutting tool, cuts away the hood as close as possible to the eye-lens outsert, voicemitter, and inlet-valve cover. Make sure nothing is left dangling below the bottom of the mask. If a replacement hood is not available, decon Buddy #2's hood in the same manner as for a MOPP-gear exchange.
- The attendant—
  - Instructs Buddy #1 on how to roll Buddy #2's hood to ensure that it is accomplished to standard.
  - Will monitor the shuffle pit and add more STB after 10 soldiers have processed through it.

STATION 3

OVERGARMENT REMOVAL

TASK: Remove contaminated overgarment before the agent penetrates material and touches the undergarments or the skin

The attendant—

- Assists the soldier in removing his overgarment.
- Cuts or removes the M9 detector paper from around the soldier's wrist.
- Unfastens the Velcro closure over the jacket zipper.
- Unties the waist cord and unfastens the Velcro straps on the soldier's jacket.
- Unfastens the back snaps.
- Cuts or removes the M9 detector paper from the soldier's trousers, if binding.
- Unfastens the Velcro straps and zipper on the cuffs of the soldier's trousers.
- Unfastens the front waist snaps and unzips the front zipper. He has the soldier lift one leg and point that foot down and bend slightly at the knees for stability.
- Grasps the cuff of the elevated boot with a hand on each side and pulls the cuff in an alternating, jerking motion until the soldier can step out of the trouser leg. The process is repeated on the other leg.

STATION 4

OVERBOOT AND GLOVE REMOVAL

TASK: Remove contaminated overboots and gloves to limit the spread of contamination (the overboots and gloves may also be decontaminated for reissue, if servicable)

- **Engineer tape is used to mark the liquid-contamination control line.**
- **The attendant unfastens or cuts the elastic closures on the soldier's overboots.**
- **The soldier stands next to and faces the liquid-contamination control line.**
- **The soldier steps back about 12 inches from the liquid-contamination control line.**
- **The attendant steps on the back of the soldier's overboot and instructs him to lift his heel and work his foot out of the overboot and step across the liquid-contamination control line. Repeat the process on the other foot.**
- **The attendant discards the overboots.**
- **The soldier holds the fingertips of his gloves and partially slides his hands out.**
- **The attendant removes the soldier's gloves.**
- **The attendants—**
  - **Put the gloves and overboots into the containers on the "dirty" side of the control line.**
  - **Submerge the gloves and overboots in the container of hot, soapy water, ensuring that no water remains inside when removed from the container.**
  - **Submerge the gloves and overboots in the STB/HTH solution and thoroughly scrub them.**
  - **Rinse the scrubbed items and check for holes, tears, or punctures. If found, discard the items.**
  - **Place the usable items on a poncho or a plastic tarp to air dry.**
  - **Place the air-dried usable items into plastic trash bags along with an M256A1 detector kit. If the kit shows contamination remaining, the attendants can recycle the items or discard them. However, if the kit shows no contaminated, they can be reused.**

STATION 5

MONITOR

TASK: Identify contamination on personnel (spot-decon capability and medical aid are provided, as required)

- The monitor checks the soldiers for contamination.
- The aidman checks the soldiers for any chemical-agent symptoms and treats as necessary.
- The monitor asks the soldiers about any damage to their MOPP gear that was identified at stations 2, 3, or 4.

NOTE: If all the liquid contamination is absorbed into the clothing, the M8 detector paper will be negative even though there is a hazard.

STATION 6

MASK REMOVAL

TASK: Remove the mask without contaminating the soldier

- The attendant pulls the hood over the front of the mask, grabs the mask by the voicemitter cover, and pulls the mask off the soldier while he holds his breath.
- The attendant holds the mask open so that the soldier can remove the inserts without touching the outside of the mask if the mask has optical inserts.
- The soldier walks upwind 16 feet, crosses the vapor-contamination control line, and then resumes breathing.
- The attendant brings the mask to station 7.

NOTE: No vapor hazard is expected beyond the vapor-hazard contamination control line.

- The chemical unit positions the M8A1 or M22 ACAA upwind of the site.
- The soldier moves straight ahead while his mask, which may still be emitting vapors, is held on the "dirty" side of the vapor-contamination control line and taken to the next station.

STATION 7

MASK DECON POINT

TASK: Remove all contamination from the protective masks

The attendants—

- **Discard the winterization kits, if used.**
- **Remove the eye-lens outserts and the hoods if the hoods were not cut off at station 2.**
- **Remove or discard the filters or canisters and dispose of them in the properly marked containers.**
- **Wash the masks, hoods, eye-lens outserts, and hoses on the M42 and M43 masks in hot, soapy water.**
- **Rinse these items in clean water.**
- **Dip them in the sanitizing solution and agitate for 5 minutes.**
- **Rinse them in clean water again.**
- **Add one tube of mask sanitizing solution (calcium hypochlorite) to each quart of water.**
- **Wipe the masks with rags until they are almost dry.**
- **Discard each gallon of mask sanitizing solution into a sump after every 10 masks.**
- **Check for contamination with a CAM.**
- **Recycle the masks if they are still contaminated and then decon their gloves.**
- **Take the unassembled masks to the reissue point if they are not contaminated.**

STATION #8

REISSUE POINT

TASK: Reissue the masks with their components to the soldiers

- The unit NBC NCO reissues the masks with their components to the soldiers.
- The soldiers will reassemble the masks in the postdecon AA.
- The unit NBC NCO provides the soldiers with replacement parts for all types of protective masks.
- The soldiers pick up individual gear and move to the postdecon AA.

Glossary

AA	assembly area
AC	hydrogen cyanide
ACAA	automatic chemical-agent alarm
AD	air defense
AFM	Air-Force Manual
AFR	Army-Force Regulation
AMT	ambulatory medical treatment
AO	area of operation
APC	armored personnel carrier
AR	Army regulation
ART	assessment and recovery team
attn	attention
AVIM	aviation intermediate maintenance facility
BAS	battalion aid station
BDO	battle-dress overgarment
BDU	battle-dress uniform
BFV	Bradley fighting vehicle
bio	biological
C^2	command and control
CAM	chemical-agent monitor
CANA	convulsant antidote for nerve agent
CARC	chemical-agent-resistant coating
CG	phosgene
cGy/hr	centigray per hour
chem	chemical
CHS	combat health support
CK	cyanogen chloride
Class II Supplies	clothing, individual equipment, tentage, organizational tool sets and kits, hand tools, maps, and administrative and housekeeping supplies and equipment

Classes of Supply the grouping of supplies by type into 10 categories to facilitate supply management and planning

CLU command launch unit

CMO chemical officer

CN chloroacetophenone

CP chemical protective

CPS collective protective shelter

CPU chemical protective undergarment

CRDEC Chemical Research Development and Engineering Center

CSS combat service support

CTA common table of allowance

CVCUS combat vehicle crewman uniform system

CX phosgene oxime

DA Department of the Army

DAD detailed aircraft decon

DAP decontaminating apparatus, portable

DD Department of Defense

DED detailed equipment decon

DISCOM Division-support command

DP diphosgene

DS2 decontaminating solution number 2

DTD detailed troop decon

DU depleted uranium

EDTA ethylene diamine tetra acetic acid

EIC end-item code

EMT emergency medical treatment

ENCU environmental control unit

EOD explosive ordnance disposal

F Fahrenheit

FARP forward arming and refueling point

FLOT forward line of troops

FM field manual

FMC field medical card

FMFM Fleet Marine Force Manual

FRAGO	fragmentary order
FSB	forward-support battalion
G4	Assistant Chief of Staff, G4 (Logistics)
G5	Assistant Chief of Staff, G5 (Civil Affairs)
GA	tabun
GB	sarin
GD	soman
GF	soman
GI	Government issue
GP	general purpose
GPM	gallons per minute
Gy	gray
HD	distilled mustard
HE	high explosive
HEMTT	heavy expanded-mobility tactical truck
HEPA	high-efficiency particulate air filter
HL	mustard-lewisite
HMMWV	high-mobility multipurpose wheeled vehicle
HN-1	nitrogen mustard
HN-2	nitrogen mustard
HN-3	nitrogen mustard
HP	horsepower
HQ	headquarters
HTB	high-test bleach
HTH	high-test hypochlorite
IEDK	individual equipment decon kit
JMAO	Joint Mortuary Affairs Office
JP8	aviation fuel
L	lewisite
LAW	light antitank weapon
LCE	load-carrying equipment
LDS	Lightweight Decontaminating System
LIN	line-item number
LO	lubrication order

LTA	launcher-tube assembly
MADCP	mortuary affairs decontamination collection point
METT-TC	mission, enemy, terrain, troops, time available, and civilian consideration
MGX	MOPP-gear exchange
MOPP	mission-oriented protective posture
MP	military police
MRE	meals, ready-to-eat
MTF	medical-treatment facility
MTOE	modification table of organization and equipment
NATO	North Atlantic Treaty Organization
NAVMC	Navy/Marine Corp
NBC	nuclear, biological, and chemical
NBC5 report	once the NBC reports are posted on the situation map, an NBC5 report showing the contaminated area is prepared by the division
NCO	noncommissioned officer
NCOIC	noncommissioned officer in charge
NSN	national stock number
OEG	operational exposure guidance
OPORD	operation order
OS	operator's spray down
PAA	peracetic acid
PALADIN	artillery howitzer
PDDA	power-driven decon apparatus
PDDE	power-driven decon equipment
PL	platoon leader
PLL	prescribed load list
PNVS	Pilot Night-Vision System
POL	petroleum, oils, and lubricants
POMCUS	prepositioning of material configured to unit sets
PPW	patient protective wrap
PSG	platoon sergeant
psi	pounds per square inch
PT	pointed
PW	personal wipe down

rad	radiological
RD	rounded
RPO	radiological protection officer
S4	Supply Officer (US Army)
SA	arsine
SAW	squad automatic weapon
SB	supply bulletin
SD	skin decon
SDK	skin decontaminating kit
SMCT	soldiers' manual of common tasks
SOI	signal operating instructions
SOP	standing operating procedure
STANAG	Standardized North Atlantic Treaty Organization Agreement
STB	supertropical bleach
STP	soldier training publication
T20	tritium oxide
TADS	Tactical Air-Defense System
TAP	toxicological apron, protective
TB	technical bulletin
TC	training circular
TGD	thickened soman
THD	thickened distilled mustard
TM	technical manual
TO	technical order
TOE	table of organization and equipment
TOW	tube launched, optically tracked, wire guided
TPU	tank and pump unit
TRADOC	United States Army Training and Doctrine Command
uGy	microgray
US	United States
USMC	United States Marine Corp
UV	ultraviolet
VDR-2	radiac meter
VR-55	soman

VW	vehicle washdown
VX	soman
WBGT	wet-bulb gradient temperature
WO	warning order

Bibliography

AR 11-9. *The Army Radiation Safety Program*. 28 May 1999.

AR 385-10. *The Army Safety Program*. 23 May 1988.

CRDEC Technical Report. CRDEC-CR-87036. *Project VALID Final Comprehensive Report. Herman Nelson Heater Utilization Study*. December 1986. AD-B108506.

CRDEC Technical Report CRDEC-TR-175. *Decontamination Effectiveness of Water Streams*. March 1991. AD-B154567.

CTA 50-909. *Field and Garrison Furnishings and Equipment*. 1 August 1993.

DA Form 2028. *Recommended Changes to Publications and Blank Forms*. 1 February 1974.

DA Form 2404. *Equipment Inspection and Maintenance Worksheet*. 1 April 1979.

DD Form 1380. *US Field Medical Card*. December 1991.

DO49 Technical Report DPB/JOD-91/025. *MOPP Exchange Evaluation, Operational*. September 1991. AB-B158128.

DO49 Technical Report DPG/TA-86-015. *Standard and Nonstandard Decontaminants, and Decontamination Efficiency (U)*. May 1987. AD-C041660.

DO49 Technical Report DPG/TA-88/018. *Effects on High Water Pressure on Composite Materials*. June 1988. AD-B124033.

DO49 Technical Report DPG/TA-88/023. *Commercial and Field Expedient Chemical Warfare Agent Decontamination*. March 1988. AD-B121231.

DO49 Technical Report DPG/TA-88/030. *Decontamination on Selected Military Equipment: US Army Ammunition Stocks*. September 1988. AD-B128325.

DO49 Technical Report DPG/TA-89024. *Decontaminative Effects of Soil*. November 1989. AD-B140334.

DO49 Technical Report DPG/TA-90/007. *Water Quality Requirements for Decontamination*. December 1989. AD-B140336.

DO49 Technical Report DPG/TA-90/014. *Mission Oriented Protective Posture (MOPP) Exchange Procedures Evaluation*. January 1990. (No DDIC Number Assigned).

DO49 Technical Report DPG/TA-90/025. *Human Factors Evaluation of the Mission Orientated Protective Posture (MOPP) Exchange Procedures*. July 1990. AD-B148421.

FM 3-3. *Chemical and Biological Contamination Avoidance (FMFM 11-7)*. 16 November 1992.

FM 3-3-1. *Nuclear Contamination Avoidance (FMFM 11-18)*. 9 September 1994.

FM 3-4. *NBC Protection (FMFM 11-9)*. 29 May 1992.

FM 3-4-1. *Fixed Site Protection*. 16 August 1989.

FM 3-6. *Field Behavior of NBC Agents (Including Smoke and Incendiaries) (AFM 105-7; FMFM 7-11-H)*. 3 November 1986.

FM 3-7. *NBC Field Handbook*. 29 September 1994.

FM 3-9. *Potential Military Chemical/Biological Agents and Compounds (AFR 355-7)*. 12 December 1990.

FM 3-100. *Chemical Operations Principles and Fundamentals*. 8 May 1996.

FM 3-101. *Chemical Staffs and Units*. 19 November 1993.

FM 8-9. *NATO Handbook on the Medical Aspects of NBC Defensive Operations*. 1 February 1996.

FM 8-10-4. *Medical Platoon Leader's Handbook*—Tactics, Techniques, and Procedures. 16 November 1990.

FM 8-10-7. *Health Service Support in a Nuclear, Biological, and Chemical Environment*. 22 April 1993.

FM 8-33. *Control of Communicable Diseases Manual*. 9 April 1996.

FM 8-285. *Treatment of Chemical Agent Casualties and Conventional Military Chemical Injuries (FMFM 11-11)*. 22 December 1995.

FM 21-10. *Field Hygiene and Sanitation*. 22 November 1988.

FM 21-11. *First Aid for Soldiers*. 27 October 1988.

FM 100-5. *Operations*. 14 June 1993.

FM 100-9. *Reconstitution*. 13 January 1992.

FM 100-10. *Combat Service Support*. 3 October 1995.

Joint Publication 3-11. *Joint Doctrine for Nuclear, Biological, and Chemical (NBC) Defense*. 10 July 1995.

Joint Publication 3-12. *Doctrine for Joint Nuclear Operations*. 18 December 1995.

Joint Publication 4-06. *Joint Tactics, Techniques, and Procedures for Mortuary Affairs in Joint Operations*. 28 August 1996.

LO 3-4230-209-12. *Decontaminating Apparatus, Power-Driven, Skid-Mounted, Multipurpose, Nonintegral, 500 Gallon, ABC-M12A1*. 21 June 1974.

LO 5-2805-259-12. *Engine, Gasoline, 20 HP, Military Standard Models 4A084-2, 4A084-3, 4A084-4, (LI-03524B-10-1A)*. 30 January 1989.

SB 10-523. *Size Tariff for Clothing, equipage, and Footwear*. 1 August 1998.

STANAG 2426 NBC. *NBC Contamination Control Policy for NATO Forces (Edition 1)*. 8 February 1995.

STP 21-1-SMCT. *Soldier's Manual of Common Tasks Skill Level 1*. 1 October 1994.

STP 21-24-SMCT. *Soldier's Manual of Common Tasks Skill Levels 2/3/4*. 1 October 1992.

TB 43-0116. *Identification of Radioactive Items in the Army*. 1 April 1998.

TC 3-4-1. *Chemical Agent Monitor Employment*. 17 December 1991.

TM 3-4230-209-20&P. *Unit Maintenance Manual Including Repair Parts and Special Tools List Decontaminating Apparatus: Power-Driven, Skid-Mounted, 500 Gallon, M12A1 (NSN 4230-00-926-9488)*. 28 June 1986.

TM 3-4230-214-12&P. *Operator's and Unit Maintenance Manual Including Repair Parts and Special Tools List for Decontaminating Apparatus: Decontaminating Apparatus: Portable, 14 Liter, M13 (Practice) (NSN 4230-01-345-5172)*. 15 August 1992.

TM 3-4230-228-10. *Operator's Manual for Decontaminating Apparatus: Power Driven, Lightweight, M17 (NSN 4230-01-251-8702)*. 15 May 1989.

TM 3-4230-229-10. *Operator's Manual for Decontaminating Kit, Skin: M291, (NSN 4320-01-251-8702) (SS010-AA-MMO-1-0; TO 11D1-1-131; TM 4230-10/1)*. 2 October 1989.

TM 3-4230-235-10. *Operator's Manual for Decontamination Kit, Individual Equipment: M295 (NSN 6850-01-357-8456) (EIC: Y67)*. 20 February 1994.

TM 3-6665-254-12. *Operator's and Organizational Maintenance Manual: Detector Kit, Chemical Agent, ABC-M18A2 (NSN 6665-00-903-4767)*. 16 February 1976.

TM 3-6665-268-10. *Operator's Manual for Sampling Kit, CBR Agent 34*. 6 September 1968.

TM 3-6665-307-10. *Operator's Manual for Chemical Agent Detector Kit, M256 (NSN 6665-01-016-8399) and M256A1 (NSN 6665-01-133-4964)*. 1 September 1985.

TM 3-6665-311-10. *Operator's Manual for Paper, Chemical Agent Detector: M9 (NSN 6665-01-226-5589 (TO 11H2-2-21)*. 31 August 1998.

TM 3-6665-312-12&P. *Operator's and Organizational Maintenance Manual Including Repair Parts and Special Tools List for M8A1 Automatic Chemical Agent Alarm (NSN 6665-01-105-5623) and Auxiliary Equipment M10 Power Supply (NSN 6665-00-859-2225); M10A1 Power Supply (NSN 6665-00-093-2739); M228 High Profile Mounting Kit (NSN 6665-00-859-2212) and M182 Low Profile Mounting Kit (NSN 6665-00-110-9492) (TO 11H2-17-1)*. 17 March 1985.

TM 3-6665-321-12&P. *Operator's and Unit Maintenance Manual (including Repair Parts and Special Tools List) for Alarm, Chemical Agent, Automatic: M22 (NSN 6665-01-438-6963) (EIC: Y14) and Auxiliary Equipment Power Supply, Chemical Agent Automatic Alarm: M28 (6130-01-438-6960) (EIC: Y40) Mounting Kit, Chemical Agent Automatic Alarm: M281 (NSN 6665-01-438-6959) (EIC: Y38) Alarm Unit, Chemical Agent Automatic Alarm: ABCA-M42 (NSN 6665-00-859-2215) (EIC: 399) (TO 11H2-23-1; MC 10434A-12&P; EE168-DB-OMP-010)*. March 98.

TM 3-6665-331-10. *Operator's Manual for Chemical Agent Monitor (CAM) (TO 11H2-20-1)*. 12 June 1992.

TM 10-277. *Chemical, Toxicological, and Missile Fuel Handlers Protective Clothing.* 1 November 1980.

TM 11-6665-251-10. *Operator's Manual for Radiac Set AN/VDR-2 (NSN 6665-01-222-1425).* 1 March 1988.

TM 43-0001-26-1. *Army Equipment Data Sheets: Chemical Defense Equipment.* 12 May 1982.

NAVMC 10245. *Equipment Maintenance Request.* February 1994.

Index

TABLE OF CONTENTS

PART I. SCOUTING

PART I. SCOUTING

CHAPTER 1. FUNDAMENTALS OF SCOUTING

Scouting involves observing terrain and/or the enemy, and accurately reporting those observations. Scouting requires proficiency in the use of weapons, cover and concealment, route selection, and skill in unobserved day or night movement.

1001. PURPOSE

When an infantry unit is not actively fighting the enemy, it should be actively searching for the enemy, which is the primary purpose of scouting. The unit attempts to keep the enemy off balance while making preparations for further attacks. Physically locating and keeping the enemy off balance are normally accomplished by small units ranging from a two-person scouting party to a squad-size patrol.

Infantrymen are sent out as scouts or as members of a patrol because the commander needs information about the enemy, terrain, and the location of friendly troops. The lives of the entire unit may depend upon the success or failure of a scout or patrol and the accuracy and timeliness of the report. The success of the scout or patrol will depend upon their training, preparation by the commander, and understanding their mission and the commander's requirements.

To wage combat successfully, a commander must have accurate, detailed, and timely information about the enemy, the terrain, and adjacent friendly units. Well-trained scouts and capably led patrols are among the most effective means the commander has for acquiring the information necessary to plan tactical actions and make decisions in execution.

1002. REQUIRED SCOUTING SKILLS

To be effective, a scout must be able to—

- Recognize terrain features.
- Read a map and determine direction.
- Practice and implement the principles of cover and concealment.
- Fully utilize movement and route selection.
- Know the enemy (estimate enemy unit composition and strength).
- Observe and report information accurately.
- Select routes and move through numerous types of terrain.

CHAPTER 2. TERRAIN, MAPS, AND DIRECTION

A scout must understand map symbols, identify elevations from contour interval lines, scale distance on a map, relate natural and man-made features shown on the map to the actual features on the ground, plot a course from one point to another, and locate his current position. To relate a map to the actual terrain and its features, a scout must be able to orient it to the ground using a compass, two points, a watch, and the sun or the stars.

2001. TERRAIN FEATURES

Since the infantry works and fights on the ground, terrain ashore information that scouts gather and report on is of great importance. Hills, valleys, woods, and streams are the forms and growths commonly referred to as natural land features. Artificial or man-made features include houses, bridges, and railroads. Figure 2-1 shows some important terrain features.

2002. THE LENSATIC COMPASS

The best method of finding direction, during both day and night, is with a compass. The lensatic and M2 are the two types of compasses issued to infantry units. Both work on the same principle.

The standard compass for general use in the Marine Corps is the pivot-mounted lensatic compass, so called because azimuths are read through a magnifying lens in the eyepiece. Figure 2-2, on page 2-2, shows the lensatic compass and its nomenclature. The meter graphic scale on the side of the compass is 1:50,000, which is the most used scale in military mapping. The graphic scale is useful in the field as a straightedge, an aid in orienting the map, and a means of reading map grid coordinates. The plastic dial is graduated in both degrees and mils. Numbers on the dial are printed in black. There are luminous markings on the bezel, floating dial, and on both ends of the sighting wire, plus a 3-degree bezel serration and clicking device that permit reading azimuths at night. The compass is carried in a nylon case that may be attached to the cartridge belt.

Magnetic compasses are affected by the presence of iron or magnetic fields. Consequently, the scout should not be within the influence of local magnetic

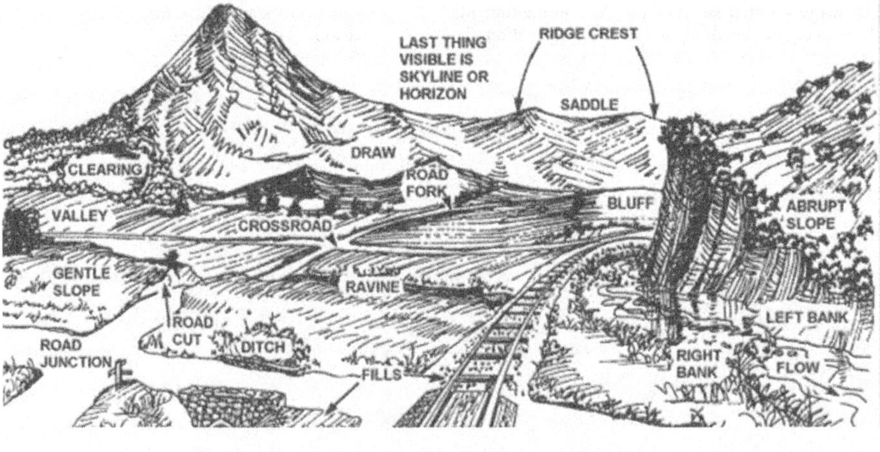

Figure 2-1. Natural and Artificial Terrain Features.

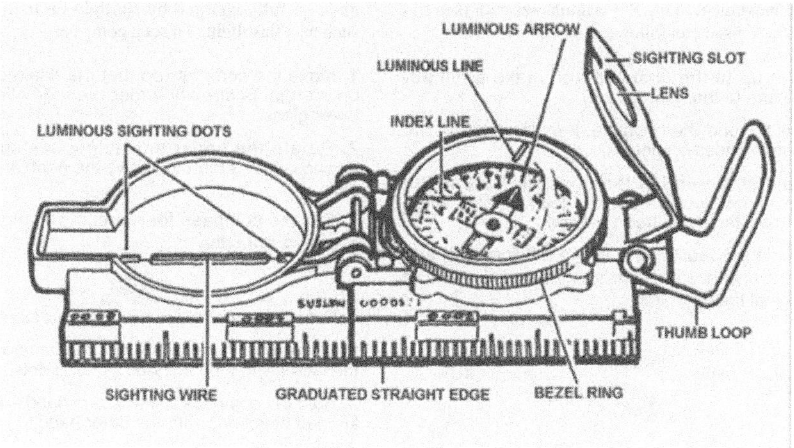

Figure 2-2. Lensatic Compass.

attraction while using a compass to determine direction. The rifle, pistol, and other metal objects must be laid aside when reading the compass. The minimum distances the scout should be from visible masses of iron and electrical fields of magnetism for the compass to provide accurate readings are provided below.

Magnetic Fields	Minimum Distances (meters)
High tension powerlines	60
Field guns	20
Vehicles (wheeled or tracked)	20
Telephone and telegraph wires	10
Barbed wire	10
Machine gun	3
Rifle, pistol	1

Reading an Azimuth

To read an azimuth to any point, the cover of the compass is raised to an angle of 90 degrees in relation to the index face, and the eyepiece is lifted to a 45-degree angle in relation to the bezel, or so the numbers on the dial can be seen. The thumb of either hand is placed in the thumb loop, the index finger extended along the side of the compass case, and the remainder of the hand closed. The closed hand and wrist are grasped with the other hand. The elbows are drawn in close to the body, forming a firm foundation for the compass. The eye is placed to the lens on the eyepiece. The compass is pointed at the object or point to which the azimuth is to be read. A sighting is taken through the sighting slot in the eyepiece, and the point is lined up with the sighting wire in the cover. The compass is held until the dial steadies; then the reading is taken through the lens of the eyepiece. This reading is the magnetic azimuth of the line from the observer to the point.

Reading a Back Azimuth

A back azimuth is the direction opposite the line of sight. If the azimuth is less than 180 degrees, the back azimuth is obtained by adding 180 degrees. If the azimuth is greater than 180 degrees, the back azimuth is obtained by subtracting 180 degrees. Back azimuths are used to determine a return route or to resection to determine a current position.

Circumventing Obstacles

When a scout is traveling on an azimuth and comes upon an obstacle—such as a contaminated area, minefield or swamp—the following steps (sometimes referred to as the *90-degree offset* method) is employed to go around or circumvent the obstacle and resume movement along the original azimuth (see fig. 2-3). The steps are as follows:

1. Move up to the obstacle and make a full 90-degree turn to the right (or left).

2. Walk beyond the obstacle, keeping track of the distance in paces or meters.

3. Stand at the end of the obstacle, face in the original direction of march, and follow that azimuth until the obstacle has been passed.

4. Make a 90-degree turn to the left (or right) and move the distance previously measured to return to the original line of march.

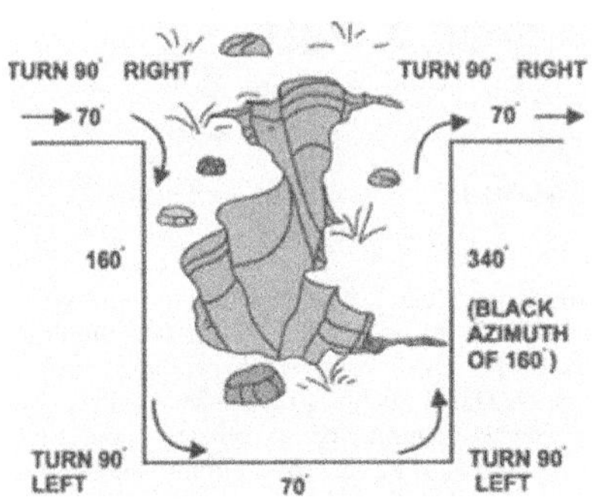

Figure 2-3. Circumventing an Obstacle.

Following an Azimuth During the Day

The eye is placed to the lens in the eyepiece and the compass moved until the desired azimuth reading is visible beneath the fixed index. Without moving the compass, the vision is shifted from the lens through the sighting slot in the eyepiece, and a sighting is taken out beyond the sighting wire in the cover. A prominent terrain feature on this line of sight is selected, the compass closed, and the landmark approached. When the scout reaches the landmark, the procedure is repeated.

Following an Azimuth at Night

It is necessary to prepare and set the compass before departing on a night movement because at night only the luminous parts of the compass can be seen. To prepare the compass for night use, the luminous parts must be fully charged by sunlight or artificial light, such as a flashlight. To set a compass—

1. Move the compass so that the desired azimuth on the dial is directly under the index line on the lower glass.

2. Rotate the upper movable glass so that the luminous line is directly above the north arrow of the dial.

3. Set the compass for marching at night on the specified azimuth.

—OR—

1. Face the general direction of movement.

2. Line up the north arrow and the luminous line on the bezel with the luminous sighting dots.

3. Hold the compass still with one hand and grip the knurled bezel ring with the other hand.

4. Turn the bezel ring the prescribed number of clicks in the proper direction, remembering that each click equals 3 degrees. For example, to set an azimuth of 21 degrees, the bezel ring would be turned seven clicks to the left.

5. Turn the whole compass until the north needle lines up with the luminous line. The compass is then set on the desired azimuth. The azimuth is the line formed by the two luminous sighting dots on the inside of the cover.

To march on a preset azimuth during night movement, open the compass and move it so the north arrow is directly below the luminous line (see fig. 2-4). Move

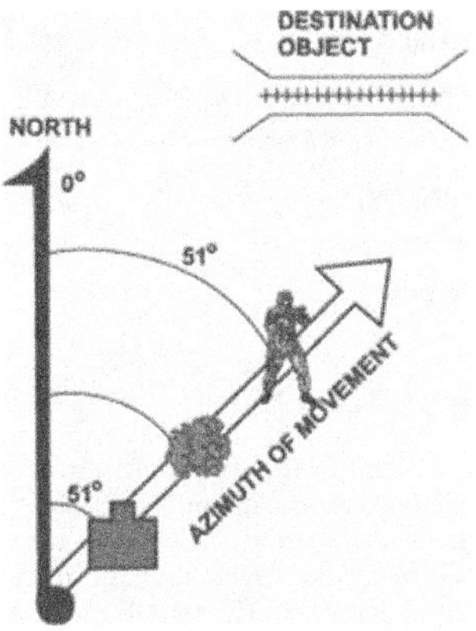

Figure 2-4. Following a Night Azimuth.

in the direction of the line formed by the two luminous sighting dots. It is necessary to refer to the compass more frequently at night than during the day. If stars are visible, find a prominent star along the azimuth of movement to use as a reference point. When the view of the sky is restricted by overcast conditions or vegetation, send a scout forward along the azimuth of movement to the limit of visibility. This scout is guided along the azimuth of movement by a stationary navigator. When the scout reaches the limit of visibility, the navigator moves to the scout's location. This process is repeated until the destination is reached.

A more rapid method for reaching the scout's destination is to equip the navigator with a compass. The navigator can set the compass as explained earlier and the scout proceeds providing security 180 degrees to the front on the specified azimuth, receiving right and left corrections from the navigator while both are on the move. The point scout must stay within visual range of the navigator. If available, a strip of white or luminous tape on the back of the point scout's helmet will assist.

Intersection

Intersection is the location of an unknown point by successively occupying at least two, preferably three known positions and sightings on the unknown point. It is used to locate features not depicted on the map or not readily identifiable. To determine an intersection, perform the following steps (see fig. 2-5):

1. Orient the map using the compass.

2. Locate and mark your position on the map.

3. Measure the magnetic azimuth to the unknown position; then convert to grid azimuth.

4. Draw a line on the map from your position on this grid azimuth.

5. Move to a second known position from which unknown point is visible. Locate this position on the map and again orient the map using the compass. The second unknown position should be a minimum of 30 degrees offset from the first position.

6. Repeat steps 4 and 5.

Figure 2-5. Intersection.

To check accuracy, move to a third position and repeat steps 1 thorough 4. Where the lines cross is the location of the unknown position. Using three lines, a triangle is sometimes formed—called the *triangle of error*—instead of an intersection. If the triangle is large, recheck your work to find the error. Do not assume that the position is at the center of the triangle.

Resection

Resection is the location of the user's unknown position by sighting on two or three known features that are identifiable on the map. To determine a resection, perform the following steps (see fig. 2-6):

1. Orient the map using the compass.

2. Locate two or three known positions on the ground and mark them on the map.

3. Measure the magnetic azimuth to a known position then convert to grid azimuth.

4. Change the grid azimuth to a back azimuth and draw a line on the map from the known position back toward the unknown position.

5. Repeat step 3 and step 4 to determine a second known position.

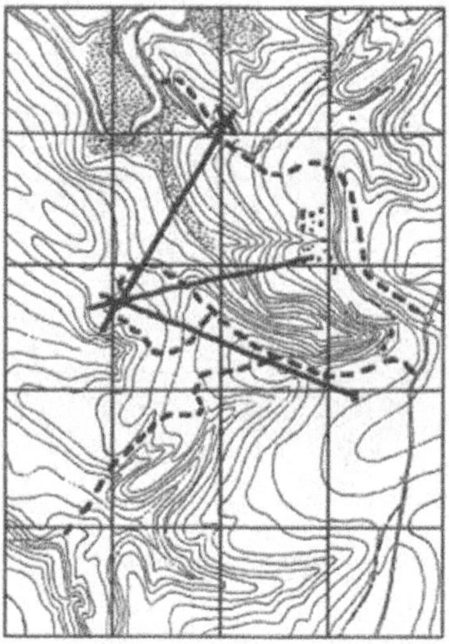

Figure 2-6. Resection.

To check accuracy, repeat the steps above for a third known position. The intersection of the lines is your location. Using three lines, a triangle of error may be formed. If the triangle is large, recheck.

2003. STARS, SUN, AND OTHER FEATURES

In rare cases when a scout is without a compass, the following examples are alternate means to determine direction. When using constellations to determine direction, identify your location's Temperate Zone. A Temperate Zone is the area between the tropics and the polar circles.

At night, the stars provide an excellent means of maintaining a line of march. In the North Temperate Zone (north of the equator), the Big Dipper constellation is one key to determining direction of true north. It is made up of seven fairly bright stars in the shape of a dipper with a long curved handle (see fig. 2-7). The two stars that form the side of the cup farthest from the handle, used as pointers, are situated in the direction of a bright star that is about five times the distance between the two stars of the dipper cup. This bright star is the North Star and is directly over the North Pole. The pointers always designate the North Star, which is the direction of true north.

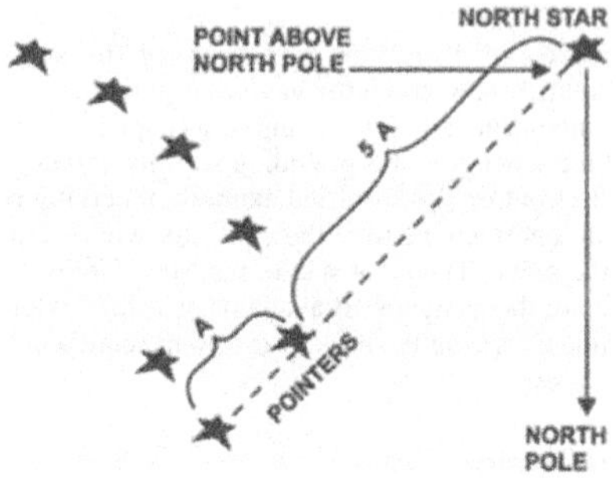

Figure 2-7. Locating the North Pole.

In the Southern Hemisphere, true south is determined in relation to the Southern Cross, a constellation composed of five stars. Two bright pointer stars in the vicinity of the Southern Cross serve as locators to help locate true south (see fig. 2-8). The outer four stars are

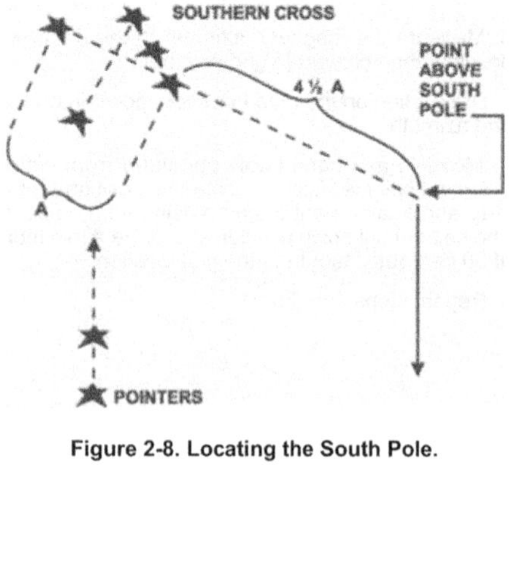

Figure 2-8. Locating the South Pole.

fairly bright and form a cross. This cross is imagined as the frame of a kite. A straight tail, four and one half times as long as the length of the kite itself, is put on the kite using finger widths for a measuring stick. The end of this tail will be close to a position directly over the South Pole. Usually, it will not be possible to see a star in the immediate vicinity, because there is no bright star visible directly above the South Pole.

During daylight hours, a watch and the sun can be used to determine direction within 8 degrees. In the North Temperate Zone, the watch is held horizontally, face up, and the hour hand pointed at the sun (see fig. 2-9). The north-south line and the direction of south can be found midway between the hour hand and the number 12, if the watch is set on standard time. If in daylight savings time, the direction of south is found midway between the hour hand and the number 1.

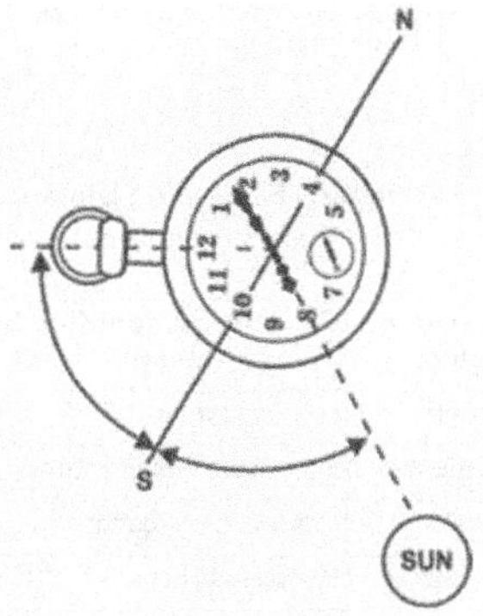

Figure 2-9. Determining Direction by Watch and Sun (North Temperate Zone).

In the South Temperate Zone, if the watch is set on standard time, the number 12 on the watch is pointed at the sun; if the watch is set on daylight savings time, the number 1 is pointed at the sun. North is midway between 12 (or 1) and the hour hand (see fig. 2-10).

When laying in a north-south line, if any doubt exists as to which end of the line is north, remember that the sun is in the east before noon and in the west in the afternoon.

In addition to the sun and stars, other methods a scout without a compass can use to determine direction include determining prevailing wind direction and

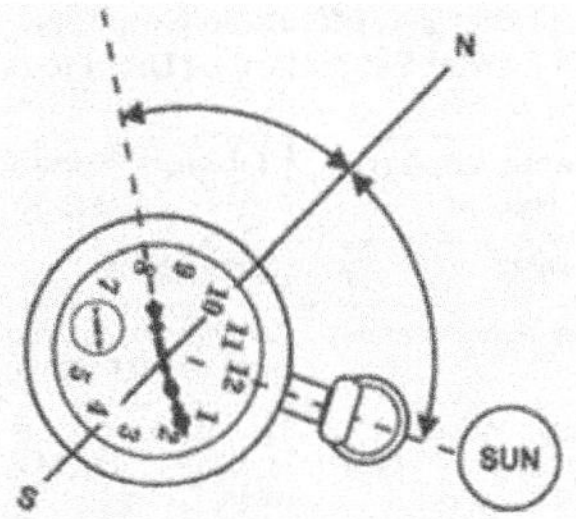

Figure 2-10. Determining Direction by Watch and Sun (South Temperate Zone).

using a mountain for orientation. By previous study of maps and photographs, a scout can keep informed of location and direction by using a distinctive edge of woods, a deep ravine or the direction of a stream's flow. A scout should constantly evaluate and memorize both the immediate terrain and general area for prominent features and landmarks.

2004. RANGE DETERMINATION

Range determination is the method of finding the distance between an observer and an enemy target or an object. By accurate range determination, the members of a given unit can set their sights correctly and place effective fire on enemy targets. The degree of accuracy is dependent on several factors, such as terrain relief, time available, and experience of the observer.

Mental Estimation

A mental distance estimate is made using a known unit of measure. Distance is estimated to the nearest 100 meters by determining the number of known units of measure between the observer's position and a target. For example, a football field, which is 100 yards, can be used as a known unit of measure for determining the distance between an observer's position and a target. For longer distances, progressive estimation may be necessary. To do this, the observer determines the number of units of measure to an intermediate point and doubles the value. The observer should consider the effects in table 2-1 in estimating distances.

Table 2-1. Effects to Consider in Mental Estimation of Distances.

Objects Appear Nearer	Objects Appear More Distant
In bright light.	In poor light or in fog.
In clear air at high altitude.	Only a small part of the object can be seen.
The background is in contrast with the color of the object.	The background is similar in color to that of the object.
The observer is looking down from a height.	The observer is looking over a depression, most of which is visible.
The observer is looking over a depression, most of which is hidden.	The observer is kneeling or sitting, especially on a hot day, when the ground is moist.
The observer is looking down a straight feature such as a road.	
The observer is looking over water, snow, or a uniform surface such as a cultivated field or desert.	

Estimating in Good Visibility

When visibility is good, distances can be estimated by using the appearance of tree trunks, branches, and foliage (as seen by the naked eye) in comparison with map data. Table 2-2 is a guide for wooded terrain. Table 2-3 is a guide for urban environments.

Estimating From a Terrain Study

The Marine should always use terrain/map analysis to assist in estimating distances. When the Marine is looking in a specific direction, the estimation of distance can be enhanced by studying the terrain and comparing it with the map. Particular emphasis should be given to color contrasts of terrain features seen along the observer-target line (OTL). For example, the distance across successive ridge lines or depressions in the distance may be identifiable by only slight changes of color to the eye. Different colors of grass might reveal a hidden terrain feature such as a stream.

Table 2-2. Estimating Distance in Wooded Terrain.

Distance in Meters	Tree Description
1,000	Trunk and main branches are visible. Foliage appears in cluster-like shape. Daylight may be seen through the foliage.
2,000	Trunk visible, main branches distinguishable, foliage appears as smooth surface. Outline of foliage of separate trees distinguishable.
3,000	Lower half of trunk visible. Branches blend with foliage. Foliage blends with adjoining trees.
4,000	Trunk and branches blend with foliage and appears as a continuous cluster, smooth in appearance. Movement of foliage due to wind cannot be deleted.
5,000 and beyond	Whole area covered by trees and appears smooth and dark.

Table 2-3. Estimating Distance in Urban Terrain.

Distance in Meters	Object Identified by the Unaided Eye
1,000	Lone tree trunk
1,500	Individuals and horsemen
3,000	Chimneys on rooftops
4,000	Windows in houses
4,000–5,000	Individual houses in populated area
8,000–9,000	Villages and individual houses
15,000–18,000	Large houses, towers, and steeples

CHAPTER 3. ENEMY ACTIVITY

A commander often acts on information furnished by scouts. Therefore, scouts must aim at absolute accuracy in reporting enemy activity. This chapter discusses estimating enemy strengths, interpreting signs and tracks, and knowing the enemy.

3001. ESTIMATING ENEMY STRENGTHS

If troops cannot be counted, their strength may be estimated by: noting the length of time it takes various types of moving columns to pass given point, the area required of a unit in camp or bivouac, or the front on which they are deployed. When the ground is dry, infantry on the march raise a low, thick cloud of dust, and motor vehicles or mechanized units raise a thick, rapidly moving cloud. Additionally, through practice, a scout may gain information as to the strength and composition of enemy forces by listening to noises and observing lights, fires, and smoke. A scout gains valuable experience in estimating enemy strengths by observing friendly forces in camp, on the march, and deployed. The knowledge scouts gain during field exercises of the appearance and tactical dispositions of squads, platoons, companies, and larger units will be of great assistance in estimating the strength and composition of enemy units observed under various conditions.

3002. INTERPRETING SIGNS AND TRACKS

In addition to estimates made through direct observation, a scout may often be able to estimate size, composition, direction, rate of movement, condition, discipline, state of training, and morale of enemy forces through signs and tracks left behind.

Signs

The examination of vacated enemy positions provides valuable information. The size of a bivouac or defense area ordinarily indicates the number of enemy occupants. Clothing, ration containers, dumps, etc., further indicate the quantity of the departed enemy force. The condition of the bivouac area and amount of material abandoned give an indication of the enemy morale, training, and discipline. A well-policed area indicates good discipline. Rubbish, ration and smoking residue, and nonessential personal items of equipment adrift indicate a lower state of morale, training, and discipline. Stores and material left behind in good condition may indicate a hasty movement or withdrawal. Burned or destroyed materials indicate a deliberate, orderly withdrawal or movement. Letters, insignia, and other articles may reveal the identity of the enemy unit.

In the case of a moving enemy, the distance between periodic halts indicates the rate of march if enemy habits relative to marches and halts are known. Condition of the halt areas indicates the state of morale, training, and discipline.

The physical condition of enemy dead and wounded and their personal equipment and weapons are reported. The general condition and state of maintenance of destroyed or abandoned vehicles should also be reported.

Tracks

A track is a mark left on the ground by the passage of a person or object. Examination of tracks reveals information about the enemy.

Troops

A few tracks overlapping each other on both sides of a road or trail may indicate a patrol in staggered formation. A large number of tracks indicates troops in column formation. A large column will wear a dry road smooth and flat. In damp terrain, a freshly made track will have sharp edges; ordinarily, signs of moisture will disappear in about 15 minutes. A runner's toes are dug into the ground; a walker's footprint is fairly even.

Vehicles

The type of track indicates whether the vehicle is wheeled or tracked. A scout acquires the necessary experience to make the proper determination by observing vehicle tracks during training.

The direction of travel can be determined by the way tracks pass across ruts, by impressions on the edges of holes in the ground, how water is splashed from puddles, or by the way grass, twigs, and branches are bent; for example—

- A vehicle (wheeled or tracked) entering a rut pushes dirt into the rut and leaves an indentation on the exit side of the rut.
- A wheel going over holes in the ground leaves a deeper impression on the edge toward the direction of travel.
- The side of a puddle with the greater splash indicates the direction of travel of the vehicle.
- When traveling cross-country, the direction in which grass is bent and/or twigs, branches, and bushes are broken indicates direction of travel.

A general rate of speed can be estimated by the amount of water or mud splattered. A fast-moving vehicle will throw larger amounts of water or mud a greater distance to the front and sides than a slow-moving vehicle, and it leaves a deeper impression on the exit edges of holes. The faster the travel, the deeper the impression.

3003. KNOWING THE ENEMY

A scout should learn as much as possible about enemy psychology, habits, organization, and tactics. The more knowledge gained about the enemy, particularly the enemy's normal security measures, the better the scout's chances are to observe and obtain accurate information with minimum risk to the mission's success. Scouts gain much of this knowledge through experience, but they also gain a great deal of their preliminary information, particularly that pertaining to enemy organization and tactics, during training and may be updated by unit commanders and intelligence officers.

CHAPTER 4. DAYLIGHT SCOUTING

A scout must be able to operate in all types of terrain and under all conditions of visibility. He must be thoroughly familiar with the principles for using cover and concealment, camouflage, individual movement, and route selection, both to and from the objective.

4001. COVER AND CONCEALMENT

Cover is protection from the fire of hostile weapons. Concealment is protection from observation or surveillance from hostile air and ground observation, but not from hostile fire. Both cover and concealment are divided into two main categories: natural and artificial. Natural cover includes small hills, ditches, rocks or vegetation. Fighting holes, bunkers, and brick walls are examples of artificial cover. Some features, such as buildings, provide both cover and concealment. In deciding whether to seek cover or concealment, a scout must make the best choice to complete the mission (see fig. 4-1).

Figure 4-1. Correct Use of Cover.

Concealment Principles

Concealment principles are as follows:

- Remain motionless while observing. Anything in motion attracts the eye.
- Use all available concealment.
- Observe from the prone position (it offers a low silhouette and makes detection by the enemy difficult.
- Expose nothing that reflects light.
- Blend with the background because contrasting colors are noticeable.
- Remain in the shade because moving shadows attract attention.
- Distort or change the regular outline of objects. Most military objects have distinctive shapes that make obvious shadows and silhouettes.
- Avoid the skyline. Figures on the skyline can be seen from great distances and are easily identified by their outlines.

Concealment Techniques

Concealment techniques are as follows:

- When observing, the scout looks around an object's side (unless it is transparent) and prepares to fire, if necessary, around the side of or, if possible, through an object.
- Looking or firing over an object can make the scout an easily visible target for the enemy. If the scout must fire over the top of concealment or cover, the outline of the head or helmet should be broken or distorted.
- Upon the approach of an airplane, the scout takes a prone position, turns face-down, and remains motionless. If surprised by an airplane, the scout remains in place and does not look up.

- The scout covers exposed body parts such as the face, back of the neck, and hands with grease paint, mud or other materials to reduce sun reflection.
- Camouflage for equipment can be improvised from garnishing or sandbags to prevent sun reflection.
- In snowy terrain, white overgarments are worn.
- The helmet cover outline should be distorted.

4002. CAMOUFLAGE

Camouflage is the use of concealment and disguise to minimize the possibility of detection and/or identification of troops, material, equipment, and installations. The purpose of camouflage is to provide concealment of military objects from enemy observation. Camouflage is also used to conceal an object by making it look like something else. A scout's mission usually requires individual and equipment camouflage. If natural camouflage is not adequate, the position is camouflaged. In using camouflage, remember that objects are identified by their form (outline), shadow, texture, and color. The principal purpose of camouflage in the field is to prevent direct observation and recognition.

Individual Camouflage

Successful individual camouflage involves the ability to recognize and take advantage of all forms of natural and artificial concealment available (vegetation, soil, debris, etc.) and knowledge of the proper use of artificial camouflage materials.

Aids to Individual Camouflage

A scout must recognize the terrain's dominant color and pattern and must change the appearance of clothing and equipment accordingly in order to blend and not contrast with the terrain (see fig. 4-2).

The helmet is camouflaged by breaking up its shape, smooth surface, and shadow. Use of a helmet cover works best. In the absence of a helmet cover, mud can be irregularly blotched on the helmet to disguise its form and dull the surface. A helmet cover may be improvised from irregularly colored cloth or burlap to blend with the background. Foliage can be draped to prevent the visor of the helmet from casting a dark shadow across the face. Foliage should not stick up like plumes because any head movement will give away the position.

A small, thin bush in the shadow of a large bush makes a good observation point. Lone trees, rocks, fence corners, and outstanding landmarks are easily picked up by the enemy as obvious observation posts.

If camouflage clothing is not available, other available clothing can be attached in irregular splotches of appropriate colors.

Exposed skin reflects light and attracts the enemy's attention. Even very dark skin will reflect light because of its natural oil. The buddy system is recom-mended when applying camouflage. Standard Marine Corps issue camouflage face paint sticks are two toned:

- Loam and light green for light-skinned troops, in all but snow regions.
- Sand and light green for dark-skinned troops.
- Loam and white for troops in snow-covered terrain.

Shiny areas (forehead, cheekbones, nose, and chin) are painted with a dark color. Shadow areas (around the eyes, under the nose, and under the chin) are painted

Figure 4-2. Avoid Contrasting Backgrounds.

with a light color. Skin that is exposed on the back of the neck and hands is painted with a two-color combination in an irregular pattern (see fig. 4-3).

When standard issue face paint sticks are not available, burnt cork, charcoal or lamp black can be used to tone down exposed areas of skin.

Mud is used only in an emergency because it changes color as it dries and may peel off, leaving the skin exposed. Since mud may contain harmful bacteria, mud should be washed off as soon as possible.

Any equipment that reflects light should be covered with a nonreflective material that aids in the concealment of the weapon (for example, black electrical tape or mud). The straight line of the rifle or other infantry weapons may be very conspicuous to an enemy observer. The barrel and hand guard should be wrapped with strips of contrasting colored cloth or

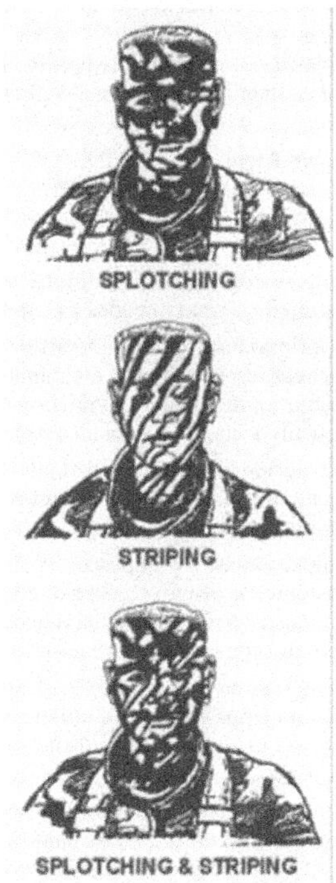

Figure 4-3. Face Camouflage.

tape to break the regular outline. Mud or dirt dulls the reflecting surface of the stock, barrel, and bayonet where coloring has been worn. Lamp black may also be used on metal parts. The function of the weapon must not impaired.

If time, material, and surroundings permit, a ghillie suit should be constructed. (Refer to MCWP 3-15.3, *Scout Sniping*.)

Aids to Camouflage a Position

To successfully camouflage a position, the scout must remember to—

- Camouflage the position as soon as it is occupied.
- Avoid using too much material for camouflage. Even though natural materials are used, too much may make the object and its shadow stand out from its surroundings, thus attracting the attention of a hostile observer.
- Inspect completed camouflage work from the enemy's point of view to check effectiveness.

Continuous Camouflage

Camouflage around and on the scout's position must be maintained in a fresh condition as wilted and dead foliage can give the position away. If the mission dictates that the position should be occupied for longer periods, wilted foliage should be replaced during periods of reduced visibility.

4003. INDIVIDUAL MOVEMENT

Principles

The principles of individual movement are as follows:

- Scouts move from one concealed position to another. When not changing positions, they remain motionless.
- The scout's head is lifted slowly but steadily, without abrupt movements, to search for a new position.
- Scouts select the next stopping place before moving and ensures it is not contained by the enemy.
- Scouts change position on the run: springs up, runs with the body bent low, zigzags, quickly drops to the ground slightly to the right or left of the objective, then rolls or crawls to the desired position.

(Remember the phrase, "I'm up—he sees me—I'm down.")

Rushing

When starting from the prone position—

- Raise the head slowly and steadily and select a new position.
- Lower the head slowly, draw arms inward, cock right leg forward, and prepare to rush.
- Use one movement to raise the body by straightening both arms.
- Spring to your feet, stepping off with the left foot.
- Bend forward as low as possible when running. Never advance directly to the next position; always zigzag.

When hitting the deck—

- Stop.
- Plant both feet in place.
- Drop quickly to the knees and slide the hand to the heel of the rifle.
- Fall forward, breaking your fall with the butt of the rifle. (To confuse the enemy, roll over after hitting the deck and roll into firing position with feet, knees, and stomach flat on the ground.)
- Keep head down if you do not intend to fire.

When rolling over—

- Hit the deck and assume the prone position.
- Bring the rifle in close to the body, placing the rifle butt in the crotch.
- Roll over swiftly to confuse any enemy observers as to final intended location. Never reappear at the same place you went down.

Low Crawl

The low crawl is used when—

- Cover and concealment are scarce.
- The enemy has good observation over the area in which the scout is moving.
- Speed is not essential.

To perform the low crawl, keep the body as flat as possible against the ground. Grasp the rifle sling at the upper sling swivel. Let the balance of the rifle rest on the forearm and let the butt of the rifle drag on the ground. Keep the muzzle off the ground.

To start forward, push arms forward and pull right leg forward. To move forward, pull with arms and push with right leg. Change the pushing leg frequently to avoid fatigue.

High Crawl

The high crawl is used when—

- Cover and/or concealment are available.
- Poor visibility reduces enemy observation
- Greater speed of movement is required.

To perform the high crawl, keep body off the ground. Rest weight on forearms and lower legs. Cradle rifle in arms, keeping the muzzle off the ground. Keep knees well behind the buttocks to stay low.

Move forward, alternately advancing right forearm and left knee; then left forearm and right knee.

Movement Aids

Aids to movement include—

- Carrying only necessities. Additional weight causes premature fatigue and impedes free movement.
- Not disturbing birds or animals whose flight would betray your presence. If birds or animals are alerted, remain motionless under cover for a few minutes, as the enemy's attention may also be attracted.
- Moving during an incident that diverts attention, such as an airplane flight, a distant disturbance or sudden bursts of fire.
- Fog, smoke, or even light haze offer concealment for movement; however, the enemy may have thermoimagery and night vision devices. Therefore, darkness and smoke cannot be used as easily.
- Following a stream or road by staying as far away from them as possible while still keeping them in sight. Keep close to the dune line when moving along a beach.
- When moving through tall grass or similar growth, move when the wind blows, changing direction frequently. A straight route will be more readily noticed.
- Whenever possible, avoid areas of soft ground so as not to leave tracks.

- When crossing a road or water obstacle, choose crossing sites where the enemy's observation is restricted (an area in shadows or near a bend) and cross rapidly.

4004. ROUTE SELECTION

Prior to Movement

A scout and the immediate commander conduct a map reconnaissance before starting on a mission. This assists them in selecting the route according to available cover and concealment and any indicated enemy activity.

Prior to and during the course of the mission, move to an observation point to visually reconnoiter the terrain for movement and select the tentative route. It may be necessary to make wide detours around open spaces or those containing enemy patrols or other enemy activity.

Carefully study the country to be traversed and pay close attention to the general features, streams, ridges to be crossed, and their relation to the general direction to be taken (see fig. 4-4).

Make notes of terrain features and landmarks along the proposed route and rely on notes for guidance (see fig. 4-5 on page 4-6). Additionally, determine the compass direction and readings for each change of direction at the start. Finally, learn the location of unit boundaries and observation/listening posts as well as general location of other friendly or scouting parties. Be sure to avoid man-made and natural obstacles as they will slow progress and overall success of the mission. If possible, use the local populous as a source of intelligence. When returning to friendly lines, avoid using the same route.

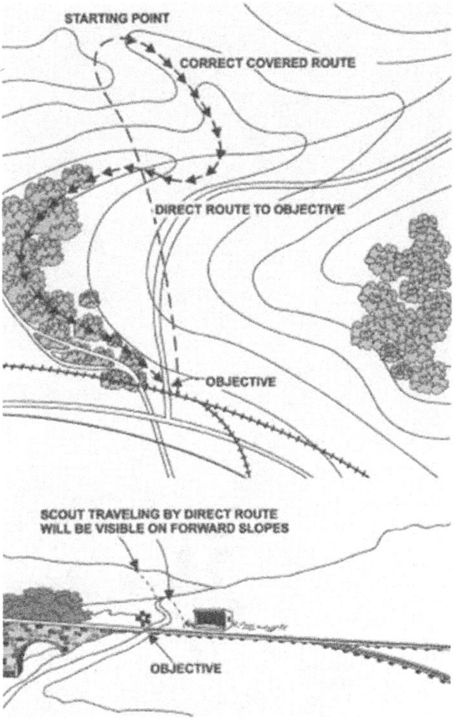

Figure 4-4. Choosing a Concealed Route of Advance from a Map.

En Route

En route, the actual advance will be a series of movements from one observation point to the next. The distance and route will depend on cover and terrain. Assess the cover, terrain, and any enemy or civilian activity to determine whether or not to modify the approach or return routes. Unless the mission requires it, avoid danger areas (for example, houses, villages, potential assembly or bivouac areas, roads, and streams) that may give away your position by being observed by the enemy. When required to reconnoiter danger areas, choose a covered approach and return, and make entry or passage as quietly and quickly as possible. If part of a larger effort, the approach and return should be covered by observation and fires of the other members of the scouting party or patrol.

Stream Crossings

When the crossing does not appear to be held by the enemy, advance upon it rapidly. If there are two or more scouts, one crosses while the other(s) provide protection. Note the length, width, depth, and approaches to a crossing. Observe the condition of the road or trail that crosses the stream, and report on the suitability of the crossing for use by tracked and wheeled vehicles. If the crossing is under observation by enemy, seek another crossing site or dash across to avoid detection.

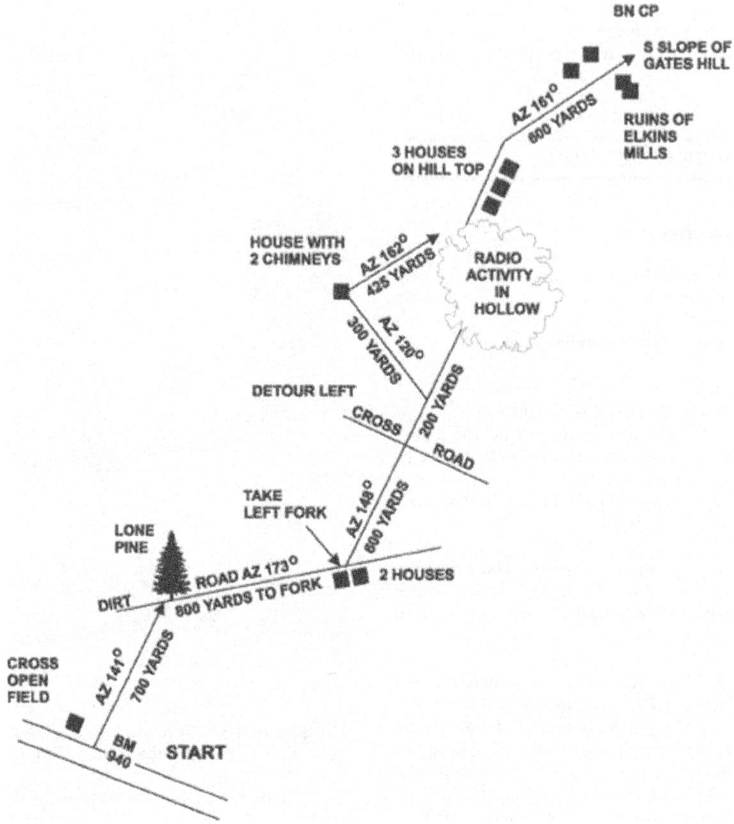

Figure 4-5. Proposed Route Sketch.

Chapter 5. Night Scouting

Night scouting presents many of the same problems encountered in day operations-such as cover, concealment, movement, and camouflage-as well as additional considerations. Knowledge of human eye construction and operation will enable maximum advantage under night conditions or poor visibility.

5001. NIGHT VISION

Certain parts of the eye correspond to parts of a simple camera (see fig. 5-1). The lens focuses light entering the eye similar to a camera lens. The iris (colored part of eye) corresponds to the diaphragm of a camera, opening and closing to regulate the amount of light entering the eye through the pupil. The retina corresponds to camera film. Light rays strike the retina, form an image, and cause an impression to be transmitted to the brain through the optic nerve. In a camera, the image is formed and fixed on film.

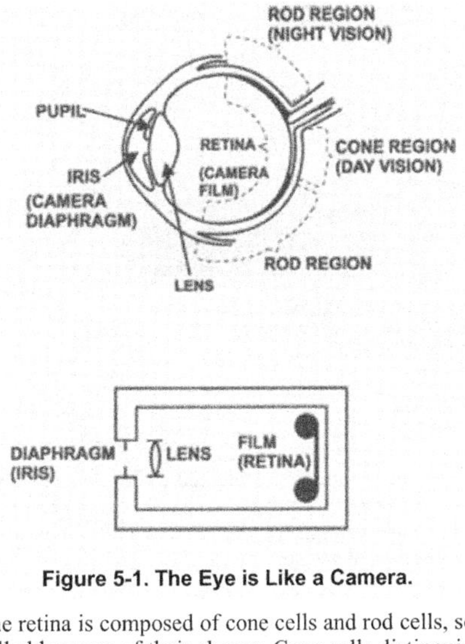

Figure 5-1. The Eye is Like a Camera.

The retina is composed of cone cells and rod cells, so-called because of their shapes. Cone cells distinguish color, shape, and sharp contrast. Because they are activated by light conditions, they are blind during periods of low illumination. Rod cells produce a chemical substance called visual purple that makes them active in darkness, low illumination or night conditions. Rod vision distinguishes black, white, shades of gray, and general outlines.

Principles

To effectively "see" at night, the principles of night vision dark adaptation, off-center vision, and scanning are applied.

Dark Adaptation

Allowing the eyes to become accustomed to low levels of illumination is called dark adaptation. It takes the rod cells about 30 minutes to produce enough visual purple to activate them and enable the eye to distinguish objects in dim light. This may also be accomplished by staying in a red-lighted area, or by wearing red goggles for 20 minutes, followed by 10 minutes in darkness (which allows the pupils to open wide). This method saves valuable time by allowing Marines to be in a lighted area to receive orders, check equipment, or perform some other function before moving into darkness.

Off-Center Vision

The technique of focusing on an object without looking directly at it is called off-center vision. When looking directly at an object, the image is formed on the cone region, which is not sensitive at night (see fig. 5-2 on page 5-2). When looking slightly to the left, right, above or below an object, the image is formed on the area of the retina containing rod cells, which are sensitive in darkness. The most sensitive area varies in individuals, but is usually found by looking 6 to 10 degrees away from an object; in effect, out of the corner of the eye (see fig. 5-3 on page 5-2).

Scanning

Off-center vision used to observe an area or an object is called scanning. When using rod vision, the visual purple in the rod cells bleaches or blacks out in 4 to 10 seconds and the object observed disappears. As the visual purple in the rod cells in one area bleaches out, the eyes must slightly shift to use fresh rod cells. Eyes

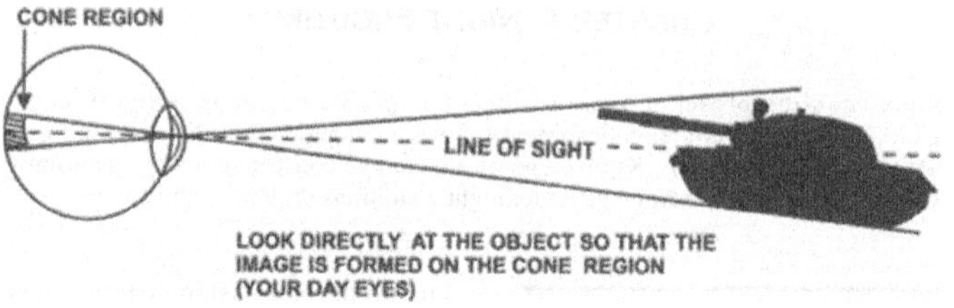

Figure 5-2. Day Vision.

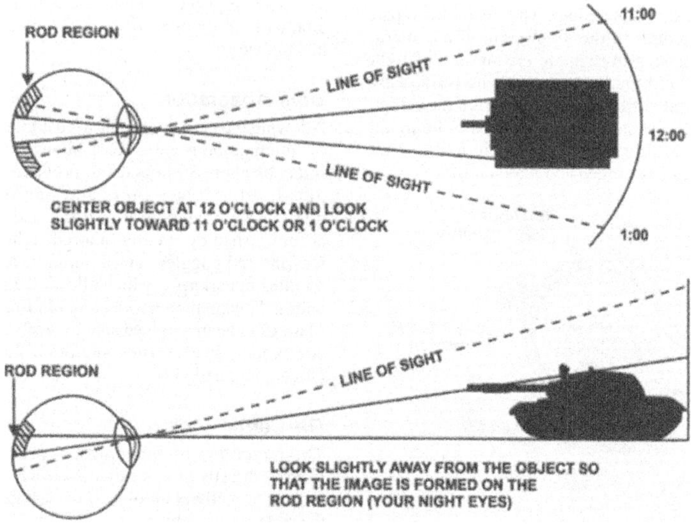

Figure 5-3. Night Vision.

should be moved in short, abrupt, irregular movements over and around the target (see fig. 5-4).

Preserving Night Vision

Night vision is quickly destroyed if bright light is allowed to enter the eye. When entering a lighted area or when observing in a temporarily lighted area (illumination, flares), one eye should be closed and covered to preserve its night vision. When the light goes off, fades or the lighted area is exited, the night vision retained by the protected eye enables it to see until the other eye adapts to the darkness. Red light helps preserve night vision, but like white light, it can be observed at long distances.

Factors that decrease night visual acuity include fatigue, lack of oxygen, long exposure to sunlight, alcohol, nicotine (within the past 48 hours), and age. When night vision has been attained, straining will not improve effectiveness; however, practice identifying objects at night will improve perception.

5002. APPEARANCE OF OBJECTS

Darkness not only makes objects difficult to see but also changes their appearance, distorts size, and blots out details. A tree visible against the night sky appears smaller than in the daytime because the twigs at the end of branches cannot be seen. A scout must train to identify objects by block outlines at night and cannot rely on details visible in daylight. Binoculars enlarge objects or parts of objects otherwise too small to be seen and help identify objects already spotted. Night observation devices increase night visibility and should be used whenever possible.

5003. SOUNDS

At night, sounds become very important. By listening, a scout gains information about the enemy and by exercising care, keeps information from the enemy. A scout stops frequently to listen. Scouts must listen for long periods in perfect silence. Hearing is amplified with the mouth open; removing the helmet will reduce sound distortion. Sounds are transmitted a greater distance in wet weather and at night than in dry weather and in the daytime. By holding the ear close to the ground sounds of people walking and vehicles moving can be heard. Sound travels approximately 370 meters a second. When a flash from a fired weapon is observed, the range to the weapons can be easily estimated by counting the time interval between the flash and hearing the report. For example, counting to three (one thousand one, one thousand two, one thousand three), indicates the distance is 1,110 meters. The cadence is determined by actual practice at known ranges.

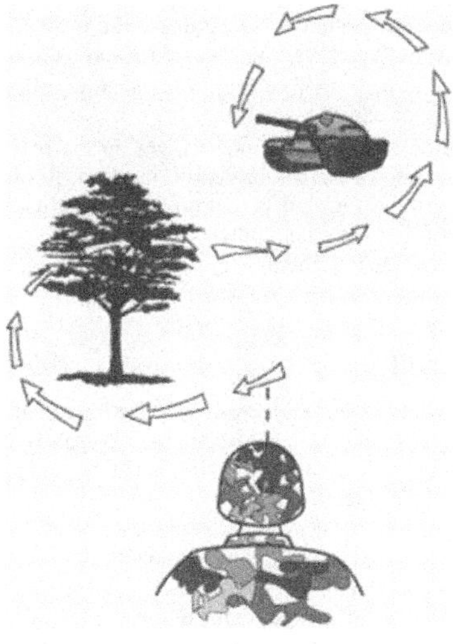

Figure 5-4. Scanning

5004. SMELLS AND TOUCH

A scout's sense of smell can warn of enemy fires, cooking, motor parks, gasoline and diesel engines, and bodies of water. A scout must feel and recognize objects in the dark, and adjust and operate equipment quietly by sense of touch.

5005. CLOTHING AND WEAPONS

All loose clothing must be secured (string or tape can be used) to prevent snagging on barbed wire, brambles, and brush. Helmet covers are worn to muffle sounds made by low branches.

The belt buckle should be turned around to the side in order to move in a prone position without scraping the buckle against stones or hard surfaces. Identification tags can be taped together to prevent rattling. Hands, face, and neck can be blackened so skin does not reflect light or appear as white spots in the darkness. (Refer to para. 4002.)

Scouts are normally armed with rifles. Rifle slings should be taped to prevent rattling. All weapons parts should be checked for glare elimination measures.

5006. CONCEALMENT

Although total darkness provides concealment, scouts must observe the same principles of concealment during moonlight conditions as in the daytime. Scouts should assume enemy employment of night observation devices and observe the principles of night movement such that presence will not be disclosed by noise when close to the enemy.

5007. AIDS TO NIGHT SCOUTING

Aids to night scouting include the following:

- Carry out scouting missions close to or within hostile positions on dark or rainy nights.
- Stifle a threatening sneeze by pressing fingers upward against the nostrils.
- Stifle a threatening cough by applying slight pressure with the finger on the Adam's apple.
- Stop a ringing sound that interferes with hearing by yawning.
- Speak softly rather than whispering if voice communication is necessary.
- Move boldly and rapidly when taking advantage of any sound-such as shelling, rustling wind or distant firing-to push forward if firing is taking place.
- Avoid shell craters and depressions in damp weather conditions if the enemy has employed chemical munitions.
- Move the eyes constantly; concentrating on one object too long will strain them.
- Take notice of the enemy's use of flares. When the enemy employs flares, few enemy patrols are apt to be out; when flares are not employed, the enemy's patrols are likely to be numerous.
- Drop to a prone position upon hearing a flare being fired and before it illuminates. Remain motionless while it is burning. If open or moving when a flare bursts in the air, freeze or drop quickly in the split second after the flare illuminates while the enemy is blinded. You are an easy target for the enemy if the flare bursts in the air or on the ground behind you. Never look at a flare. If you activate a trip flare, drop to the ground and crawl away from the illuminated area.
- Consider all patrols or individuals encountered as hostile until proven friendly. If encountering someone, crouch low, silhouetting the approaching person against the sky. At the same time, make yourself an indistinct target in case the person encountered is an enemy.
- Return fire only to avoid capture if fired on when close to enemy positions.

5008. AIDS TO NIGHT MOVEMENT

- Aids to night movement include the following:
- Move silently.
- Advance in stealthy legs. Each leg should follow some terrain feature that serves as a guide. When there are no terrain features to serve as guides, move in a straight or nearly straight line from one defined point to another, or maintain direction by using a compass.
- Avoid running, except in an emergency.
- Take advantage of sounds that may distract the enemy.
- Fall silently without making an outcry.

Walking

When walking at night—

- Place the heel down first. Balance the weight of the body on the rear foot until a secure spot is found.
- Lift the forward foot high to clear any stiff grass, brush, or other obstruction.
- Continue to balance body weight on the rear foot, lower the forward foot gently, toe first, to explore the ground for objects that might make noise. Step over fallen logs and branches, not on them.
- Lower the heel of the forward foot slowly; gradually transferring body weight to that foot.

Creeping

The low crawl and high crawl are not suitable at night when very near the enemy because an easily heard shuffling noise results. Creeping is the recommended method of movement:

- Creep at night on the hands and knees.
- Use your hands to feel for twigs, leaves or other substances that might make a noise. Clear a spot to place your knee. Keeping your hand at that spot,

bring your knee forward until it meets your hand. Then place your knee on the ground and repeat the action with the other hand and knee.

- Lay the rifle on the ground at your side and clear an area for it. Lift the rifle up and move it forward. Movement is slow and tedious, since it must be done silently.

Hitting the Deck at Night (Right-Handed Shooter)

To safely hit the deck at night from the standing position—

- Advance your left leg, place the butt of the rifle in your right armpit with the hand remaining on the pistol grip, and grasp it with the right hand at the balance.
- Quietly drop down on the right knee and left hand.

- Move the left leg carefully to the rear, and then move the right leg to the rear.

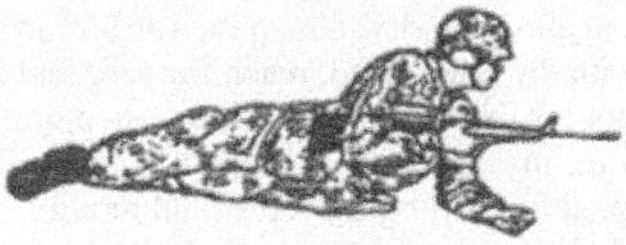

- Lie flat on the ground, or take up a firing position if necessary.

Wire Obstacles

A mission often requires a scout to pass through and work behind enemy positions. To accomplish this, the scout must be able to quietly pass through enemy wire obstacles and cross trenches. Cutting a gap in wire is time-consuming. If possible, walk over the low bands of enemy wire and crawl under the high bands (see fig. 5-5). Avoid movement along wire barriers, as enemy

Figure 5-5. Crossing Wire Silently at Night.

covering fires are generally planned parallel to them to take advantage of canalization and enfilade fire.

To step over low wire at night, crouch low to view the strands against the sky. Grasp the top strand with one hand; with the other hand, reach forward and feel for a clear spot for foot placement without stepping on other strands or any object apt to make a noise. Raise the body up, still grasping the top strand of wire. To avoid catching the foot in another strand, lift the foot up and over, passing it close to the hand grasping the wire.

If a high wire obstacle is encountered at night and wire cutters are not available, pass under the wire with your back on the ground. Grasp the lowest strands in your hands and hold them clear of the body while you slide under them.

When cutting wire and working solo, cut a wire near a post (see fig. 5-6), then dispose of all but one loose end. Grasp the wire close to a post and cut between your hand and the post, muffling the sound and keeping the loose wire in your grasp. When cutting wire in tandem, one firmly holds the wire with the hands positioned close to the cutters, in order to muffle the sound and prevent the loose ends from flying back, while the other one cuts. In both instances, the loose ends of the wire are bent back to form a passage.

Figure 5-6. Cutting Wire Silently at Night.

Wrap a sandbag around the wire cutters and wire to deaden the sound.

Do not cut a complete gap in the wire; cut only the bottom wire(s). Leave the top wire(s) intact to lessen the chance of discovery by the enemy.

Crossing Trenches

Before approaching a trench, wait outside the trench for awhile and listen. Do not enter or cross a trench near its junction with a communication trench. Crawl silently up to the edge of the trench and look into it. Remove all loose dirt and rocks from the edge. If it is a narrow trench, spring up and jump across, sinking quietly to the ground on the other side and remaining there a moment to listen before proceeding. If the trench is wide, climb silently and slowly down into it and out the other side, using the revetment for support (see fig. 5-7). Do not enter enemy trenches unless it is absolutely necessary in order to accomplish the mission. Ordinarily, work is better accomplished from outside the trench. Sentries usually pay more attention to sounds in front of them; therefore, if it is necessary to enter a trench, cross it first at the place where enemy observation is restricted, then approach from the rear.

5009. LOCATING AND PLOTTING THE ENEMY AT NIGHT

For night work, a scout must understand the use of a lensatic compass. (Refer to para. 2002.) Using the lensatic compass, the scout can guide the platoon into position, locate adjoining elements of the command, keep direction when on patrol, determine the location of gaps in the enemy wire and the position of enemy out guards.

Locating Gaps in Enemy Wire

When searching for gaps in enemy wire, at least two lensatic compasses are needed: one to register the gap in the wire and the other for navigation. When a gap in enemy wire is located, lie outside the gap, keeping a distance of 10 meters from the barbed wire. Sight with the lensatic compass on a prominent point on the skyline in line with the gap. In selecting the prominent point in the skyline, pick one that appears on the map (i.e., hill mass, house, road junction). If the only prominent point available is one not identifiable on a map (i.e., a tree, destroyed vehicle, enemy position),

Figure 5-7. Crossing Trenches Silently at Night.

use it. The exact location of this point can be fixed the next day by visual reconnaissance of the area from an observation point. When the needle rests—

- Clamp it in place by lowering the eyepiece to the closed position.
- Rotate the movable bezel ring until the luminous line is directly over the north end of the needle. The azimuth of the gap from the prominent point is now registered.

The compass is carried back without further adjustment. The azimuth setting can be recorded later on a map (see fig. 5-8).

Locating Enemy Out Guards

At night, locate enemy guards by their sounds and failure to maintain light discipline. As sounds of the enemy are heard, and/or observations of the enemy made, shoot an azimuth with one compass. When the needle comes to rest, clamp the compass by lowering the eyepiece to the closed position.

Figure 5-8. Locating a Gap in Enemy Wire at Night.

Note the time and nature of each sound, the estimated distance, and which compass was used to fix the location. Plot this data on a map then wait until the debrief to turn in the notes and compasses. See figure 5-9.

Figure 5-9. Locating Enemy Outguards at Night.

50010. ROUTES OF MOVEMENT

Prior to beginning a night missions, a scout studies the ground in detail from an observation point, air photos, and a map during daylight. The route of advance should be below the skyline. Avoid becoming a silhouette (see fig. 5-10).

Unless the moon is bright, avoid, if possible, passing through woods, ditches, ravines, and brush, because noises of movement may lead to discovery. If the enemy is known to have night observation device capability, avoiding these kinds of terrain may not be possible. To avoid enemy ambushes, return by a different route from the advance and change routes on successive nights.

Figure 5-10. Difference Between Correct Routes by Day or Night.

CHAPTER 6. OBSERVING AND REPORTING

When sent out on a mission, a scout's duties are to observe and report. the first five chapters covered scout protection measures and movement in enemy territory. This chapter discusses performing the assigned mission and reporting the mission after its completion.

6001. OBSERVATION POSTS

Positions

The following guidelines apply to observation posts. See figure 6-1.

- When selecting observation posts, scouts should chose the least prominent position.
- Scouts may occupy one or more observation posts.
- An observation post should not be manned for more than 24 hours.
- A selected observation post should be observed for 10 to 15 minutes to ensure it is not occupied.
- Scouts move to the chosen observation post by a concealed route.
- If the post is located on a hill, crawl to a position where the skyline is broken.
- If a tree is used, the position should have a background so as not to be silhouetted against the sky while climbing or observing.
- When leaving the observation post, a different route from that of the approach should be used.
- If a radio is used, its antenna should be located to provide clear communication to the controlling commander but masked from enemy observation and direction-finding equipment. Upon departure, scouts should remove the antennae from the observation post so as not to give away the position.

Observing

Using all senses available, be particularly alert for movement, objects, sounds, and smells inappropriate to the surroundings. While observing, avoid all unnecessary movement. If observing from a building, keep back from doors and windows.

In daylight, look first at the ground nearest you. Begin observing close to your post and search a narrow strip 50 meters or less deep, going from right to left parallel to your front; then search from left to right a second and similar strip farther away but overlapping the first. Continue to observe until the entire field of view has been searched (see fig. 6-2 on page 6-2).

At night, use a night observation device. If one is not available, search the horizon with short, jerky movements, and short pauses. Look a little to one side of an object and then to the other. Lower the head close to the ground to view the object more clearly. Use low-powered field glasses to increase sight range.

Figure 6-1. Method of Approaching an Observation Post.

Figure 6-2. Method of Searching Ground.

6002. REPORTING

It is imperative that the scout accurately and completely report who, where, when, and what was observed to the proper authority upon mission completion. In most cases, a scout will not be equipped with a radio.

Verbal Reports

Verbal reports should be made when writing is impractical, when the information is not complicated or when the enemy is likely to intercept a messenger carrying a written message.

Written Messages

Written messages, preferred to verbal reports, are recorded in message book blanks issued for that purpose (see fig. 6-3) and delivered to a higher authority as soon as possible. The NATO spot report (SPOTREP) should be reviewed as the written message is a NATO format.

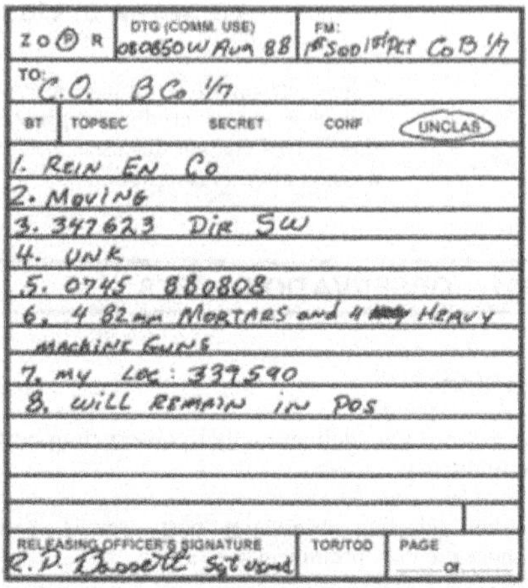

Z O P R | DTG (COMM. USE) 080850W AUG 88 | FM: 1st SQD 1st PLT Co B 1/7

TO: C.O. B Co 1/7

BT | TOPSEC | SECRET | CONF | UNCLAS

1. REIN EN CO
2. MOVING
3. 347623 DIR SW
4. UNK
5. 0745 880808
6. 4 82mm MORTARS and 4 HEAVY MACHINE GUNS
7. MY LOC: 339590
8. WILL REMAIN IN POS

RELEASING OFFICER'S SIGNATURE R.P. Bassett Sgt USMC | TOR/TOD | PAGE ___ OF ___

Figure 6-3. Sample Field Message.

The message body is brief, accurate, and clear; facts and opinions are distinguished. If secondhand information is reported, its source is included. Reports include all information of value, first about the enemy, and then about the originator. Information about the enemy should cover—

- Size and/or strength.
- Actions or activity.
- Location and direction of movement.
- Unit identification. (The designation of the enemy unit may be derived from unit markings, uniforms worn or through prisoner interrogation.)
- Time of observation.
- Equipment and weapons.

Messages are printed in block letters. Individual items of information are numbered and separated into paragraphs. If doubt exists as to message receipt by the commander, a summary of its contents is included in the next message. Information about the originator or writer should cover—

- Location at the time of enemy observation (reference to an important terrain feature, by map coordinates, by the back azimuth from each of two definitely located points, or the back azimuth and distance from one known point).
- Intentions. (Remain in position? Continue on the mission? Take other action?)

The message is carefully reread and if possible read by another person to ensure understandability. If a messenger is used, the messenger must read and understand the message in order to answer any questions the commander might have.

Sketch

Information difficult to describe may be given accurately on a simple sketch. The sketch may give all the necessary information or it may be used to supplement a written message. A military sketch is generally one of two types: simple or panoramic. The simple sketch is easily made and read.

Figure 6-4 shows a simple sketch that has been included in the message itself. The sketch may be on a separate sheet of paper, but all of the necessary information must be contained in either the sketch, the message or both.

The panoramic sketch is a picture of the terrain's elevation in perspective, as seen from one point of observation. Although a panoramic sketch is not difficult to create, skill and training are necessary to enhance usefulness. Figure 6-5 on page 6-4 illustrates how to make a panoramic sketch.

Overlay

The same information sent back to higher headquarters on the sketch may be sent on an overlay, if the sender and the person to whom the message is to be sent have copies of the same map. Figure 6-6 on page 6-6 illustrates a simple overlay. The overlay is drawn on transparent paper as follows:

- Orient the map and place it on a hard, flat surface.
- Place the transparent paper over the part of the map of the object or information to be transmitted and hold the paper in this position.
- Orient the overlay to the map by tracing in the intersecting grid lines at two opposite corners of the overlay. Write the correct number designation on the overlay. The cross made by the intersection is called a *tick mark* and enables the receiver to locate the exact area on the map covered by the overlay.
- Sketch the object seen or the information to be transmitted on the tracing paper (the sheet on top of the map) in the exact location it would appear on the map (the sheet underneath the tracing paper).

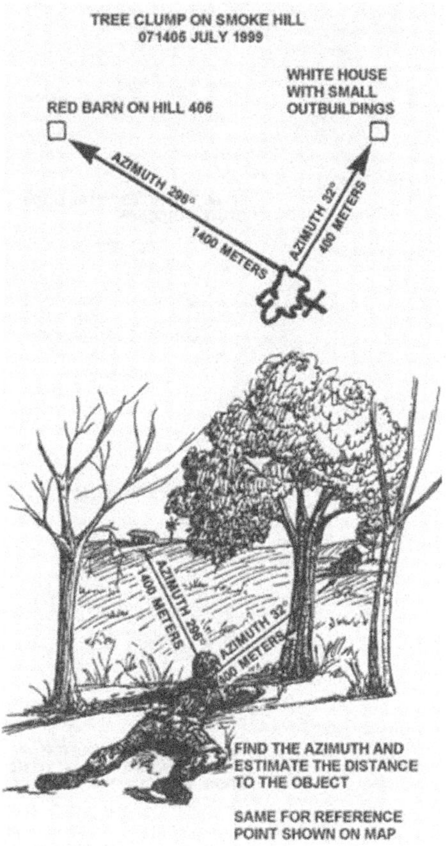

Figure 6-4. Making a Simple Sketch.

Explanatory notes are annotated in the overlay's margin, arrows point to the objectives.

- Indicate with an "X" and an appropriate explanation the position from which the observer saw the object or obtained information.
- Include title and scale of the map from which the overlay was made, date and hour the information was obtained, and signature of the observer on the overlay in the lower right-hand corner.

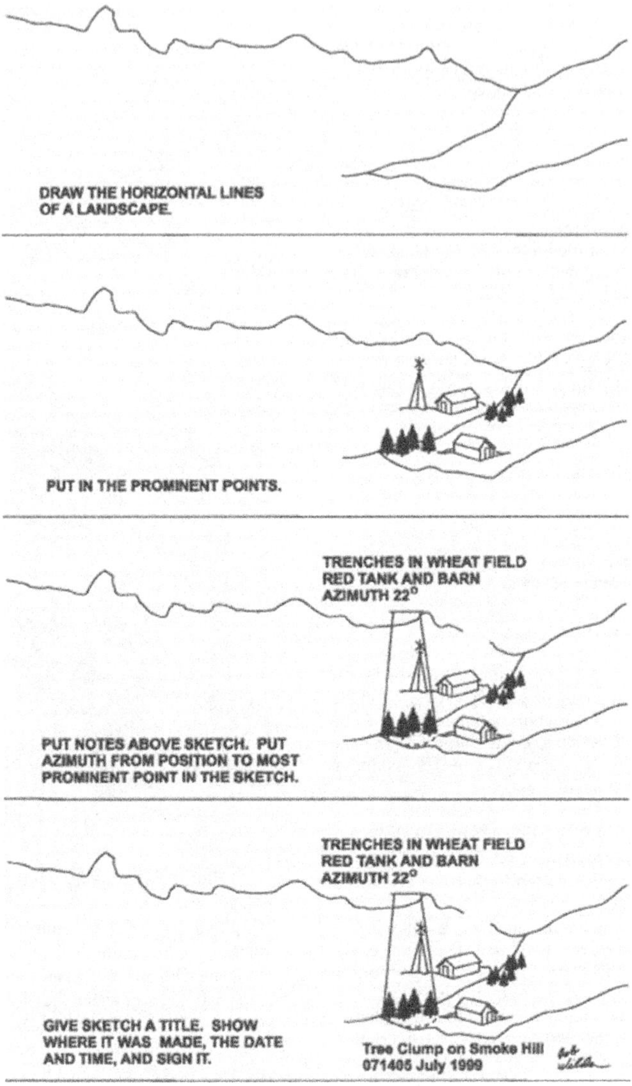

Figure 6-5. Panoramic Sketch.

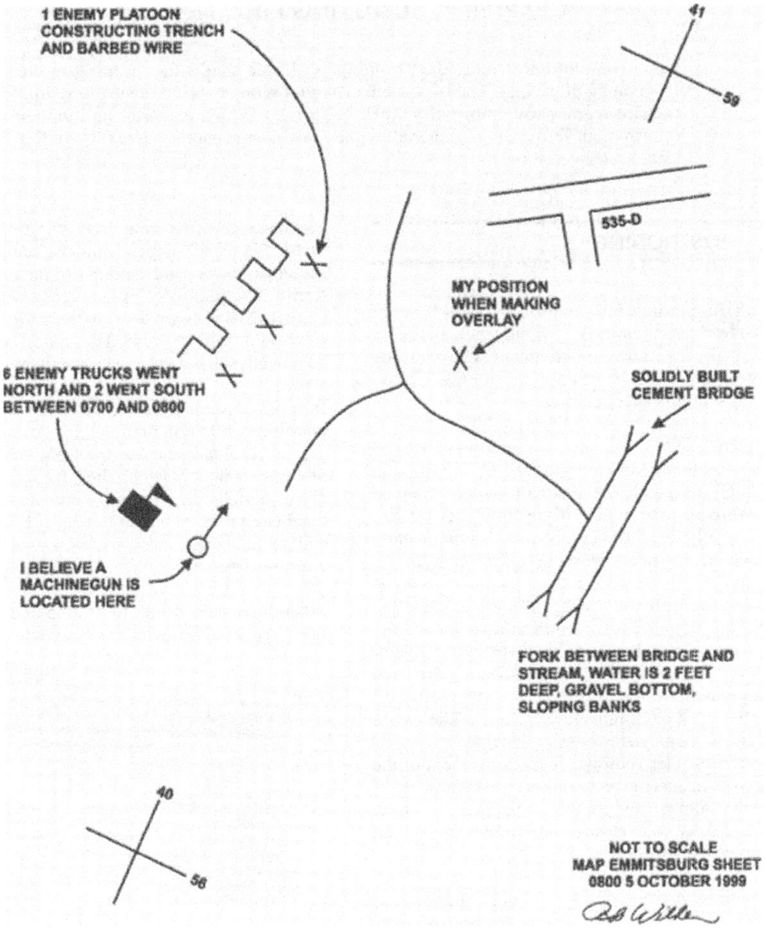

Figure 6-6. Simple Overlay.

CHAPTER 7. SCOUTING FIRE TEAMS

Upon completion of individual training, a Marine adapts this training to the functioning of the unit. The basic tactical unit for scouting is the fire team. Within the fire team, scouts normally work in pairs to furnish security and gather information. Information gathered by the individual scouts is passed to the fire team leader.

7001. POSITIONING

When a rifle platoon in the approach march is not preceded by adjacent units (left, right, front, and rear), it employs its own scouting elements. The scouting element for a platoon is usually one fire team; however, an entire squad may be used.

A fire team used as a scouting element is called a scouting fire team and is controlled by the platoon commander, assisted by the squad leader. A squad leader whose squad is providing the scouting fire teams normally marches near the platoon commander to assist in the control of the scouting fire teams.

A scouting fire team moves aggressively to cover the front of the advancing platoon and to locate the enemy's position(s). It generally moves in a wedge or skirmisher's formation. Normally, a scouting fire team is deployed on a frontage of 50 to 75 meters (10 to 17 meters between each individual scout). The entire squad may be employed to cover a wider frontage. The platoon commander coordinates the movement of the scouting fire team(s) so as to protect the main body of the platoon from enemy fire from points within 400 to 600 meters away, or in close terrain from points within the limits of enemy observation.

Scouting fire team(s) should have enough firepower to overcome resistance from small enemy advanced posts and patrols; the intent is to make enemy riflemen and machine gunners open fire and disclose their positions. Without scouts in advance, the platoon may move into areas where enemy fire may prevent further advance or maneuver and inflict heavy casualties (see fig. 7-1).

Scouting fire teams are covered by the platoon or, when the platoon is masked, the fire team leader ensures individual scouts are maneuvered and coordinated so that the fire team covers its own advance. The fire team leader constantly watches for signals from the platoon commander and remains in visual contact at all times.

The distance between the scouting fire team and the platoon is terrain-dependent. The scouting fire team should not be beyond visibility of the platoon. In open terrain, the platoon commander usually directs the scouting fire team to move by bounds along a succession of locations designated by the platoon commander as intermediate objectives.

Individual scouts should advance as stealthily as possible, while remaining consistent with their mission of reconnaissance to the front, taking advantage of cover without delaying the advance. An occasional glimpse of scouts constantly advancing over a wide front can make the enemy uneasy. It is this activity, and not the target the scouts offer, that may cause the enemy to open fire and disclose its location.

When fired upon, scouts must drop to cover and return fire only when necessary to complete the mission. If

Figure 7-1. Position of Scouts Preceding an Attacking Platoon.

necessary, one scout reports back to the fire team leader who informs the platoon commander.

7002. LOCATING ENEMY POSITIONS

The key terrain of defensive positions are those points that afford extended observation over the ground where the attack must advance. The enemy will place machine guns and infantry to defend critical points.

Members of a scouting fire team preceding an attacking platoon identify the probable enemy infantry and machine gun positions (see fig. 7-2). They use concealment and cover to conduct their advance in order to discover the exact location of enemy positions.

7003. ACTION WITH AN ATTACKING PLATOON

The scouting fire team reconnoiters to the front of the advancing platoon. As soon as the scouting fire team leader indicates the area is secure, the platoon advances and the scouting fire team moves forward. Squads within the platoon advance by bounds; at least one squad is positioned to support the other(s) by fire. Successive positions along the line of advance are selected and designated by the platoon commander as intermediate objectives, and reconnoitered by the scouting fire team before occupation. By conducting proper reconnaissance, surprise by the enemy or movement in the wrong direction may be prevented.

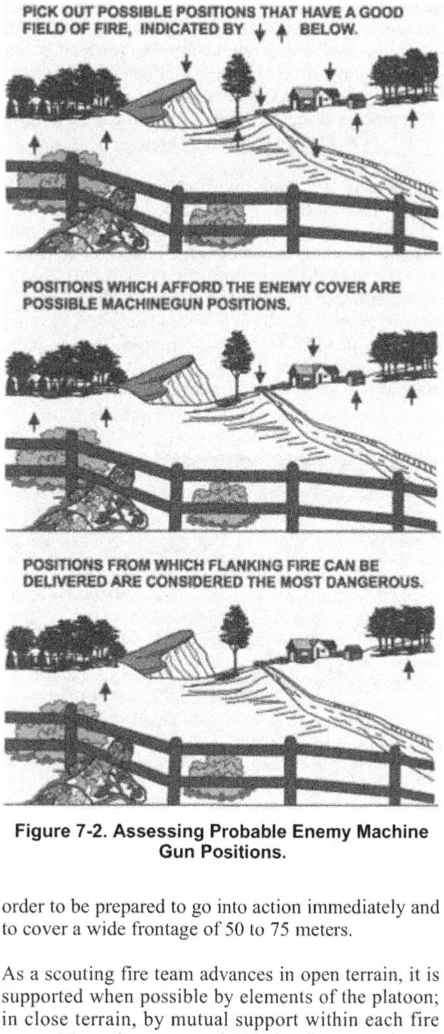

Figure 7-2. Assessing Probable Enemy Machine Gun Positions.

Movement

The distance between the scouting fire team and the front of the main body of the platoon is dependent on the mission, enemy, terrain and weather, troops and support available, and time available (METT-T). In close terrain, such as dense woods, the scouting fire team's movements closely resemble those used for night operations. In approaching houses, woods, and villages, one scout of each pair covers the other while the latter reconnoiters (see fig. 7-3).

A scouting fire team moves forward aggressively to cover the front of the advancing platoon, usually adopting either skirmisher's or wedge formation in order to be prepared to go into action immediately and to cover a wide frontage of 50 to 75 meters.

As a scouting fire team advances in open terrain, it is supported when possible by elements of the platoon; in close terrain, by mutual support within each fire team. Mutual support within the fire team is accomplished by the fire team leader and the automatic rifleman forming a team that supports by fire the advance of the rifleman and assistant automatic rifleman until they reach a location designated by the fire team leader. The rifleman and assistant automatic rifleman then support by fire the movement of the team leader and automatic rifleman. These successive points to which the sub elements of the fire team move are normally designated as fire team intermediate objectives by the fire team leader. The fire team leader sets as many fire team intermediate objectives as necessary to maintain mutual support within the team. This process is repeated until the team can be covered by other elements of the rifle platoon.

A scouting fire team takes advantage of available cover and concealment without delaying its advance. The orders of the platoon commander govern the distance at which it precedes the platoon. The terrain and the probable position of the enemy affect the scouting fire team's distance in front of the platoon. It may be as much as 400 to 600 meters in advance of the platoon. In open terrain, the platoon commander usually directs that the scouting fire team move by bounds to a succession of intermediate objectives. In close terrain or conditions of limited visibility, the scouting fire team is normally ordered to precede the platoon at the limit of visibility, maintaining visual contact with the platoon commander.

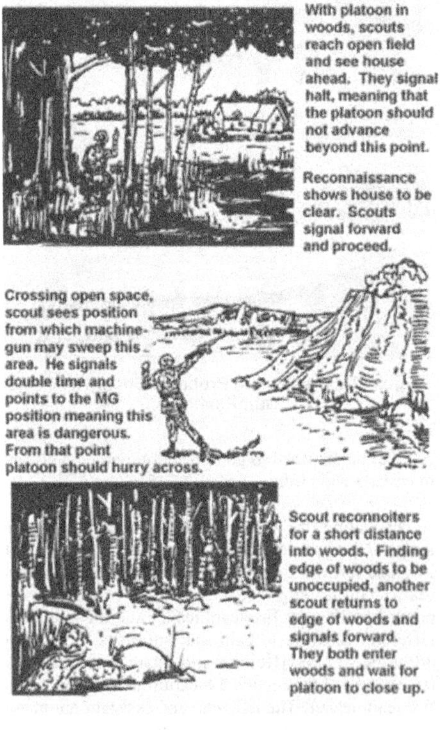

Figure 7-3. Conduct of Scouts Preceding a Platoon.

Action in Woods

When a scouting fire team is directed to advance over open ground to the edge of a woodline, two members of the team, preferably the rifleman and assistant automatic rifleman, reconnoiter inside the woodline while the remainder of the fire team covers them. It is not recommended they separate until finished with their reconnaissance of the far side of the danger area. Both members staying together can cover the same area using a zigzag reconnaissance and they are better equipped to overcome any opposition.

In heavy underbrush and/or poor visibility, the rifleman and assistant automatic rifleman proceed into the woodline together for 50 to 60 meters. The two then separate, searching out either flank to the first high ground or limits of observation, probably 50 to 100 meters (see fig. 7-4 on page 7-4). After the initial search and out posting the limit of advance, the remaining scout signals the fire team forward.

In light underbrush and/or good visibility, the assistant automatic rifleman remains at the edge of the woodline while the rifleman searches the woodline. (see fig. 7-5 on page 7-5). The rifleman searches the woodline in a zigzag pattern, reports back to the assistant automatic rifleman, then moves to an outpost position at the limit of advance. In turn, the fire team leader signals the platoon commander that it is safe for the platoon to move forward.

The fire team leader then moves the remainder of the fire team into the woods, joining up with the forward scout manning the outpost. The scouting fire team occupies and holds a line 50 to 75 meters within the woods and observes toward the direction of movement until the platoon closes up. The scouting fire team leader awaits further word from the platoon commander before moving the team further into the woods.

When directed, the scouting fire team leader moves the team forward until they reach the far edge of the woods. The team is held at the edge of the woods and the fire team leader notifies the platoon commander of the situation. The platoon commander moves the platoon to a position where it can cover the scouting

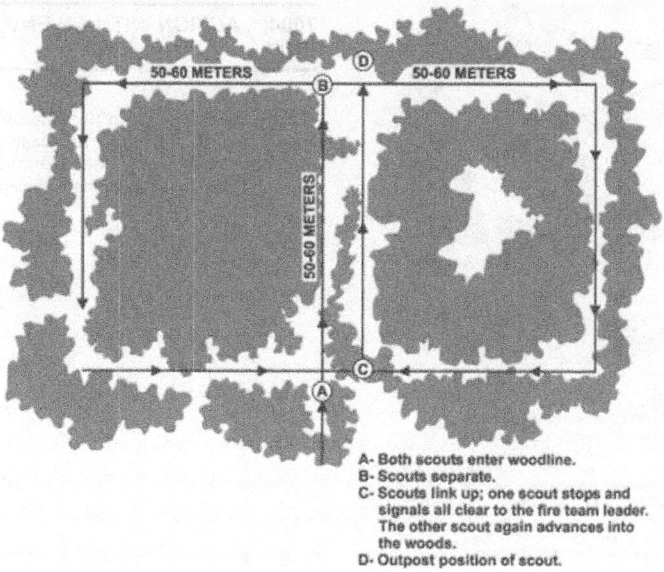

Figure 7-4. Searching Edge of a Woodline (Dense Underbrush/Poor Visibility).

fire team as it exits the woods and directs the team leader to move out and continue the scouting mission.

A scouting fire team passing through woods ahead of its unit maintains a distance allowing visual and oral communications. If an obstacle is encountered, reconnaissance to its front and flanks must be carried out. When advancing along a road or path, scouts precede the platoon to provide necessary protection and to prevent surprise fire on the platoon. When crossing a road or path, they reconnoiter well to the flanks before signaling "all clear" to the platoon.

The scouting fire team will not exit the woods until the arrival of the platoon commander, who will then be given an opportunity to alter the disposition or direction of march. The point where the platoon exits the woods is considered a danger area where the platoon is vulnerable to enemy fire. The scouting fire team is sent ahead to reconnoiter the danger area, as well as the next area to be occupied by the platoon. They signal back whether conditions require a halt, an advance or a quick rush across the open area. The scouting fire team leader must be continually on the lookout for signals from the rear.

Action Under Fire

When a scouting fire team is fired upon, they immediately take cover, locate targets, and return fire. The scouting fire team leader then determines—

- Location of enemy (range and reference points).
- Extent of position (location of flanks).
- Types of positions (obstacles, bunkders, fighting holes, etc.).
- Number of enemy.
- Enemy weapons (machine guns, mortars, tanks, etc.).

The platoon commander assesses the situation as quickly as possible based on the limited information obtained. Usually the platoon commander brings up the remaining squads, sets up a base of fire, and assaults the enemy position. Should the enemy position prove too strong for the platoon, the platoon remains engaged with the enemy as a base of fire until the remainder of the company is committed to clear the enemy resistance.

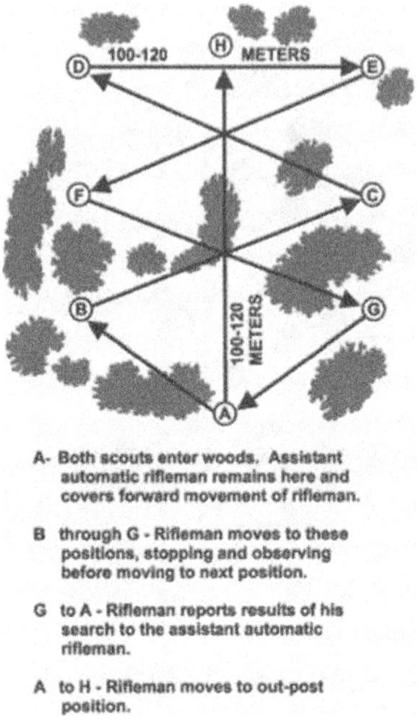

Figure 7-5. Searching Edge of a Woodline (Light Underbrush/Good Visibility).

7004. ACTION WITH AN ENVELOPING UNIT

When a platoon is given the mission to envelop an enemy position, a scouting fire team is employed for protection and reconnaissance in the same way as when the platoon is advancing in the approach march.

PART 2. INFANTRY PATROLLING

CHAPTER 8. FUNDAMENTALS OF INFANTRY PATROLLING

This chapter begins Part 2 and provides basic information about infantry patrols; specifically, their purpose, types, and missions. Infantry patrol training is also addressed and keys to successful patrolling are presented. Subsequent chapters of this part cover patrol organization, preparation, movement, and reconnaissance actions.

8001. DEFINITIONS

A patrol is a detachment of ground, sea or air forces sent out for the purpose of gathering information or carrying out a destructive, harassing, mopping-up or security mission (Joint Publication [JP] 1-02). The mission to conduct a patrol may be given to a fire team, squad, platoon or company.

8002. RELATION OF PATROLLING TO SCOUTING

Each patrol member must be knowledgeable in the principles of scouting and maintain membership of a larger team. To develop the teamwork skills required among the members of a patrol, additional training beyond the basic principles is necessary to become a well-trained scout. A patrol member must respond quickly to the decisions and orders of the patrol leader. There must be complete confidence among all members of the patrol and the confidence that they, as a team, will be successful in their mission.

8003. PURPOSE

A commander must have current information about the enemy and the terrain in order to employ the unit effectively. Patrols are an important means of gaining this information and are used to destroy enemy installations, capture enemy personnel, perform security missions or prevent the enemy from gaining information. Modern warfare places a high premium on effective patrolling because units have larger areas of operations and can be threatened from all directions. As distances between units increase, more patrolling becomes necessary to prevent infiltration by guerrillas or small enemy units, as well as to maintain contact with friendly adjacent units. Active patrolling by numerous small groups is needed to locate the enemy and gather information on the enemy's disposition, strength, morale, and weapons, as well as gather and confirm information about the terrain.

8004. TYPES OF PATROLS

Classification as to Mission

Reconnaissance Patrol

Reconnaissance patrols gather information about the enemy, terrain or resources. Relying on stealth rather than combat strength, they gather this information and fight only when necessary to complete the mission or to defend themselves. The distance covered by reconnaissance patrols varies based on the terrain and mission. The squad is ideally suited for reconnaissance patrol missions because of its relative small size and its experience of working together.

Combat Patrol

A combat patrol is a fighting patrol assigned missions that require engagement with the enemy in combat. Larger and more heavily armed than reconnaissance patrols, combat patrols have a mission to capture enemy documents, provide security, and capture or destroy enemy equipment and installations. Such action is ordinarily followed by a return to friendly positions. Regardless of the mission, the patrol reports any information concerning the enemy and terrain acquired during the accomplishment of the assigned mission. There are four types of combat patrols: raid, contact, ambush, and security (normally conducted by a Marine rifle platoon). A rifle platoon reinforced with crew-served weapons is normally considered the minimum size for contact, economy of force or ambush patrols. In some situations, such as the capture of a small enemy outpost, a rifle platoon could conduct a raid. However, a raid is a complex mission and, due to the organization of a raid force (command, reconnaissance, assault, support, security, and reserve elements), a rifle company is normally the smallest force assigned to a raid.

Classification as to Means of Movement

Foot Patrols

Movement by foot is the most common means; however, there are inherent disadvantages. Foot patrols travel slowly and carry limited quantity and types of equipment and supplies. Range and area coverage is relatively restricted. Foot patrols also have apparent advantages in that they have fewer restrictions as to terrain that can be covered; are more difficult for the enemy to detect; provide thorough coverage within limits of range; and are generally not inhibited by weather.

Motorized Patrols

Where terrain and road networks permit, a motorized patrol overcomes the inherent disadvantages of the foot patrol. Mechanized forces require patrolling units that can keep pace with them. However, motorized patrols are restricted to certain types of terrain, and tend to bypass areas that may be advantageous to and occupied by enemy infantry.

Waterborne Patrols

Waterborne patrols move over seas, lakes, rivers and streams, canals, and other inland waterways. The water is either used as a medium of entry to an objective area or is the actual patrol route. Waterborne patrols are limited by the location of water routes in the terrain and tend to bypass areas that may be advantageous to and occupied by the enemy.

Helicopterborne Patrols

Where terrain is extremely difficult or the enemy situation precludes the use of vehicle or motorized patrols, helicopterborne patrols are a method or means to conduct a patrol.

8005. TRAINING

Training is essential to successful patrolling. Premature and unordered actions by members of the patrol destroy coordination and control. Leaders are trained to issue their orders calmly to inspire confidence and discipline, and to avoid misunderstanding. Patrol members must work together and fight as a team. Training should develop the following skills:

- Expertise in handling individual and special weapons, and familiarity with enemy weapons that may be captured.
- Recognize camouflaged personnel, equipment, and defensive positions; ability to pick up fleeting targets fire the rifle from any firing position.
- Understand fire discipline and, after weapon firing, immediately change position (see fig. 8-1).
- Quick and accurate observation skills, and the ability to recall and transmit clearly and briefly, both orally and in writing.

Figure 8-1. Changing Position After Firing.

- Recognize and quickly respond to improvised signals, visually or by sound.
- Ability to swim with weapon and equipment.
- Use issued or improvised camouflage suits and garnish helmet in order to blend with the surroundings. Smudge face, hands, and any bright surfaces of weapons and equipment with some substance, such as mud or charcoal, to prevent the reflection of light.
- Silence self, equipment, and weapon.
- Use antimalarial and water purification tablets.
- Acclimation to temperature extremes.
- Develop a sense of direction and learn how to follow a course by compass, stars, sun, flow of streams, prominent terrain features, and by observing other natural phenomenon. Learn to determine the distance traveled from a known point and to keep a record of azimuths and the distance traveled on each azimuth (dead reckoning).
- Call for and adjust indirect fire assets.
- Familiarization with all communications assets and the use of field expedient antennae.

8006. KEYS TO SUCCESSFUL PATROLLING

Regardless of the category or means of conducting a patrol, the keys to successful patrolling are—

- Detailed planning. Every portion of the patrol must be planned, all possible contingencies considered.
- Productive, realistic rehearsals. Each phase of the patrol is rehearsed, beginning with actions in the objective area. Similar terrain and environmental conditions are used when conducting rehearsals.
- Thorough reconnaissance. Ideally, the patrol leader will physically conduct a reconnaissance of the route and objective. Photographs and/or maps will be used to supplement the reconnaissance.
- Positive control. The patrol leader must maintain positive control, this includes supervision during patrol preparations.
- All-around security. Security must be maintained at all times, particularly near the end of the patrol where there is a natural tendency to relax.

CHAPTER 9. PATROL ORGANIZATION

Organizing a patrol is a two-step process: the general organization of the entire patrol and the task organization of various patrol elements. Normally, the nature of patrolling does not permit long preparation periods and rehearsals to specifically build a unit for each mission. Accordingly, the patrol leader must combine unit integrity considerations with proven concepts of patrol organization.

9001. GENERAL ORGANIZATION

The patrol leader establishes a patrol headquarters and elements to accomplish the mission.

The headquarters is composed of the patrol leader and the personnel who provide support for the entire patrol, such as a forward observer, corpsman, and radio operator.

The major subdivisions of reconnaissance and combat patrols are elements. The existing infantry structure (squads and fire teams) is reinforced as required. For example, a reinforced platoon tasked to conduct a combat patrol that will raid an enemy outpost could be organized as follows:

- Platoon headquarters (command element):
  - Patrol leader (platoon leader).
  - Assistant patrol leader (platoon sergeant).
  - Navigator.
  - Radio operator (company tactical net).
  - Radio operator (patrol tactical net).
  - Corpsman.
- The first squad (security element) provides security en route to the objective area (the point, flank security, and rear security) and at the objective area (the flanks and the objective rally point).
- The second squad (support element) supports by fire for the attack, covering fire for the withdrawal, and supporting fires to cover the crossing of danger areas.
- The third squad (assault element) provides the assault force to attack and seize the objective; searchers to clear the objective; pacers, compass man, navigator, and the assistant patrol leader en route and back from the objective area.

Any attachments a patrol may have (i.e., demolition team, scout snipers, and machine gun squad) will be added to the element that supports its function. For example, the demolition personnel should go with the unit conducting the attack, and scout snipers and machine gunners should stay with the support squad.

9002. TASK ORGANIZATION

The preceding paragraphs described the elements necessary for a patrol to accomplish its mission. These elements reflect the internal functions or tasks required for the patrol to succeed. Depending on the METT-T, there are various methods of grouping these elements together. Task-organization is the further subdivision of patrol elements into teams that are required to perform essential tasks. In creating teams, unit integrity of infantry units should be maintained.

The patrol is organized so each individual, team, and element is assigned a specific task, but capable and prepared to perform other tasks. This may not be possible for certain specialist tasks requiring a trained technician.

CHAPTER 10. PATROL PREPARATION

For a patrol to succeed, all members must be well trained, briefed, and rehearsed. The patrol leader must have a complete understanding of the mission and a thorough understanding of the enemy and friendly situations. The patrol leader should make a complete reconnaissance of the terrain to be covered (either visual or map), and must issue an order to the patrol, supervise preparations, and conduct rehearsals.

10001. MISSION

The mission assigned to a patrol must be clear and oriented toward one objective with a specific task and purpose. More than one primary objective or indefinite missions invites confusion, casualties, and failure.

10002. FACTORS INFLUENCING PATROL SIZE

The circumstances under which patrolling by infantry units is conducted make it necessary for combat patrols to be able to fight offensively, security patrols to defend themselves, and reconnaissance patrols to move quickly and only fight if necessary to break contact or defend themselves. The size of a patrol depends on METT-T.

Generally, a patrol should contain the least number of members needed to accomplish the mission. Combat missions ordinarily require larger patrols than reconnaissance missions.

10003. COMMANDER'S DUTIES

Determining Patrol Requirements

The need for conducting patrols derives from the commander's stated mission (issued by higher level commander) and other specified and implied tasks (secondary or supporting efforts necessary to accomplish the stated mission). This requires analyzing the unit's mission and determining the necessary reconnaissance and/or combat tasks that must be performed. By considering the mission, along with time available, the commander develops the overall concept of operations to include the patrol plan and the specific tasks for each patrol.

Assigning Units

When assigning patrol missions, maintaining the integrity of the existing unit while considering the skills and experience of the unit and its leader are critical factors to the infantry commander. To provide operational depth and equitable apportionment of hazardous assignments, the commander ensures that each of the subordinate leaders and units develops the skills and experience necessary to conduct successful patrols. Assignment of patrol units must consider the commander's concept of operations as a whole and the plans for subsequent employment of assigned forces after completion of the assigned patrolling mission.

Providing Adequate Time

The commander must allow the patrol sufficient preparation time by completing the mission analysis, estimate, and preparation of orders. The commander should use warning orders to alert subordinates to possible requirements and afford them the opportunity for concurrent planning.

Providing the Patrol Leader Information

The following information should be provided:

- A simple, straightforward explanation of the mission, particularly for night patrols.
- General routes (defined by checkpoints) or exact routes (defined by avenues of approach or other terrain features) to follow.
- Enemy composition, disposition, and strength.
- Location and activities of friendly troops.
- Outposts or other security elements through which the patrol is to pass.
- Terrain conditions.
- Missions and routes of other patrols.

- Time patrol is to depart and return.
- Method of reporting information while on patrol (radio, messenger), place where messengers are to be sent, and place where the patrol leader is to report upon completion of the patrol.
- The challenge and password to be used for exiting and reentering friendly lines.
- Special instructions such as locations to be avoided and essential elements of information higher headquarters is seeking.
- Report barrier/obstacle plan of friendly units if known; support available from friendly unit (e.g., medical evacuation [MEDEVAC], react force).
- Fire support available.

Providing Special Skills and Equipment

There will be situations when the unit assigned to conduct a patrol does not have the necessary technical skills or equipment organic to it to successfully accomplish the patrol mission. In these cases, additions (attachments) are made to the unit, such as—

- Machine gun and/or shoulder-launched, multipurpose assault weapon (SMAW) team and/or squad.
- Forward observers (mortar/artillery).
- Radio operator(s).
- Combat engineers.
- Tracked vehicle crewman to assess the ability to traverse the terrain.
- Corpsman.
- Other personnel (snipers, translators) as required.

Providing Miscellaneous Support

The commander must ensure that the patrol leader is provided with the food, water, ammunition, radios and batteries, maps, special clothing, and any other items required by the unit (includin attachments) needed for the mission. Post-patrol support such as debriefings by intelligence personnel must also be planned.

Reviewing the Patrol Leader's Plan and Preparations

Once the patrol leader receives the mission, conducts visual and/or map reconnaissance, and develops the plan, the commander confirms the patrol leader's understanding of the mission and plan for accomplishing it. This discussion between the patrol leader and commander ensures that the patrol leader understands the commander's intent and is properly prepared to provide the patrol leader an opportunity to ask for clarification or additional support, if required.

Debriefing the Patrol

Upon return of the patrol, the commander receives the patrol report at a debriefing attended by the patrol leader and all patrol members. The debriefing should be conducted as soon as possible following the patrol's return, while information is still fresh in the minds of the patrol members. A patrol report, based on the information collected during the debrief, is generated and forwarded to the next higher com-mander.

10004. PATROL LEADER DUTIES

The patrol leader organizes and prepares the patrol by using the six troop-leading steps to make the best use of resources available. These steps are to—

1. Begin planning.
2. Arrange for reconnaissance and coordination.
3. Make reconnaissance.
4. Complete the plan.
5. Issue the order.
6. Supervise.

Begin Planning

The patrol leader begins by evaluating all factors affecting the mission. He looks for possible courses of action that lead to a decision, and then transforms this decision into an order. The first step includes making an initial assessment and decision on using available time, issuing a warning order and initial preparatory tasks, and initiating his estimate.

The initial planning effort assesses the time, assistance, and information available, and plans the proper use of each. Time allowances include reconnaissance; completion of the estimate and order; troop preparation; and such briefings, rehearsals, and inspections as required before beginning the patrol.

The patrol leader reviews the mission and the attachments and/or support available and decides what preparatory efforts must begin immediately. Proper use of subordinates to manage these initial tasks during this period reduces preparation time and frees the patrol leader for proper planning and reconnaissance. A warning order is issued using a modified five-paragraph order format (situation, mission, execution, administration and logistics, and command and signal [SMEAC]) as a checklist. (Refer to app. A for specific information that a patrol warning order should contain within each checklist item.

Once the initial preparations are set in motion, the patrol leader begins estimating by analyzing the mission; considering the friendly, enemy, and operating environments; considering each course of action available against what the enemy might do; comparing the courses of action in terms of mission accomplishment, capabilities, and probable casualties; and choosing one that becomes the basis for concept and order. While the patrol leader should use and organize notes, estimates must be done quickly and accurately, particularly for immediate situations. To organize thinking, the patrol leader uses METT-T, which consists of:

- Mission—the mission assigned to the patrol and how it relates to the mission of the commander who is sending the patrol.
- Enemy—what is known or suspected of enemy presence and capabilities, habits and characteristics, and fighting techniques.
- Terrain and weather—including ground, vegetation, drainage, weather, and visibility.
- Troops and support available—friendly situation and support available.
- Time available—the constraints and impact of time on preparation and mission accomplishment.

The estimate begins with mission analysis, which is the most important part of the entire planning process. Here, the patrol leader considers the specified tasks of the mission assigned and identifies other significant actions (specified and implied tasks) that must be undertaken to accomplish the stated mission. These tasks are arranged in sequence of accomplishment. The tasks and sequence create the framework for developing courses of action for the patrol concept of operations.

Arrange for Reconnaissance and Coordination

The patrol leader arranges a personal reconnaissance to observe as far forward as possible and also coordinates with the appropriate commanders for the patrol's "passage of lines" (see paragraph 11001 for complete definition) and supporting fires. The patrol leader also coordinates with other patrol leaders who may be operating in the same or adjacent areas and requests that the commander assigning him the patrol mission coordinate the patrol action with adjacent commanders, local security, and night defensive fires, as appropriate. The patrol leader may delegate any or all of these arrangements to the assistant patrol leader if the patrol leader requires the time for planning.

Make Reconnaissance and Complete the Estimate

The patrol leader uses personal reconnaissance to answer questions that arise from the map reconnaissance and METT-T evaluation. Specific points include passage points, lanes through obstacles, locations of friendly listening posts and observation posts, possible approach and return routes, enemy positions (if any), and intermediate observation points on the way to the objective.

In selecting approach and return routes, the patrol leader chooses routes that best use concealment and avoid opposition and obstacles. To lessen the chances of ambush by the enemy, the return trip is planned along a different route. In addition to personal reconnaissance and review of the map and aerial photographs, the advice of other patrol leaders who already are familiar with the terrain and the objective area should be considered.

After compiling information about the situation and possible time constraint, the patrol leader completes an estimate. The first step is developing courses of action, each of which will provide for movement to the objective area, mission accomplishment, and the return, based on the tasks and their sequencing identified in mission analysis. While the eventual concept of operations is presented in order of occurrence, the patrol leader must develop the courses of action by either backward or forward planning. In situations where the objective is well defined and there is sufficient information to plan the action for mission accomplishment (reconnaissance or combat), the patrol leader begins the scheme for accomplishing the mission at the objective and then, planning backwards, considers the options for getting there and back.

The following sections on movement to and return from the objective area, reconnaissance missions, and combat missions discuss methods and options available to the patrol leader in developing the courses of action. The principal variables between courses of action will be who, where, and how in the following:

- Patrol task organization.
- Routes to the objective area.
- Observation point(s) (reconnaissance patrol), ambush site, form of maneuver, type of ambush (combat patrol), fire support plan.
- Return routes.

The patrol leader then mentally considers the progress of each course of action (a map or simple sketch is a useful aid) against expected and unexpected enemy action. By comparing the options against each other and prospective enemy opposition, the patrol leader chooses the course of action that has the best chance of success. Includes in this mental preview process is the time to determine the patrol plan for unexpected contingencies (enemy attack/counterattack, casualty handling). These contingency actions, together with the selected course of action, become the patrol concept of operations.

Once the patrol leader determines the scheme of maneuver, the fire support required to accomplish the mission is addressed including the fire power organic to the patrol and what additional indirect fire support will have to be provided by other units. When planning for indirect fire support, the patrol leader considers the following questions:

- Will artillery, mortar, or close air support be required at the objective area (combat patrols)?
- What artillery and mortar targets exist along the routes to and from the objective area that can be employed by the patrol if it encounters the enemy during movement (reconnaissance patrols and combat patrols)?
- What additional fire support will be required to cover the patrol's movement from the objective area back to the friendly area once the enemy is aware of the patrol's actions at the objective area (combat patrols)?

The effect that casualties have upon the patrol depends upon many factors. Generally, more casualties can be expected in a combat patrol than in a reconnaissance patrol. A patrol may continue on to the objective carrying its casualties, send them back with a detail of Marines, abort the mission and return the entire patrol with the casualties, or call their parent unit for assistance.

Some factors that determine what action the patrol leader takes are: patrol's mission; unit's standing operating procedure for handling wounded; number of casualties and nature of their injuries; availability of aid, helicopters or other means of casualty evacuation. Helicopter evacuation should only be used for the most serious casualties. For infantry units conducting patrols in proximity to the enemy, helicopter evacuation of casualties may compromise the patrol's mission and force the patrol to return to friendly positions before the mission is completed.

The patrol leader determines the requirement for nuclear, biological, and chemical (NBC) defense equipment. Gas masks should always be carried due to the availability of riot control agents (RCAs) to the enemy. If chemical or biological agents have been employed in the area that the patrol must pass through, protective garments will have to be worn by patrol members for part of or the entire patrol. Wearing extra clothing and carrying extra equipment affects the speed of the patrol's movement. A contingency plan for post patrol decontamination must be developed.

Complete the Plan

At this point, the patrol leader has completed the basic thinking necessary for accomplishing the assigned mission. The patrol leader prepares the patrol order to spell out the details, assign tasks to subordinates, and explain the entire endeavor for ease of understanding by the other members of the patrol.

Prepare the Order

The patrol leader's order contains more detailed information than discussed in the warning order. Orders follow the prescribed five-paragraph order format but contain greater detail. The patrol order is a modified 5-paragraph order; the major modifications are to paragraphs 3a and 3c. The format for the order is contained in appendix B.

Issue the Order

The patrol leader asks for a status report on the initial preparatory tasks assigned to subordinate leaders and specialists when the warning order was issued. When the patrol leader has completed planning and initial preparations have progressed to the point where the patrol order may be issued, the members of the patrol are assembled. Roll call is taken to ensure all patrol members are present, then the prepared order is issued.

This will be the only opportunity for the patrol leader to issue detailed instructions. The mission, in particular, must be unmistakably clear so that once the patrol is committed, all subordinate leaders can act with unity of purpose.

Whenever possible, the patrol leader should have a Marine, such as the navigator, build a terrain model using dirt, sand, twigs, etc., explaining the concept of operations for movement to the objective area, actions at the objective area, and the return. Terrain models provide patrol members with a clear and simple layout of the area of operations and key terrain.

Supervise

Inspections and rehearsals, vital to proper preparation, are conducted even though the patrol leader and patrol members are well experienced in patrolling. Inspections determine the patrol's state of physical and mental readiness.

The patrol leader inspects before rehearsals to ensure completeness and correctness of uniform and equipment. The following areas are checked:

- Camouflage.
- Identification tags, Geneva Convention cards.
- Prescribed equipment, weapons, and ammunition are available and serviceable.
- Tape and other items are used to "silence" equipment (prevent noise produced during movement).
- Items that could provide information to the enemy (e.g., letters and papers) remain behind.
- Unnecessary equipment and excess weight remain behind.

The patrol leader questions each patrol member to ensure the following is known:

- The mission, planned routes (primary and alternate), and the fire support plan of the patrol.
- The individual's role: what to do and when to do it.
- What others are to do and how their actions impact.
- Challenges and passwords, codes, reporting times, radio call signs, frequencies, and any other pertinent details.

There is usually a period of time between final rehearsal and departure. The patrol leader reinspects just before departure to ensure all equipment is still in working order and the unit is ready to embark on the mission.

Rehearsals ensure the operational proficiency of the patrol. Plans are checked and needed changes are made. The patrol leader verifies the suitability of equipment. It is through rehearsals that patrol members become thoroughly familiar with the actions to take during the patrol.

If the patrol will operate at night, both day and night rehearsals are conducted. Terrain similar to that over which the patrol will operate is used. All actions are rehearsed. If time is limited, the most critical phases are rehearsed. Action at the objective is the most critical phase of the patrol and is always rehearsed.

An effective method is to talk the patrol through each phase, describing the actions and having each member perform individual duties. When satisfied, the patrol leader walks the patrol through all phases of the patrol using only the signals and commands to be used during the actual conduct of the patrol. Rehearsals continue until the patrol is thoroughly familiar with the plan. The rehearsal is also used to test the soundness of the patrol order and patrol organization.

After the rehearsal, the patrol leader makes final adjustments to the plan and patrol organization based on what was learned during the rehearsal and from other sources, such as the S-2 and adjacent patrols. When this is completed, the patrol leader issues final instructions to subordinate leaders noting any changes made in the patrol organization or plan. While the subordinate leaders are briefing the remainder of the patrol members, the patrol leader reports to the commander stating that the patrol is ready to begin the mission. The patrol leader also coordinates the location and time that the patrol can test fire all weapons prior to departure.

Chapter 11. Movement to and Return from the Objective Area

This chapter provides guidance to patrol leaders for movement to and return from the objective area. Action in the objective area depends on whether the patrol is assigned a reconnaissance mission, combat mission or security mission. Chapters 12 and 13 provide detailed guidance on these types of missions.

11001. PASSAGE OF LINES

A passage of lines is an operation in which a force moves forward or rearward through another force's combat positions with the intention of moving into or out of contact with the enemy. (JP 1-02)

During the initial preparation for the patrol, the patrol leader selects a patrol assembly area and reconnoiters the area of passage designated by the commander. In coordination with the unit commander responsible for the area of passage, the patrol leader identifies gaps or lanes in minefields and wire obstacles and locates local security elements through which the patrol will pass. The patrol leader also checks the route from the patrol assembly area to the passage point or contact point where the patrol will depart friendly lines. If possible, both the route to the passage point and the route through the frontlines should be concealed from the enemy's view.

The patrol leader also reconnoiters the area for return passage of lines and coordinates with the unit commander responsible for the area of passage for passage points and lanes as necessary. The patrol leader observes these points from the direction that the patrol will use upon return to friendly lines, if possible, to aid in recognition upon return. The patrol leader provides the forward unit with information about the size of the patrol, general route, and expected time of return. The manner of challenge and recognition of the returning patrol should be coordinated in detail.

Upon return to friendly forward local security squads and/or frontlines, the patrol leader leaves the patrol in a covered position and moves forward with a radio operator and at least one Marine for security to make contact with the friendly unit in the manner agreed. After contact is made and recognized, the patrol leader rejoins the patrol and takes them to the passage point, personally checking in each member.

11002. ORGANIZATION FOR MOVEMENT

The patrol's task organization establishes the elements and teams needed to accomplish the mission in the objective area and to and from the objective. The patrol leader determines the formation(s) in which the patrol moves to the objective area.

Formations

The proper use of patrol formations is critical to the patrol's success. The squad and fire team formations described in FMFM 6-5 (proposed MCWP 3-11.2), *Marine Rifle Squad*, also apply to infantry patrolling formations. Because the movement of the patrol must be concealed from the enemy, the patrol normally moves through terrain that provides concealment. Control of the patrol in this type of terrain is difficult; thus the column formation, which is easily controlled, is normally used. However, as various types of terrain are encountered, the patrol leader uses the same considerations in determining the appropriate formation used in other infantry operations.

The standard squad and fire team formations are adaptable to a patrol. The patrol may change formations en route to match the situation and terrain. The patrol leader may have to sacrifice some control for better dispersion or give up some speed for greater stealth and more security. Other considerations include—

- Visibility, weather, terrain, and vegetation will influence dispersion and control of individuals and units. These factors may also affect the enemy; if visibility is good for the patrol, it is also good for the enemy. Two pieces of luminous tape worn on the back of the collar will aid in control and movement on dark nights. The collar is turned down when near the enemy. The tape can also be worn on the back of the cap, but should be covered or removed when near the enemy.

- Preserving the integrity of fire units (fire teams and, if attached, machine-gun teams) is of primary importance. If team members are detached from a machine-gun team, the unit integrity is lost and effectiveness is reduced. The patrol leader must also position fire units so as not to mask their fires.
- All-around defense of the patrol must not be sacrificed under any circumstances. The conventional squad and platoon formations provide adequate firepower in any direction required. When attachments are made to the patrol, the attachments are positioned within the formation to enhance the firepower of the patrol. If a fire unit, such as a machine-gun team or squad, is attached, it is incorporated into the all-around defense of the patrol by modifying the conventional formation (see fig. 11-1). It is permissible to employ machine-gun teams individually during movement. However, assault rocket squads and teams should be employed primarily as rifle fire units for movement. Firing the SMAW or AT-4 from within a patrol formation can present a serious back-blast danger to patrol members and should only be fired on the patrol leader's direction.
- Time allotted for mission accomplishment is also a major consideration. In selecting the formations, the patrol leader must consider the speed of the movement required to meet the time constraints (if any) imposed on the patrol. If required to meet a time schedule, a formation that permits rapid movement should be used. Speed, however, must never be permitted to force the patrol leader to make rash tactical decisions.

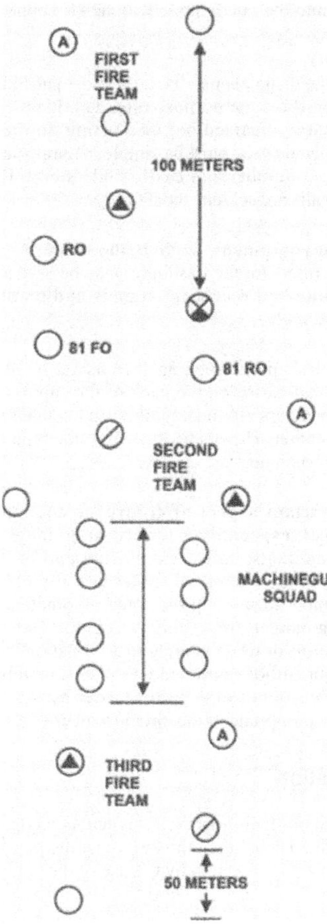

Figure 11-1. Squad Patrol with Attached Mortar Forward Observer Team and Machine Gun Squad.

Exercise of Control

The patrol leader is positioned for best control over the patrol. The assistant patrol leader moves at or near the rear of the patrol. Other subordinate leaders move with their elements. All patrol members assist by staying alert and passing on signals and orders. A signal to halt may be given by any patrol member, but the signal to resume movement is given only by the patrol leader.

Arm-and-hand signals are the primary means of communication within a patrol and should be used exclusively when near the enemy. All members must know the standard infantry signals (refer to FMFM 6-5), as well as any special signals required, and be alert to receive and pass them to other members.

The patrol leader should speak just loudly enough to be heard. At night, or when close to the enemy, the patrol leader halts the patrol and has subordinate leaders come forward. They speak in a low voice and then pass the information to their subordinates by moving from member to member.

Radios provide a means of positive control within a large patrol; however, radios should be used only when arm-and-hand signals or face-to-face contact between the patrol leader and subordinate leaders is impractical. When close to the enemy, words are spoken into the microphone with hands cupped over it in a low voice.

Other sound signals may be used if the patrol leader is sure they serve the purpose intended. Planned sound signals are rehearsed before starting on the patrol. Sound signals used must be simple, natural sounds that are few in number and easily understood. Bird and animal calls are seldom satisfactory.

Infrared equipment, such as the sniper scope and infrared filters for the flashlight, may be used as means of sending and receiving signals and maintaining control at night.

Luminous tape may be used to assist in control at night. Small strips on the back of the cap or collar of patrol members aid in keeping visual contact with the front member. However, the luminous tape must be covered when near the enemy.

An important aspect of control is accounting of personnel, especially after crossing danger areas, enemy contacts, halts, and exiting and re-entering friendly lines. The patrol leader may arrange for the last member to send up the count automatically after crossing danger areas, enemy contact, and halts. In large patrols or when moving in a formation other than a column, subordinate leaders check members and report the count to the patrol leader by the quickest method appropriate to the circumstances.

Navigation

One or more members are assigned as navigators for the patrol. Their function is to assist the patrol leader in maintaining direction by the use of the compass.

The patrol leader assigns at least two members as pacers to keep track of the distance from point to point. The average of their count is used to approximate the distance traveled. Pacers are separated so they do not influence each other's count. Pacers must know how to convert their own paces to meters.

The route is divided into increments, with each increment starting at a recognizable point on the ground. The pacers begin their counts from zero at the beginning of each leg. This makes the pace count easier to keep and checks for accuracy.

The pace count is sent forward when the patrol leader turns to the member behind and in a low voice says, "send up the pace" or uses the arm-and-hand signal of tapping his boot. This signal is passed to both pacers, who in turn send up the pace count in meters; for example, "two-hundred" or "one-seven-five."

Patrol members must understand that the counts of both pacers are sent forward. The patrol leader must know the counts of both men in order to check them.

Security

The patrol is dispersed consistent with control, visibility, cover, and concealment. Scouts are employed to the front, flanks, and rear of the patrol to provide security. For the patrol members traveling in the main body of the patrol, areas of responsibility are assigned to the front, flanks, and rear. Scouts are the eyes and ears of the patrol leader. They move when and where directed by the patrol leader and maintain contact with the patrol leader at all times, except when momentarily obscured by vegetation or other terrain features.

Front

Small patrols (squad size) may employ from one scout up to a fire team as the point, depending on the enemy situation, terrain, and route being followed by the patrol. Normally, squad-size patrols will use two scouts as the point. The point is responsible for investigating the route of advance immediately to the front of the patrol.

The point moves as far ahead of the patrol as visibility and terrain permit. When visibility is good, the point may precede the main body by as much as 100 meters. The point travels right and left ahead of the patrol, searching the area over which the patrol will pass.

The point maintains direction by knowledge of the general route to be followed and visual contact with the patrol leader. The patrol leader or the navigator ensures that the point is proceeding correctly.

The point, which stays far enough ahead of the patrol to provide security, is not a trail breaker for the patrol. If the point loses contact with the patrol, the point waits for the main body to catch up or moves rearward if contact is not quickly regained.

One of the navigators may be positioned with the point. One or more members works as the point while the other is the navigator.

Flank

Flank security for a patrol of squad size or less may be provided by using one or two members on either flank. If two scouts are assigned to a flank, one is positioned to observe the patrol leader and the other works farther out from the patrol. The scout who must observe the patrol leader remains within a maximum distance of 100 meters. The scout farther out remains in sight of the inside scout but normally does not move more than 20 or 25 meters away and remains prepared to relieve flank security regularly. Moving through dense woods or jungle may render the use of flank security impractical because of reduced visibility. In such cases, it moves with the patrol itself, but maintains observation to its assigned flanks.

Rear

A small patrol normally has only one rifleman assigned as rear security. An interval between the member assigned as rear security and the last Marine of the patrol is maintained at the limit of visibility, up to 50 meters. This member maintains rear security for the patrol by constantly observing to the rear.

Halting

Speed of movement is slower at night than in day patrols and reduces the danger of a Marine becoming separated from the patrol. The patrol occasionally halts to observe and listen for enemy activity; this is called a security halt. Upon signal, when reaching a danger area and periodically throughout movement en route, every member freezes in place, remains quiet, observes, and listens. It may be necessary to call a security halt just after departing friendly areas and just before entering friendly areas.

The patrol may halt briefly to send a message, eat, rest, check direction, or make a reconnaissance. The area selected should provide adequate concealment and cover, as well as favor the defense. All-round security is established and the patrol leader ensures all members move out when the patrol resumes movement. (For extended halts, see paragraph 11005.)

Infiltration

The disposition of enemy forces may sometimes prevent a patrol from entering the enemy occupied area as a unit; however, pairs of scouts or fire teams may slip through without being discovered. (Refer to FMFM 6-5 for infiltration techniques and procedures).

11003. CONTROL MEASURES FOR MOVEMENT

Checkpoints

A checkpoint is a predetermined point on the surface of the Earth used as a means of controlling movement, a registration target for fire adjustment or reference for location (JP 1-02). Checkpoints are a means of control between the parent unit and the patrol. These locations are decided upon and coordinated before the patrol leaves, so that both the patrol members and parent unit will know the patrol's location when it reports in. The parent unit can follow the progress of the patrol without transmitting coordinates over the radio that the enemy could monitor.

Rally Points

A rally point is an easily identifiable point on the ground where units can reassemble and reorganize if they become dispersed (MCRP 5-12C). It should provide cover and concealment, be defensible for at least a short time, and be easily recognized and known to all patrol members. All rally points are considered and identified as tentative rally points until they are occupied, found to be suitable, and designated as rally points by the patrol leader.

Initial Rally Point

The initial rally point is within friendly lines where the patrol can rally if it becomes separated before departing the friendly area or before reaching the first en route rally point. It may be the assembly area where the patrol waits while the patrol leader contacts the last friendly position through which the patrol will pass. The location of the initial rally point must be coordinated with the forward unit commander.

En route Rally Points

En route rally points are between the initial rally point and the objective rally point; and from the objective rally point back to the point where the patrol reenters friendly lines. They are determined as the patrol passes through a likely area that is suitable for a rally point.

Objective Rally Point

The objective rally point is located nearest the objective where the patrol makes final preparations prior to approaching the objective. It also serves as a location where the patrol reassembles after completing actions on the objective. The objective rally point must be suitable to accommodate those activities accomplished prior to actions on the objective. This position must provide the patrol concealment from enemy observation and, if possible, cover from enemy fires. It may be located short of, to a flank or beyond the objective. It should be out of sight, sound, and small-arms range of the objective area. The patrol leader's reconnaissance of the objective is made from this position; it is the release point from which patrol elements and teams move to the objective to accomplish the mission.

Rally Point Selection

The patrol leader selects likely locations for tentative rally points during reconnaissance or map study. A tentative rally point must be confirmed and announced after examination proves suitability. A tentative initial rally point and a tentative objective rally point are always selected and identified in the patrol order. If necessary, the patrol leader selects additional rally points en route as suitable locations are reached. When the patrol reaches a danger area that cannot be bypassed, such as an open meadow or stream, the patrol leader selects a rally point on both the near and far side. If good locations are not available, the patrol leader designates the rally points in relation to the danger area. The patrol leader will say, for example, "50 meters this side of the trail" or "50 meters beyond the stream."

Rally Point Use

The initial rally point and en route rally points are designated to enable the patrol to reassemble if it is unavoidably separated or dispersed. Identifying features are pointed out. The patrol leader ensures that the information is passed to all patrol members. When crossing a danger area, a near side rally point and a far side rendezvous point are designated. These rally points should only be used when all other methods of retaining control of the patrol have failed. The success of the patrol may be jeopardized if it is dispersed and cannot rally expeditiously.

If the patrol has left the friendly area and becomes dispersed, patrol members return to the last designated rally point (the initial or an en route rally point) unless the patrol leader gives other instructions.

As previously noted, the patrol leader selects two rally points at the near and far sides of danger areas that cannot be bypassed. If the patrol becomes separated or dispersed at a danger area, and there has been no enemy contact, the patrol should reassemble at the rally point on the far side of it. If, however, the patrol is separated or dispersed at a danger area as a result of enemy contact, members who have already crossed the danger area assemble and reorganize at the rally point on the far side; members who have not crossed assemble and reorganize at the rally point on the near side. In the absence of the patrol leader and unless directed otherwise, the senior member at the rally point on the near side takes charge, attempts to move the rallied patrol members to the rally point on the far side, and rejoins the remainder of the patrol.

Rally Point Actions

The patrol leader plans the actions to be taken at rally points and instructs the patrol accordingly in the patrol order. Planned actions at the initial rally point and en route rally points must provide for the continuation of the patrol as long as there is a reasonable chance to accomplish the mission. Plans for actions at rally points should provide for—

- Recognition signals for assembly at rally points.
- Minimum number of members and maximum amount of waiting time required before the senior member at the rally point moves the rallied patrol members onward toward the objective or returns to friendly lines.
- Instructions for patrol members who find themselves alone at a rally point.

11004. PRECAUTIONS AT DANGER AREAS

A danger area is any place where the patrol is vulnerable to enemy observation or fire (open areas, roads, trails, and obstacles such as barbed wire, minefields, rivers and streams, and lakes). Any known or suspected enemy position the patrol must pass is also a danger area. The patrol leader plans for crossing each danger area and includes these plans in the order.

The patrol reconnoiters the near side of a danger area first, then the patrol leader sends scouts to reconnoiter the far side. Once the scouts report that the far side is clear of the enemy, the remainder of the patrol crosses the danger area. As each individual or group crosses the danger area, they are covered by those remaining and by those who have successfully crossed. Enemy obstacles are avoided since they are usually covered by fire.

In crossing a river, the near bank is reconnoitered first; then the patrol is positioned to cover the far bank. Scouts are sent across to the far bank. After the far bank has been reconnoitered and the scouts report that it is clear of the enemy, the patrol crosses as rapidly as possible. This may be done individually or in pairs. If crossing the river requires swimming, the patrol uses improvised rafts to float equipment, weapons, and ammunition across. (Refer to MCRP 3-02C, *Water Survival Handbook.)*

A road or trail is crossed at or near a bend or where the road is narrow. Observation is restricted and, if the enemy is present, the patrol is exposed as short a time as possible. The near side is reconnoitered first, then scouts are sent across to reconnoiter the far side. This includes reconnoitering the tentative rally point on the far side. Once the scouts report "all clear," the remainder of the patrol crosses rapidly and quietly.

If the patrol must pass close to an enemy position, it takes advantage of battlefield noises to cover the sounds of movement. If supporting fires are available, the patrol leader can call for them to divert the enemy's attention as the patrol passes.

11005. HIDE

When a patrol is required to halt for an extended period in an area not protected by friendly troops, the patrol moves into a location which, by the nature of the surrounding terrain, provides passive security from enemy detection. Such an assembly area is termed a *hide*. To establishment a hide—

- Cease all movement during daylight hours to avoid detection.
- Hide the patrol for an extended period while the patrol leader conducts a detailed reconnaissance of the objective area.
- Rest and reorganize after extended movement.
- Reorganize after a patrol has infiltrated the enemy area in small groups.

The patrol leader's plan must include tentative hide locations when the patrol's mission dictates an extended halt within enemy areas. These tentative locations must be confirmed by actual ground reconnaissance prior to occupation by the patrol. The plan for a hide includes both passive and active security measures.

Passive security measures are—

- Avoid built-up areas.
- Select an area remote from all human habitation.
- Avoid known or suspected enemy positions.
- Avoid ridge lines, topographic crests, valleys, lakes, and streams.
- Avoid roads and trails.
- Avoid open woods and clearings.
- Select areas offering dense vegetation, preferably bushes and trees that spread out close to the ground.

Active security measures—

- Establish security covering all likely avenues of approach into the site.
- Establish communications (wire, radio, signal, runner) with posted security to provide early warning of enemy approach.
- Select an alternate area for occupation if the original hide is compromised or found unsuitable.
- Plan for withdrawal in the event of discovery.
- Establish an alert plan with a certain percent of the personnel awake at all times.
- Organize the elements of the patrol so necessary activities can take place with a minimum amount of movement.

The size of the area physically occupied by a patrol in a hide and the number of security posts required are governed by the terrain, quantity and quality of cover and concealment, and size of the patrol.

If the situation permits, a hide can also be used as the final preparation position and/or objective rally point.

11006. IMMEDIATE ACTIONS UPON ENEMY CONTACT

A patrol may make contact with the enemy at any time. Contact may be through observation, a meeting engagement or ambush. Contact may be visual, in which the patrol sights the enemy but is not itself detected. When this is the case, the patrol leader can decide whether to make or avoid physical contact, basing his decision on the patrol's assigned mission and capability to successfully engage the enemy unit.

When a patrol's assigned mission prohibits physical contact (except that necessary to accomplish the mission), its actions are defensive in nature. Physical contact, if unavoidable, is broken as quickly as possible and the patrol, if still capable, continues its mission.

When a patrol's assigned mission permits or requires it to seek or exploit opportunities for contact (as in the case of a combat patrol), its actions are offensive in nature, immediate, and positive.

In patrolling, contacts (visual or physical) are often unexpected at very close ranges, and short in duration. Effective enemy fire often provides leaders little or no time to fully evaluate situations and issue orders. In these situations, immediate action provides a means for swiftly initiating positive offensive or defensive action, as appropriate.

Two types of physical contact with the enemy are meeting engagement and ambush. Meeting engagement is a combat action that occurs when a moving force, incompletely deployed for battle, engages an enemy at an unexpected time and place. It is an accidental meeting where neither the enemy nor the patrol expect contact and are not specifically prepared to deal with it. An ambush is a surprise attack from a concealed position.

Immediate Actions

Immediate actions are designed to provide swift and positive small unit reaction to visual or physical contact with the enemy. They are simple courses of action in which all Marines are well trained. Minimal signals or commands are required and they are developed as needed for the combat situation. The signals can, in many cases, be initiated by any member of the unit. It is not feasible to attempt to design an immediate action drill to cover every possible situation. It is better to know the immediate action drill for each of a limited number of situations that may occur during a patrol. Arm-and-hand signals associated with immediate actions—such as FREEZE, ENEMY IN SIGHT, and HASTY AMBUSH RIGHT or LEFT—are contained in FMFM 6-5 (proposed MCWP 3-11.2).

Immediate Halt

When the patrol detects the enemy but is not itself detected, the situation requires the immediate, in-place halt of the patrol. The first member visually detecting the enemy gives the silent signal for FREEZE. Every member halts in place, weapon at the ready, and remains absolutely motionless and quiet until further signals or orders are given.

Air Observation and/or Attack

These actions are designed to reduce the danger of detection from aircraft and casualties from air attack.

When an enemy or unidentified aircraft that may detect the patrol is heard or observed, the appropriate immediate action drill is FREEZE. The first member hearing or sighting an aircraft that may be a threat signals FREEZE. Every member halts in place until the patrol leader identifies the aircraft and gives further signals or orders. Members of the patrol must not look up at the aircraft as sunlight can reflect off their faces even when camouflaged.

When an aircraft detects a patrol and makes a low level attack, the immediate action drill air attack is used. The first member sighting an attacking aircraft shouts, "AIRCRAFT," followed by the direction of the incoming attack: FRONT, LEFT, REAR or RIGHT. The patrol moves quickly into line formation, well spread out, at right angles to the aircraft's direction of travel. As each member comes on line, the member hits the ground, using available cover, then positions the body perpendicular to the aircraft's direction of travel, to present the shallowest target possible (see fig. 11-2 on page 11-8). Between attacks (if the aircraft returns or if more than one aircraft attacks), patrol members seek better cover. Attacking aircraft are fired upon only on command of the patrol leader.

Meeting Engagement

Hasty Ambush. This immediate action is used to avoid contact and to prepare to initiate an unplanned ambush on the enemy. It may often be a subsequent action after the command freeze. When the signal HASTY AMBUSH is given (by the point member, patrol leader or another authorized patrol member), the entire patrol moves quickly to the right or left of the line of movement, as indicated by the signal, and takes up the best available concealed firing positions (see fig. 11-3 on page 11-9). The patrol leader initiates the ambush by opening fire and shouting, "FIRE"; thus ensuring initiation of the ambush if the weapon misfires. If the patrol is detected before this, the first member aware of detection initiates the ambush by firing and shouting. The patrol leader may decide not to initiate the ambush in order to avoid contact unless the patrol is detected. When used as an offensive

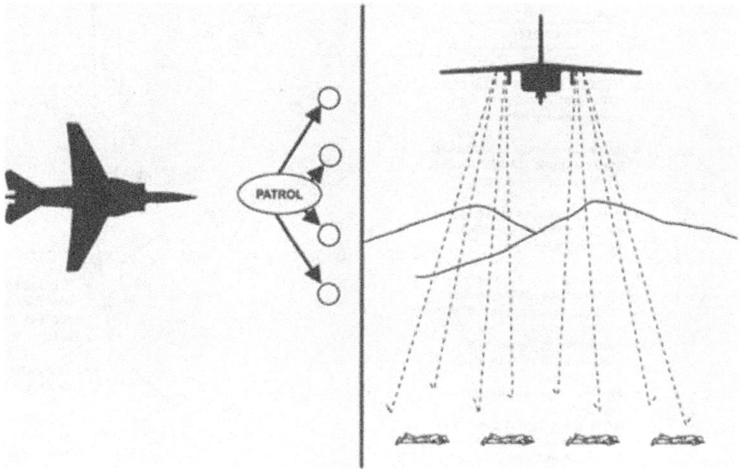

Figure 11-2. Immediate Action, Air Attack.

measure, the enemy is allowed to advance until he is in the most vulnerable position before the ambush is initiated. An alternate means for initiating the ambush is to designate an individual (for example, point or last member) to open fire when a certain portion of the enemy unit reaches or passes that member.

Immediate Assault. This immediate action drill is used defensively to make and quickly break undesired but unavoidable contact (including ambush) and offensively to decisively engage the enemy (including ambush). When used in a meeting engagement, members nearest the enemy open fire and shout, "CONTACT," followed by the direction of the incoming attack: FRONT, LEFT, REAR or RIGHT. The patrol moves swiftly into line formation and assaults (see fig. 11-4).

Defensive Measures

When used defensively, the assault is stopped if the enemy withdraws and contact is broken quickly. If the enemy stands fast, the assault is carried through the enemy positions and movement is continued until contact is broken.

Offensive Measures

When used offensively, the enemy is decisively engaged. Escapees are pursued and destroyed until orders to break contact are given by the patrol leader. If the patrol is fired upon from beyond 50 meters, the patrol must break contact as quickly as possible and continue the mission. If it engages the enemy any longer than necessary to break contact, it may put the mission in jeopardy.

Fire and Maneuver

Fire and maneuver is one means to break contact. One portion of the patrol returns the enemy fire while another portion moves by bounds away from the enemy. Each portion of the patrol covers the other by fire until contact is broken by all.

Clock System

The clock system is another means to break contact. Twelve o'clock is the direction of movement of the patrol. The patrol leader shouts a direction and a distance. For example: "TEN O'CLOCK-TWO HUNDRED," means the patrol should move in the direction of ten o'clock for 200 meters. Patrol members keep their same relative positions as they move so the original formation is not disrupted. Subordinate leaders must be alert to ensure that the members of their elements and teams receive the correct order and move as directed.

Counter Ambush

When a patrol is ambushed, the immediate action drill used depends on whether the ambush is a near ambush

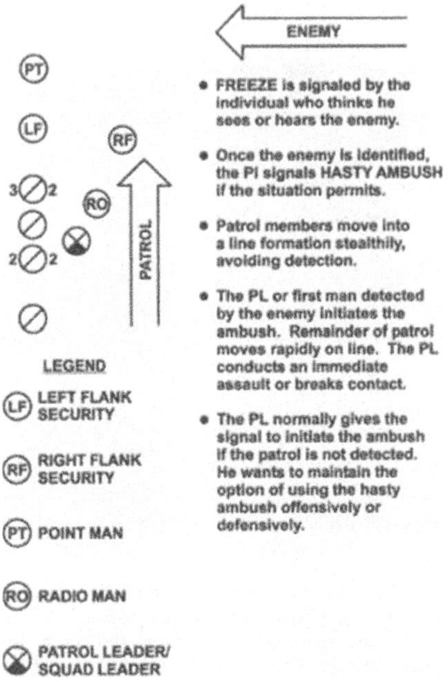

Figure 11-3. Immediate Action, Hasty Ambush.

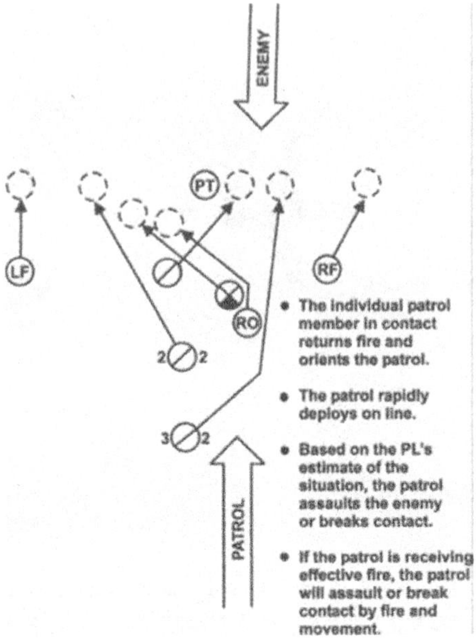

Figure 11-4. Immediate Action, Immediate Assault.

(the enemy is within 50 meters of the patrol) or a far ambush (the enemy is beyond 50 meters of the patrol). Fifty meters is considered the limit from which the ambushed patrol can effectively launch an assault against the enemy.

Near Ambush. In a near ambush, the killing zone is under very heavy, highly concentrated, close range fires. There is little time or space for members to maneu-ver or seek cover. The longer they remain in the killing zone, the greater the chance they will become casualties. Therefore, if members in the kil-ling zone are attacked by a near ambush, they immediately assault without order or signal directly into the ambush position, occupy it, and continue the assault or break contact, as directed. This action moves them out of the killing zone, prevents other elements of the ambush from firing on them without firing on their own members, and provides positions from which other actions may be taken (see fig. 11-5 on page 11-10). Members not in the killing zone maneuver against the enemy as directed. The assault continues until all patrol members are outside of the killing zone.

Far Ambush. In a far ambush, the killing zone is also under very heavy, highly concentrated fires, but from a greater range. This greater range provides members in the killing zone maneuver space and some opportunity to seek cover at a lesser risk of becoming a casualty. If attacked by a far ambush, members in the killing zone, without order or signal, immediately return fire, take the best available positions, and continue firing until directed otherwise. Members not in the killing zone maneuver against the ambush force, as directed (see fig. 11-6 on page 11-10). The assault is continued against the enemy or until the order to break contact is given.

In each situation, the success of the counter ambush employed depends on the members being well trained in quickly recognizing the distance from which an ambush is initiated and well rehearsed in the proper reaction.

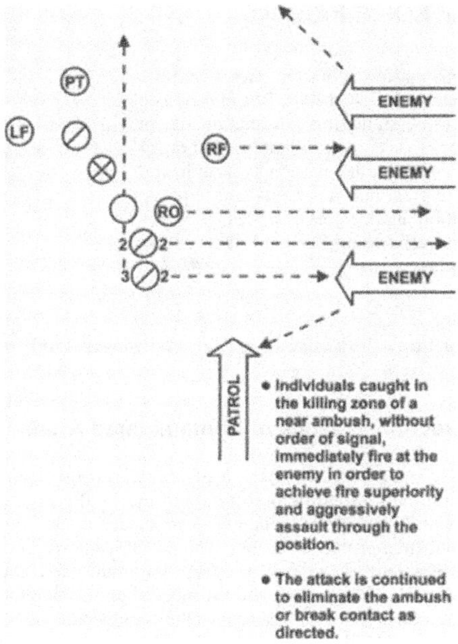

Figure 11-5. Near Ambush.

11007. PATROL LEADER'S ACTION IN A DEVELOPING SITUATION

While good patrolling depends on good planning, the patrol leader's plan must be flexible. Every combat situation develops differently than expected, and the patrol leader must be ready to quickly adapt to the situation as it develops. One of the most difficult tasks in battle is to recognize the correct moment for making a decision. Generally, it is more difficult to determine the moment for making a decision than it is to formulate the decision itself. When the situation demands, decisions must be made promptly without waiting for more information. In a developing situation, the patrol leader should use the following questions as a guide for battlefield decisionmaking:

- How has the situation changed?
- How does the change affect mission accomplishment and the immediate superior's mission?
- Must a decision be made now?
- What are the options?
- Which option best serves the mission and the unit as a whole?
- Which option offers the greatest chance of success?

11008. RETURN FROM OBJECTIVE AREA

After performing actions in the objective area, the patrol reassembles at the objective rally point. This phase of the patrol is perhaps the most difficult and dangerous. Patrol members are experiencing fatigue, emotional letdown, and wounds; they may be low on water and ammunition. Above all, the enemy was likely alerted if not in pursuit. At this point, the patrol leader must move the patrol rapidly but carefully and maintain patrol security at a high level. In returning to friendly lines, the patrol neither uses nor travels near the same route used to get to the objective area. The enemy may have that route covered. For information on reentry of friendly lines, refer to paragraph 11001.

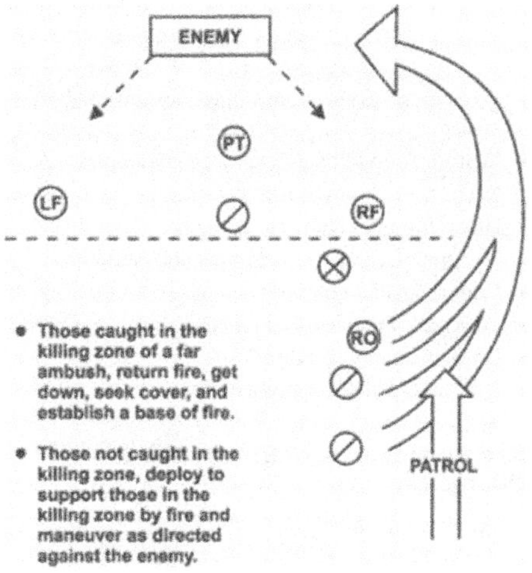

Figure 11-6. Far Ambush.

Chapter 12. Reconnaissance Patrols

The commander needs accurate, timely information about the enemy and the terrain the enemy controls to assist in making tactical decisions. Reconnaissance patrols are one of the most reliable means for obtaining this information. These patrols engage the enemy only when necessary to accomplish their mission or for protection. They generally avoid combat, accomplish the mission by stealth, and do not maintain contact with the unit that sent them out. A reconnaissance patrol is capable of carrying the search for information into the area occupied by enemy forces—usually beyond the range of vision of friendly ground observation posts—and examining objects and events at close range.

12001. GENERAL MISSIONS

General missions for reconnaissance patrols include gathering information about location and characteristics of friendly or hostile positions and installations, terrain (routes, stream crossings), and obstacles.

12002. SPECIFIC MISSIONS

Locate the Enemy

Reconnaissance patrols try to determine the—

- Location of enemy forces, installations, and equipment.
- Identification of enemy units and equipment.
- Strength of enemy forces.
- Disposition of enemy forces.
- Movement of enemy personnel and equipment.
- New or special types of weapons.
- Presence of mechanized units.
- Unusual enemy activity.
- Presence of NBC equipment.

Reconnaissance of Enemy Wire Obstacles

A patrol with the mission of investigating enemy-emplaced wire obstacles employs a formation providing for all-around security and takes precautions against being observed by the enemy. The patrol leader and one Marine inspect each gap and establish its location by means of compass bearings to prominent objects in the rear of enemy or friendly positions or through the use of global positioning satellite receivers.

Reconnaissance of Contaminated Areas

A patrol with the mission of investigating a contaminated area reconnoiters and marks the area's boundaries. Patrol members wear protective clothing and field protective masks. The patrol leader reports the extent of the area, the ttype of agent used, the terrain and vegetation and the method of marking the contaminated area. A sketch of the contaminated area, prepared by the patrol leader, should be included in the patrol report.

Reconnaissance of Enemy Minefields

Patrols assigned to reconnoiter enemy minefields are often composed of specially trained personnel. The type of mine is identified and the possibility of bypassing the area determined.

Reconnaissance of Terrain

The reconnaissance patrol is particularly suited for gathering information about the terrain within their area of operations. (Refer to MCRP 2-15.3B, *Reconnaissance Reports Guide*.) Commanders must know the location and condition of bridges, streams, and roads in order to make proper plans for the movement of troops. The following information should be obtained by a reconnaissance patrol assigned such missions:

- Bridges:
  - Maximum load capacity.
  - Material used in construction (wood, stone, concrete or steel).

  - Material used for abutments (dirt, wood, masonry, concrete).
  - Type and number of supporting members of the bridge.
  - Condition and dimensions of bridge flooring.
  - Presence or absence of mines and demolition charges on the bridge and at entrances/exits.
  - Terrain crossed (ravine, stream, road).
  - Primary use (rail, auto, foot).
  - Location and type of an alternate crossing if bridge is unusable.
- Streams and fords:
  - Width and depth of stream.
  - Composition of the bottom (mud, sand, gravel, rock).
  - Speed of current in miles per hour.
  - Composition and gradient of banks.
  - Surrounding terrain and vegetation cover.
- Roads:
  - Width.
  - Composition of surface.
  - Condition.
  - Road blocks.
  - Gradient/degrees of slopes.
  - Curves (location and length).
  - Pitch and culvert locations.
  - Primary use and frequency of use.

12003. TYPES OF RECONNAISSANCE

Route

Route reconnaissance is a directed effort to obtain detailed information of a specified route and all terrain from which the enemy could influence movement along that route. (Refer to MCRP 5-12C.) It is also described as a form of reconnaissance focused along a specific line of communications—such as a road, railway, or waterway—to provide new or updated information on route conditions and activities along the route. (Refer to MCRP 5-12A, *Operational Terms and Graphics*.)

A route reconnaissance normally precedes the movement of forces. It provides detailed information about a specific route and the surrounding terrain that could be used to influence movement along that route. Considerations include trafficablility, danger areas, critical points, vehicle weight and size limitations, and locations for friendly obstacle emplacement.

The objective area is normally defined by a line of departure, a route, and a limit of advance.

Area

An area reconnaissance is a directed effort to obtain detailed information concerning the terrain or enemy activity within a prescribed area such as a town, ridge line woods or other features critical to operations. (Refer to MCRP 5-12C.) An area reconnaissance could also be made of a single point, such as a bridge or installation. (Refer to MCRP 5-12A.)

Emphasis is placed on reaching the area without being detected. Enemy situations encountered en route are developed only enough to allow the reconnoitering unit to report and bypass.

Zone

A zone reconnaissance is a directed effort to obtain detailed information concerning all routes, obstacles (to include chemical or radiological contamination), terrain, and enemy forces within a zone defined by boundaries. A zone reconnaissance is normally assigned when the enemy situation is vague or when information concerning cross-country trafficability is desired. (Refer to MCRP 5-12C.)

The commander specifies specific routes or areas of interest within the zone. The zone to be reconnoitered usually is defined by a line of departure, lateral boundaries, and a limit of advance.

12004. TASK ORGANIZATION

A reconnaissance patrol is organized around the current structure of the Marine rifle squad with one or more of the squad's fire teams assigned as the reconnaissance element to reconnoiter or maintain surveillance over the objective. At least one fire team acts as a security element whose functions are to—

- Secure the objective rally point (see para. 11003).
- Give early warning of enemy approach.
- Protect the reconnaissance unit.

Reinforcing the squad is based on METT-T. If two squads are required, then the mission should be assigned to a platoon.

A company may send a reconnaissance patrol to specifically reconnoiter a given area (a riverbed or a bridge), or to maintain surveillance on a particular streambed for the next four nights. Reconnaissance patrols perform three basic functions en route to and from the objective:

- Provide control.
- Provide security while moving and in the objective area.
- Conduct reconnaissance or surveillance.

Depending on the size of the reconnaissance patrol, an element must be tasked to reconnoiter the area or zone. In the case of a surveillance mission, two teams are recommended to share the same task (so one could lay up and rest, while the other lays low and hides to maintain strict silence and no movement while surveillance is being maintained on the objective). Whatever the case, each unit that has a surveillance mission or reconnaissance mission must be thoroughly briefed as to what essential elements of information are to be collected as well as their location in the objective area.

The patrol should be organized with one or two fire teams to actually conduct the reconnaissance mission and the remaining fire team to provide security. Each fire team should be prepared to assume either mission. A small area reconnaissance patrol needs only one fire team for the assigned mission. A patrol with a wider area reconnaissance mission should use two fire teams to physically conduct the mission and one fire team for cover and/or security.

The security element for a reconnaissance patrol should be organized to cover the likely avenue of approach into the objective area, to protect the units conducting the reconnaissance, and to cover the objective rally point.

The variations of the special organization for reconnaissance patrols illustrate the patrol leader's flexibility in organizing the patrol to meet mission requirements. The patrol leader is not limited to the variations reflected herein but can choose any special organization as long as that task organization accommodates the requirement for command and control, reconnaissance, and security.

12005. SIZE OF RECONNAISSANCE PATROLS

A reconnaissance patrol should be kept to the minimum number of personnel required to accomplish the mission. A mission requiring a patrol to remain away from its unit for a considerable period of time, or one requiring a patrol to send back information by messenger, increases the size of the patrol. Reconnaissance patrols seldom exceed a squad in strength. Unit integrity should be preserved whenever possible. Intelligence personnel, interpreters, and other specialists, such as radio operators or engineers, are assigned to a patrol if the particular mission requires.

12006. RECONNAISSANCE EQUIPMENT

Patrol members are armed and equipped as necessary for accomplishing the mission. The automatic rifle in each fire team provides a degree of sustained firepower in case of enemy contact. The patrol should have at least two: pairs of binoculars, pairs of wire cutters, maps, compasses, and watches. Night observation equipment may be used. Pencils and small notebooks are carried so notes and sketches can be made. A message book with message blanks and overlay paper is mandatory.

12007. RECONNAISSANCE PATROL ACTIONS AT THE OBJECTIVE AREA

Route

The patrol leader halts and conceals the patrol near the objective area in the final preparation position. The patrol leader then conducts leader's reconnaissance to confirm the plan for positioning the security teams and employing units assigned to the reconnaissance mission. The patrol leader returns to the patrol and positions the security to provide early warning of enemy approach and secure the objective rally point. The reconnaissance unit(s) then reconnoiters the objective area (route). The reconnaissance unit may move to several positions, along or adjacent to the specific route, in order to conduct a thorough reconnaissance. After completing the reconnaissance, each reconnaissance team moves to the objective rally point and reports to the patrol leader. The patrol then returns to friendly lines and the leader makes a full report.

Area

The patrol leader halts and conceals the patrol near the objective area in the final preparation position. The patrol leader then conducts leader's reconnaissance to pinpoint the objective and confirm the plan for positioning the security teams and employing units assigned the reconnaissance mission. The patrol leader returns to the patrol and positions security to provide early warning of enemy approach and secure the objective rally point. The reconnaissance unit(s) then reconnoiters the objective area. The reconnaissance unit may move to several positions, perhaps making a circle around the objective area, in order to conduct a thorough reconnaissance. When the reconnaissance is completed, the patrol leader assembles the patrol and tells members what has been observed and heard. Other patrol members contribute observations. The patrol then returns to friendly lines and the patrol leader makes a full report.

Zone

The patrol leader halts the patrol at the final preparation position, confirms the plan, and conducts leader's reconnaissance. The patrol leader positions the security team and sends out the reconnaissance team. When the entire patrol is used to reconnoiter the zone, it provides its own security. After completing the reconnaissance, each reconnaissance team moves to the objective rally point and reports to the patrol leader. The patrol then returns to friendly lines.

Chapter 13. Combat Patrols

Combat patrols are assigned missions that usually include engaging the enemy. They are fighting patrols. Every combat patrol has a secondary mission: gaining information about the enemy and terrain. Combat patrols are employed in both offensive and defensive combat operations and they assist the parent unit in accomplishing its mission by inflicting damage on the enemy; establishing and/or maintaining contact with friendly and enemy forces; denying the enemy access to key terrain; and probing enemy positions to determine the nature and extent of enemy presence.

13001. Task Organization

A combat patrol is organized around the current structure of the Marine rifle platoon. A combat patrol leader should use the unit's normal organization (fire team, squad, and platoon) in assigning functions, patrol missions, and chain of command. Combat patrols must be able to perform the following four basic functions en route to and from the objective:

- Provide control.
- Provide security.
- Provide support by fire.
- Attack or assault the objective.

A rifle platoon could task-organize as a combat patrol as follows:

- Platoon headquarters (patrol headquarters).
- First squad (security).
- Second squad (support).
- Third squad (assault).

Every combat patrol must—

- Provide a control mechanism in the form of a headquarters.
- Designate a unit (a fire team or squad) to provide security while moving en route to the objective or while at the objective. At the objective area, this unit isolates the objective area, secures the objective rally point, and covers the withdrawal.
- Designate a unit to act as support. This unit provides the base of fire in the attack or covers withdrawals or advances.
- Designate a unit(s) to conduct the attack or assault. This unit(s) engages the enemy at the objective area by fire and maneuver or movement. It also operates immediately in the objective area (searching, demolition, prisoners of war, etc.). Paragraph 9001 outlines the general organization of combat patrols. As in the case with reconnaissance patrols, the task organization of a combat patrol depends on the specific mission assigned. If any special requirements are generated because of the specific mission, the patrol is task-organized to fit the needs of the mission.

13002. Equipment

Combat patrols are armed and equipped as necessary for accomplishing the mission. In addition to binoculars, wire cutters, compasses, and other equipment generally common to all patrols, it usually carries a high proportion of automatic weapons and grenades. Communications with higher headquarters is important as success of the mission may depend on being able to call for supporting fires. Also, internal radio communications with the units and teams may be useful. However, the patrol must not be so overburdened with equipment as to impede movement or mission accomplishment.

13003. Raid Patrols

A raid is a surprise attack on an enemy force or installation with the attacking force withdrawing after accomplishing its mission. Raids destroy or capture enemy personnel or equipment, destroy installations, or free friendly personnel who have been captured by the enemy. Patrolling techniques are used in planning and when moveing to and from the objective. (Refer to MCWP 3-41.2, *Raids*.) Surprise, firepower, and violence of action are the keys to a successful raid. Patrols achieve surprise by attacking—

- When the enemy is least prepared (e.g., during periods of poor visibility such as darkness, rain, fog, or snow).
- From an unexpected direction. (This might be accomplished by approaching through a swamp or other seemingly impassable terrain.)
- With concentration of firepower at critical points within the objective.

Planning

A successful raid requires detailed planning. The leader of a combat patrol engaged in raiding must anticipate probable situations and decide upon definite courses of action to meet them. Rehearsals are imperative.

A raid patrol conducts such missions as destroying an enemy outpost or seizing prisoners from an observation post or lightly defended position.

While preparing for the mission, the patrol leader requests fire support required for the accomplishment of the mission. If practical, artillery and mortars should be employed to isolate the objective to prevent movement of enemy reinforcements into the area.

Execution

The leader's plan must be detailed and complete. All of the considerations outlined in chapter 11 must be covered. Patrol formations must provide for ease of control and all-around security while moving to and from the objective area and provide for rapid and coordinated deployment of the various units once the objective area is reached. The leader's plan usually includes the encirclement of the hostile position—either physically or by fire—in order to isolate it during the assault.

The final simultaneous assault against the objective develops when enemy defensive fires at the objective are suppressed by either friendly fire superiority or surprise. The assault is covered by the fire of the unit assigned the function of support by fire.

Grenades, SMAWs, and demolitions are most effective for clearing bunkers.

Security units are posted to isolate the objective. The patrol leader signals them when the withdrawal begins. As a minimum, security is on each flank and to the rear (at the objective rally point).

Actions at the Objective

The patrol leader halts the patrol near the objective at the final preparation position. Security is established and the leader's reconnaissance is made with appropriate subordinate leaders. When the leaders return to the patrol, they confirm previous plans or announce any changes. Movements are arranged so all units reach their positions simultaneously. This improves the patrol's capability for decisive action, if prematurely detected by the enemy.

The teams of the security element move to positions to secure the objective rally point, give early warning of enemy approach, block avenues of approach into—and prevent enemy escape from—the objective area. As the assault element moves into position, the security element informs the patrol leader of all enemy activity, firing only if detected or on the patrol leader's order. Once the assault element commences action, the security element prevents enemy entry into or escape from the objective area. The security element covers the withdrawal of the assault element (and support element, if employed) to the objective rally point, withdrawing only upon order or prearranged signal.

As the assault element approaches the objective, it deploys early enough to permit immediate assault if detected by the enemy. Each team uses stealth while moving into proper position. On command, or if one or more of the assault element is detected and fired upon by the enemy, the support element opens fire to neutralize the objective, then ceases or shifts fire according to prearranged plans and signals. As supporting fires cease or shift, the assault element assaults the objective. Demolition, search, and other teams are protected by the assault element while they work. On order, the assault element withdraws to the objective rally point.

If a support element is employed, its leader deploys teams to provide fire support for the assault element. Each member of the support element must know the scheme of maneuver to be used by the assault element, specific targets or areas to be neutralized by fire, and the signals that will be employed to commence, shift, and cease fires. The support element withdraws on order of the patrol leader. At the objective rally point, the patrol leader quickly reorganizes the patrol and begins the return movement to friendly lines.

13004. CONTACT PATROLS

General

Contact patrols establish and/or maintain contact to the front, flanks or rear by—

- Establishing contact with an enemy force when the definite location of the force is unknown.
- Maintaining contact with enemy forces through direct and/or indirect fires, or observation.
- Avoiding decisive engagement with the enemy.

Task Organization and Equipment

Task organization and equipment depend on the known enemy situation and anticipated enemy contact. A patrol sent out to establish contact with an enemy force is organized, armed, and equipped to overcome resistance of light screening forces in order to gain contact with the main enemy force. It is not organized and equipped to engage the main enemy forces in combat. Communication is paramount; radios must be reliable over the entire distance covered.

Actions at the Objective

The patrol leader selects a series of objectives. Once an objective is reached, the patrol leader initiates a planned set of actions in order to establish and maintain contact with the enemy. These plans and actions are guided by the missions to establish or maintain contact—not to engage in decisive combat. Contact with the enemy is maintained for the purposes of surveillance, applying pressure, and preventing seizure of the initiative. If the contact patrol becomes decisively engaged with the enemy, many of the tasks originally assigned to the patrol cannot be accomplished, since the enemy has seized the initiative and friendly forces are not forced to react.

13005. AMBUSH PATROLS

General

An ambush is a surprise attack from a concealed position upon a moving or temporarily halted target. It is one of the oldest and most effective types of military actions. Ambush patrols conduct ambushes of enemy patrols, resupply columns, and convoys. The intent of an ambush is to place the enemy in a dilemma where staying in the kill zone or attempting to move out of it prove equally lethal. The ambush may include an assault to close with and decisively engage the enemy, or the attack may be by fire only.

Purpose of Ambushes

Ambushes are executed for the general purpose of reducing the enemy's over-all combat effectiveness and for the specific purpose of destroying its units. The cumulative effect of many small ambushes on enemy units lowers enemy troop morale and harasses the enemy force as a whole.

Destruction is the primary purpose of an ambush because loss of men killed or captured, and loss of equipment and supplies destroyed or captured, reduces the overall combat effectiveness of the enemy.

Harassment, though less apparent than physical damage, is a secondary purpose of ambushes. Frequent ambushes force the enemy to divert men from other missions to guard convoys, troop movements, and carrying parties. When enemy patrols fail to accomplish their mission because they are ambushed, the enemy is deprived of valuable information. A series of successful ambushes causes the enemy to be less aggressive and more defensive minded. His men become apprehensive, overly cautious, reluctant to go on patrols, seek to avoid night operations, are more subject to confusion and panic if ambushed, and in general, decline in effectiveness.

Classification of Ambushes

A *deliberate ambush* is one in which prior information about the enemy permits detailed planning before the patrol departs for the ambush site. Information needed to plan a deliberate ambush includes the size, composition, and organization of the force to be ambushed; how the force operates; and the time it will pass certain points or areas. A deliberate ambush may be planned for such targets as—

- Any force if sufficient prior information is known.
- Enemy patrols that establish patterns by frequent use of the same routes or habitually depart and re-enter their own areas at the same point.
- Logistic columns.
- Troop movements.

An *ambush of opportunity* is conducted when available information does not permit detailed planning before the patrol departs. This is the type of ambush that an infantry unit normally conducts. An ambush of opportunity should not be confused with a hasty ambush. An ambush of opportunity is a planned ambush; a hasty ambush is an immediate action. In planning for an ambush of opportunity, the patrol must be prepared to execute any of several courses of action based on the types of targets that may be ambushed and must rehearse prior to departure. The course of action taken is determined when the opportunity for ambush arises.

The patrol leader may be directed to reconnoiter an area for a suitable ambush site, set up at the site selected, and execute an ambush against the first profitable target that appears.

The patrol may depart just after dark, move to a specific point, observe until a designated time, ambush the first profitable target after that time, and return before daylight.

A *hasty ambush* is an immediate action where the patrol makes visual contact with an enemy force and has time to establish an ambush without being detected. The actions for a hasty ambush must be well rehearsed and accomplished through the use of hand and arm signals given from the patrol leader.

Types of Ambushes

There are two types of ambushes: point and area. The *point ambush* is one where forces are deployed to attack along a single killing zone. The *area ambush* is one where forces are deployed as multiple related point ambushes.

A point ambush, whether independent or part of an area ambush, is positioned along the enemy's expected route of approach. Formation of the forces conducting the ambush is important because, to a great extent, the formation determines whether a point ambush is able to deliver the heavy volume of highly concentrated fire necessary to isolate, trap, and destroy the enemy.

The ambush formation to be used is determined by careful consideration of possible formations and the advantages and disadvantages of each in relation to—

- Terrain.
- Visibility.
- Forces.
- Weapons and equipment.
- Ease or difficulty of control.
- Target to be attacked.
- Combat situation.

For a detailed discussion of ambush formation, see appendix D.

Ambush Operation Terms

The *ambush site* is the location where an ambush is established.

The *killing zone* is that portion of an ambush site where fires are concentrated to trap, isolate, and destroy the target. On little-traveled roads, an obstacle placed in a defile, in the woods, on a bridge or on a steep upgrade can be used effectively to force vehicles to halt, and thus render the occupants vulnerable to attack. Antitank mines may be emplaced and the occupants of the wrecked vehicle killed or captured while still dazed by the explosion.

A *near ambush* is a point ambush where the attacking force is located within reasonable assaulting distance of the killing zone (50 meters is a guide). A near ambush is most often conducted in close terrain, such as a jungle or heavy woods.

A *far ambush* is a point ambush where the attack force is located beyond reasonable assaulting distance of the killing zone (beyond 50 meters is a guide). A far ambush may be more appropriate in open terrain offering good fields of fire or when the target will be attacked by fire only.

Factors for a Successful Ambush Patrol

There are many factors that give the ambush its best chance of success. The ideal situation would be to position the ambush on favorable terrain and have detailed planning completed beforehand.

The patrol leader of an ambush looks for the most favorable terrain in which the enemy is canalized between two obstacles with limited opportunity to attack or escape. Suitable areas include defiles, small clearings, bends in trails, and steep grades. Dense undergrowth adjacent to the ambush site permits observation from concealed positions. The ambush patrol should have maximum cover and concealment, not only for the firing positions, but for the routes of withdrawal. The enemy should be in an area offering as little protection from fire as possible. Favorable fields of fire include stretches of road, trail or open ground of at least 100 meters for machine guns and 15 meters for rifle fire and grenades. The ambush site can be improved by constructing obstacles—such as felled trees, wire, land mines, or booby traps—to impede the enemy.

Planning

A deliberate ambush or an ambush of opportunity requires thorough planning.

A deliberate ambush plan is based on extensive knowledge of the enemy and terrain, and is planned and rehearsed in great detail. A physical reconnaissance of the ambush site is made during the preparation phase and information gained is incorporated into the plan. All likely immediate actions of the enemy when ambushed are examined. Planned counteractions are developed and rehearsed.

In planning an ambush of opportunity, any available information on the enemy and terrain is used. A tentative plan for the ambush that incorporates all anticipated actions is developed and rehearsed. However, the bulk of planning is done concurrently during the patrol leader's reconnaissance of the prospective ambush site. In a rapidly developing situation, hasty ambush immediate action is employed.

The route and ambush site considerations apply to both deliberate ambushes and ambushes of opportunity. A primary route that allows the patrol to enter the ambush site from the rear is planned. Entering the prospective killing zone is avoided. If the killing zone must be entered to place mines or explosives, care is taken to remove any tracks and signs that might alert the enemy and compromise the ambush. If mines or explosives are to be placed on the far side of the ambush site, or if the appearance of the site from the enemy's viewpoint is to be checked, a wide detour is made around the killing zone. Care is taken to remove any tracks that might reveal the ambush. An alternate route from the ambush site to the objective rally point, as in other patrols, is planned.

Maps and aerial photographs are used to carefully analyze the terrain. When possible, an on-the-ground reconnaissance of the ambush site is made prior to occupation. Obvious ambush sites are avoided as the element of surprise is even more difficult to achieve in these areas. An ambush site must provide for—

- Favorable fields of fire.
- Occupation and preparation of concealed positions.
- Canalization of the target into the killing zone. (An ideal killing zone restricts the enemy on all sides, confining him to an area where he can be quickly and completely destroyed. Natural obstacles, such as cliffs, streams, embankments, or steep grades, are used whenever possible to force vehicles to slow down. Man-made obstacles, such as barbed wire, mines, and craters in the roads, are used to supplement natural obstacles.)
- Covered routes of withdrawal that enable the ambush force to break contact.
- Avoidance of enemy pursuit by fire.

Occupation of Ambush Site

The surrounding area is searched for enemy patrols prior to occupation of the ambush site. Ambush formations are used to physically deploy the patrol in a manner to inflict maximum destruction upon the enemy and to provide maximum security to the patrol. Ambush formations are contained in appendix D.

Positions

The patrol is moved into the ambush site from the objective rally point. Security is positioned first to prevent surprise while the ambush is being established. Automatic weapons are then positioned so each can fire along the entire killing zone. If this is not possible, overlapping sectors of fire are provided to cover the entire killing zone. The patrol leader then selects his position, located so he can tell when to initiate the ambush. Riflemen and grenadiers are positioned and sectors of fire are assigned to cover any dead space left by the automatic weapons. The patrol leader sets a time by which positions are to be prepared. Patrol members clear fields of fire and prepare positions in that order, with attention to camouflage for both.

Suitable Objective Rally Point

An easily located objective rally point is selected and made known to all patrol members. The objective rally point is located far enough from the ambush site so that it will not be overrun if the enemy assaults the ambush. Routes of withdrawal to the objective rally point are reconnoitered. Situation permitting, each man walks the route he is to use and picks out checkpoints. When the ambush is to be executed at night, each man must be able to follow his route in the dark. After the ambush has been executed, and the search of the killing zone completed, the patrol is withdrawn quickly but quietly, on signal, to the objective rally point where it reorganizes for the return march. If the ambush was not successful and the patrol is pursued, withdrawal may be by bounds. The last group may arm mines, previously placed along the withdrawal route, to further delay pursuit.

Local Security

Security must be maintained. Security elements do not usually participate in the initial attack, but protect the rear and flanks, and cover the withdrawal.

Patience

The Marines of the ambush force must control themselves so that the ambush is not compromised. Patience and self-discipline are exercised by remaining still and quiet while waiting for the target to appear, particularly if the patrol occupies the ambush site well ahead of the arrival of the enemy. Patience is necessary so as not to alert the enemy to the presence of the ambush.

Surprise

Surprise must be achieved, or the attack is not an ambush. If complete surprise cannot be achieved, it must be so nearly complete that the target is not aware of the ambush until too late for effective reaction. Surprise is achieved by careful planning, preparation, and execution so that targets are attacked when, where, and in a way for which they are least prepared.

Coordinated Fires

Properly timed and delivered fires contribute heavily to the achievement of surprise, as well as to destruction of the enemy. The lifting or shifting of fires must be equally precise; otherwise, the assault is delayed and the enemy has an opportunity to recover and react. All weapons, mines, and demolitions are positioned and all fires, including those of available artillery and mortars, are coordinated to achieve the following results:

- Isolation of the killing zone to prevent the enemy's escape or reinforcement.
- Surprise delivery of a large volume of highly concentrated fires into the killing zone.

Control

Close control of the patrol is maintained during movement to, occupation of, and withdrawal from the ambush site. This is best achieved through rehearsals and establishment and maintenance of good communications. When the enemy approaches, the temptation to open fire before the signal is given is resisted. The patrol leader must effectively control all elements of the ambush force. Control is most critical at the time the enemy approaches the killing zone. Control measures must provide for—

- Early warning of enemy approach.
- Fire control. Withhold fire until the enemy has moved into the killing zone, then open fire at the proper time.
- Initiation of appropriate action, if the ambush is prematurely detected. Individual patrol members must be prepared to react if detected by the enemy prior to the initiation of the ambush.
- Timely and orderly withdrawal of the ambush force from the ambush site and movement to the objective rally point.

It is important to remember that an ambush patrol should have four distinct signals: one to open fire (with an alternate signal to open fire to be used at the same time as the primary); a signal to cease fire or shift fire; a signal to assault or search the killing zone; and a signal to withdraw. The signal to open fire should meet two criteria: first, it should be the firing of a weapon that will kill the enemy; secondly, it should be a weapon reliable in any weather condition. A good primary signal is a Claymore mine, and an alternate signal would be a closed bolt weapon (M16A2). Open bolt weapons (M240G, M249) should not be relied upon to initiate an ambush.

Execution of an Ambush

The manner in which the patrol executes an ambush depends primarily on whether the ambush's purpose is harassment or destruction. To a lesser degree, the execution of the ambush is determined by whether the ambush is deliberate or an ambush of opportunity.

When the primary purpose is harassment, the patrol seals off the area with security teams to prevent enemy reinforcement and escape. Maximum damage is inflicted with demolitions and automatic weapons fire. The patrol delivers a very heavy volume of fire for a short time and withdraws quickly and quietly. The patrol avoids being seen by the enemy.

When the primary purpose of the patrol is destruction, the area is sealed off with security units. Maximum damage is inflicted with demolitions, antitank weapons, and automatic weapons fire from the support team or element. When these fires cease or shift, an assault is launched into the killing zone with heavy fire and violence to complete destruction. The assault unit provides security, while designated teams search and/or capture personnel and destroy vehicles and equipment. On the patrol leader's command, or by prearranged signal, all units withdraw to the objective rally point and move out quickly.

When the patrol's primary purpose is to obtain supplies or capture equipment, security units seal off the area. Demolitions and weapons are used to disable vehicles. The assault unit must use care to ensure its fire does not damage the desired supplies or equipment. Designated teams secure the desired items; other teams then destroy enemy vehicles and equipment.

The most successful ambush is one where the attacker is deployed and concealed in such a way that the enemy will unknowingly be surrounded by fire. The usual method is for the attackers to deploy themselves along a trail or route the enemy will travel. The enemy is permitted to pass by the center of the attacker's force so that the attack can be made from the front, flank, and/or rear. One or two men are posted well forward and to the rear along the route to prevent any enemy from escaping. All fires should be delivered simultaneously on a prearranged signal.

An effective method of luring the enemy is for an ambushing patrol to cut communication or electrical wire. The patrol then deploys and ambushes the line repair crew when it arrives. Since the line crew may be protected by riflemen, the attackers must be careful to engage the entire party.

Vehicles and foot personnel moving on well-established transportation routes can sometimes be captured by altering or moving directional signs so as to divert the enemy into an area where he can be more readily attacked. The attack can best be accomplished at an obstacle, such as a stream or gully, that forces the enemy to stop or slow down.

After the enemy has been ambushed and destroyed, the unit quickly withdraws over a prearranged route to the objective rally point. Speed is very important, since the noise of the ambush could alert other nearby enemy units.

13006. SECURITY PATROLS

General

Security patrols are assigned missions that may or may not require them to engage the enemy. They are used in proximity to defensive positions, on the flanks of advancing units or in rear areas. Purposes of security patrols are to detect infiltration by the enemy, destroy infiltrators, and protect against surprise and ambush.

In any situation where there is a threat of attack, such as a rear area threatened by guerrillas or a facility that is under threat of a terrorist attack, all Marines, not just the infantry, must know how to conduct a security patrol.

In just the offensive operations, infantry units provide security patrols to screen their flanks, areas, and routes. Whereas, in defensive operations, security patrols are used to prevent the enemy from infiltrating an area, detect and destroy infiltrators, and prevent surprise attacks. In rear areas, particularly when there is guerrilla or terrorist threat, the requirement to conduct security patrols increases for all Marine air-ground task force (MAGTF) units ashore, particularly aviation and combat service support units.

Task Organization and Equipment

Generally, a Marine rifle squad or similar sized organization is considered ideal for security patrols. Communications are important to higher headquarters so that they receive information from the patrol; and communications are important to the patrol to request fire support, etc. The radio the patrol carries must have the range necessary for higher headquarters to be able to receive transmissions from anywhere along the patrol route, and the patrol must have a secondary means of reporting (i.e., flare signals upon contact).

Patrol Procedures

All of the procedures presented in previous sections are to be used in security patrols.

Patrol Planning

Security patrol planning includes—

- Rehearsing prior to departing friendly lines.
- Maintaining communications.
- Support by organic weapons.
- Reinforced if necessary.
- Using varied routes and never establishing a routine pattern.
- Staying within proximity of friendly units.

Patrol Techniques

Within rear areas, an irregular pattern of patrol is established and changed daily. Outside of friendly lines it would be prudent to establish a definite preplanned route for the patrol, of which all adjacent units know the route. The parent unit commander establishes frequent checkpoints for control. If checkpoints are designated, the patrol leader treats them as individual objectives to be searched and cleared.

The patrol has a definite plan as to what to do if contact with enemy is made, how to break contact, how to defend itself, and how to call for supporting fires. It is imperative that patrol members know what to do if they become split or separated; i.e., location of rally points and how to be recovered.

13007. URBAN PATROLS

General

As national strategy continues to focus on regional vice global conflicts, the Marine Corps will continue to conduct urban operations in various operational environments. Cities and towns are often the center of economic and political power and are therefore extremely vulnerable to urban insurgent activities and violence. The lessons learned from recent operations in Somalia, as well as experiences gained by British forces in Northern Ireland, provide the foundations of urban patrolling.

Enemy Forces

Enemy forces in urbanized areas range from organized military forces to low intensity engagements with insurgents, such as terrorists or local gangs. For Marine Corps doctrine and tactics, techniques, and procedures (TTP) for dealing with the higher intensity threat in urbanized areas, refer to MCWP 3-35.3, *Military Operations on Urbanized Terrain.*

Common Threat Tactics

Cities provide cover and concealment for both friendly forces and enemy forces. However, enemy forces generally find active support only in certain areas of the town or city. The urban-based insurgent or terrorist usually lives in a friendly community or in one where the people are too frightened to withhold support or inform anyone about the situation. The insurgent or terrorist normally maintains close contact with leaders and others friendly to the cause. The enemy will often have efficient communication and intelligence systems, sometimes involving women and children to provide cover for its activities.

The urban insurgent normally cannot, like his rural counterpart, establish bases and recruit large military units. He is generally an individual or a member of a relatively small group. He relies on the cover afforded by the city's populace and terrorizing them to coerce loyalty or support.

Urbanized areas tend to give the insurgent and/or terrorist many opportunities to initiate action and gain advantage. The normal presence of large numbers of people in cities provides the foe an opportunity to mass crowds quickly and manipulate demonstrations easily. The presence of women and children during mass demonstrations may restrict the courses of action available to friendly forces. Major incidents stemming from overreaction or excessive use of force by friendly forces may provide the insurgent with propaganda material. Publicity is easily gained in an urbanized area because major incidents can't be completely concealed from the local population. Insurgent successes can be exploited to discredit the ability of host nation police, friendly forces, and the civil government, and gain recognition for the insurgents' cause.

The urban insurgent or terrorist can usually be expected to operate more boldly than his rural counterpart. This is reflected in the enemy's tactics. A single sniper or bomber may be the norm in the urbanized area, whereas the rural threat is generally the more conventional ambush. In urbanized areas, explosive devices can be easily emplaced and used effectively against large groups or select individuals. Enemy forces may be expected to employ the following tactics in urbanized areas:

- Using local communications, such as radio and newspapers, for propaganda purposes.
- Disrupting industry and public services through strikes and sabotage.
- Generating widespread disturbances designed to stretch the resources of the security force.
- Creating incidents or massing crowds in order to lure the patrol or reaction force into a trap.
- Provoking security forces in the hope that they may react improperly, therefore discrediting the security force by means of propaganda.
- Sniping at roadblocks, outposts, sentries, and patrols.
- Attacking friendly bases with rockets and mortars.
- Planting explosive devices, either against specific targets or indiscriminately, to cause confusion and destruction, and lower public morale and confidence.
- Using ambush patrols.
- Firing on friendly helicopters.

Principles of Urban Patrolling

Patrolling in an urban environment often presents conditions considerably different and often more complex than those encountered in rural and less inhabited areas. While the principles of patrolling are still relevant in an urban situation, the nature of urban patrolling has led to the development of six specific urban patrolling principles. They are—

1. **Depth.** The restrictive, canalizing nature of urbanized terrain usually limits a patrol's ability to disperse laterally. To prevent the patrol from bunching up, patrols normally maintain dispersion along the length of a patrol formation.

2. **Mutual Support.** The positioning of units in-depth within the patrol enables one unit to cover another unit's movement and facilitates immediate action during various situations. Aircraft, vehicles, and snipers also provide good mutual support.

3. **Deception and Pattern Avoidance.** Deception and pattern avoidance are normally a planning consideration of the headquarters directing the entire patrolling plan. By varying patrol routes, durations, and departure times, hostile actions commonly used against urban patrols—such as enemy ambushes, and roadblocks—are more difficult to plan and may be preempted.

4. **Intra-patrol Communication.** Elements of an urban patrol must have the means to communicate with each other. Ideally, each element will possess a radio enabling it to remain in continuous communication that facilitates rapid response and reporting to higher headquarters, reaction force coordination, and coordination of actions with other patrols or fire support agencies.

5. **Establishment of a Reaction Force.** The nature of urbanized terrain (its compartmentalization) makes urban patrols more vulnerable to a wider range of hostile actions. The requirement for immediate, coordinated reinforcement of a patrol is best satisfied by employment of an established reaction force. The reaction force requires superior mobility (relative to the enemy's) and fire support to be effective for this task.

6. **Three-Dimensional Threat**. Patrolling in an urban environment requires constant attention to its three-dimensional aspect; hostile actions can originate from rooftops, streets, subsurface levels or combinations of all levels at once.

Classification of Urban Patrols

Mission

The vast majority of urban patrols are overt in nature, with their presence readily apparent to the local populace. Most urban patrols are combat vice reconnaissance patrols. The vast majority of urban patrols are security patrols. However, units may be assigned secondary tasks of reconnoitering specific or general areas along the patrol routes.

Raids normally involve a swift penetration of an objective to secure information, confuse the enemy or destroy installations. Raids include a planned withdrawal upon completion of the assigned mission.

Movement

Means of movement are as follows:

- Dismounted: movement on foot.
- Vehicular: movement by motorized, mechanized or armored vehicle.
- Helicopterborne: movement by helicopter, however, helicopterborne patrols will usually involve dismounted or vehicular movement after patrol insertion.
- Combination: movement using a combination of methods.

Task-Organization

Units task-organize to meet the specific requirements of the mission and situation. The Marine rifle squad is ideally suited for urban patrolling and can easily integrate attached specialists required by specific missions. Specialists who may accompany urban patrols include—

- Interrogator-translator team (ITT) and counterintelligence team (CIT) Marines.
- Explosive ordnance disposal (EOD) personnel.
- Members of host nation or allied military forces.
- Interpreters.
- Local community leaders.
- Local law enforcement officers.
- Public affairs personnel escorting media representatives.

Dismounted Patrol Organization

Patrols should maintain unit integrity (fire team and/or squad) in organizing elements for an urban patrol. Similar to reconnaissance patrols, the vulnerability of urban patrols necessitates that all elements must provide for their own security in addition to the combat aspect of the mission. Combined assault and security teams are an effective method to organize for all-around security.

Urban Patrol Base Operations

Urban patrols may operate from an established patrol base that may be located within the unit's assigned area of operations or an area designated for a patrol. The patrol base should be located in a building used exclusively for this purpose. Patrol bases may also be located within a larger site that houses other agencies, such as a higher headquarters' command echelon.

If located within a larger site, the patrol base will be included in the overall facility security plan. If isolated, the patrol base must consider the following security factors:

- External security:
  - Barrier plan.
  - Sentry posts.
  - Local security patrols.
- Internal security:
  - Covered positions for all Marines.
  - Contingency plan for hostile actions against patrol base.

The Reaction Force

The high probability of an urban patrol becoming involved in a hostile or volatile incident requires the establishment of a dedicated reaction force for rapid reinforcement, support or extraction of the patrol. Ideally, the reaction force is—

- Large enough and task-organized in a manner that it can meet and quickly defeat the expected threat. Reaction forces are normally tiered with a lead element (normally one-third of the unit's size) and a main body force (the other remaining two-thirds of the force). Reaction force response times routinely are determined in advance by higher headquarters.
- Ready to respond immediately.
- Motorized or mechanized and supported by close air support and other fire support.
- Familiar with the area of operations.
- Briefed on the patrol's plans and monitors the status of patrols in progress.
- Task-organized to be multimission capable.
- Able to communicate with the higher headquarters, fire support assets, patrol base, and the patrol.
- Controlled by higher headquarters, once employed.

Patrol Preparation

Planning

Higher headquarters will—

- Designate the area for patrol.
- Provide intelligence briefs and updates.
- Ensure liaison with allied forces and the civilian populace.
- Provide special equipment and personnel required for the mission (scout snipers, public affairs officer, interpreters, etc.)
- Provide urban maps, photos, terrain models as required.
- Consider deception and pattern avoidance when issuing mission.
- Prescribe rules of engagement (ROE).

Intelligence Brief

An intelligence brief is conducted by the S-2 officer or representative prior to a patrol conducting its mission. The brief adresses the situation relevant to the specific patrol (e.g., routes, areas, updated enemy situations).

Coordination

Higher headquarters will effect liaison with adjacent and allied forces, as well as civil authorities and other agencies, having a possible effect on the patrol. The patrolling unit generally follows the same procedures as those used during patrol planning and execution in a jungle or forest environment.

The Urban Patrol Order or Warning Order

The Urban Patrol or Warning Orders use the same format and considerations as noted in this publication for patrolling. They rely heavily on a detailed terrain model, photographs, and subterranean construction to ensure complete understanding of the plan.

Rehearsals

The limited size of the patrol base usually precludes the need for full-scale rehearsals. Immediate action drills, such as crossing danger areas, are rehearsed in as much detail as possible, despite the limited available space.

Inspections

Initial and final inspections are conducted in the same manner as other patrols. Attached personnel must be fully integrated into the patrol and familiar with the plan and unit standing operating procedures (SOPs).

Conducting an Urban Patrol

Movement

Individual and unit movement considerations are generally the same as those for other patrols. However, urban environments require consideration of additional factors. Because of these factors, an urban patrol leader should—

- Ensure that each movement within a patrol takes place under the observation or cover of another individual or element of the patrol.
- Know where cover can be taken in the event of a hostile incident or action.
- Be prepared for contact with civilians, especially children, during the patrol and be aware that they may intentionally attempt to distract patrol members.
- Expect the presence of vehicles (both moving and stationary) along the patrol route.
- Expect members of the patrol to be approached by dogs and what action to take if threatened.

Patrol Formations

Squad-sized Patrols. The need for immediate fire power outweighs the dangers of becoming canalized. In contrast to other types of patrols, the headquarters element of an urban patrol will normally locate at the lead of the patrol column. This allows the patrol leader greater flexibility through control of two combined assault and security (A&S) teams. The leaders of these elements tactically stagger their members on each side of the road (see fig. 13-1 on page 13-12).

A&S teams follow in trace of the headquarters element and maintain unit integrity on separate sides of the street. One unit will remain slightly to the rear to create a staggered interval between Marines on either side of the street. This allows A&S teams to take lateral routes in support of headquarters element without having to cross a street to do so.

Platoon-sized Patrols. Squads will generally travel abreast of each other, moving along parallel routes. The interval between squad-sized units and/or teams is situation-dependent, but is usually between 100 and 150 meters (roughly two city blocks; this often prevents visual contact between the units). The intent is to create less of a target to an aggressor, yet still allow the patrol to quickly react to an incident. Individuals within units or teams will move in a staggered column as in a squad-sized patrol. See figure 13-2, which is located on page 13-13).

Night

Night patrols will generally be at least squad-sized and will generally use the same formation as that for day patrols. At night, it may be necessary to close distances between individuals or elements to maintain control. Consideration should be given to the use of night vision devices and thermal weapons sights.

Navigation, Control, and Security Measures

The patrol leader is ultimately responsible for the navigation. The headquarters element normally functions as the base unit during movement. The designated navigator is normally assigned from within the headquarters element. City maps are often inaccurate or outdated; however, when used with aerial photographs and other navigational aids, they can be effective for urban navigation.

Checkpoints and phase lines should be related to major streets (alleys, buildings, bridges) for easy identification. Arrival at checkpoints and crossing of phase

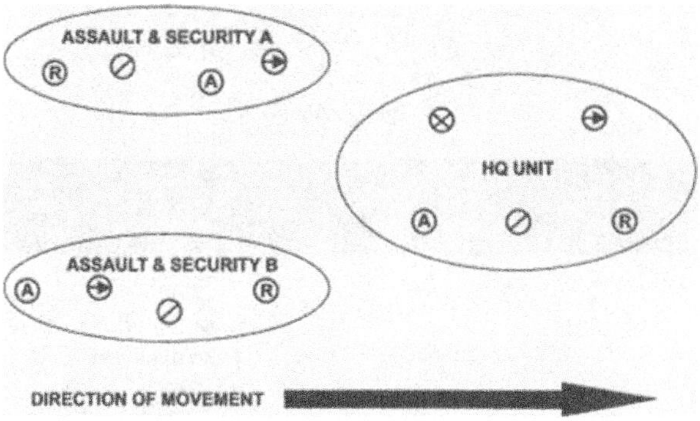

Figure 13-1. Squad-sized Dismounted Urban Patrol Formation.

lines should be relayed to higher headquarters using established brevity or codewords. A detailed patrol overlay is issued to both the reaction force and higher headquarters to keep them advised of routes and control measures used. A terrain model should be maintained at higher headquarters to aid in supervision and control of the patrol.

The use of camouflage should be limited to avoid frightening and confusing the local populace. The often overt nature of urban patrols may negate the need for camouflage. Patrols are sometimes deployed to show force presence and usually move on the urban street in plain view.

During daylight, patrols will routinely vary their rate of movement ranging from short halts to brief periods of double-timing. The British term for this urban patrolling technique is *hard-targeting*, meaning it makes the patrol harder for an enemy to target. Altering the rate of movement is intended to frustrate the enemy's ability to coordinate an attack or ambush against a targeted patrol.

Patrols should use short security halts, with Marines taking up mutually supporting firing positions. Marines must always work in pairs, ensuring mutual support. The last Marine in the element will provide rear security, but stays in his buddy's sight.

Individual Tasks

Individuals may be assigned collateral tasks performed throughout the patrol that may increase the patrol's situational awareness. Individual tasks may include—

- Vehicle spotter: looks for suspicious or known insurgent vehicles.
- Personnel spotter: observes and attempts to recognize previously identified enemy in crowds.
- Talker: attempts to gain information from casual conversations with the local populace. (Talkers are usually subordinate leaders or Marines with foreign language skills.)
- Searcher: conducts physical searches of vehicles and personnel while other patrol members provide cover and security.
- Marksman: engages point targets when the tactical situation does not permit massed or high volume fire. Other patrol members provide security to cover the marksman's engagement.

Departure of Friendly Lines

Urban patrols must vary their departure times to prevent being ambushed while exiting the patrol base.

Individual elements will usually depart exits at staggered times and at different movement rates, especially where sentries cannot provide cover. (This technique is used when the environment adjacent to the patrol base is dangerous.)

Once an element has exited friendly lines, a short halt is conducted in a predetermined, covered initial rally point 50 to 100 meters from the base. This ensures all elements are in position before the patrol continues.

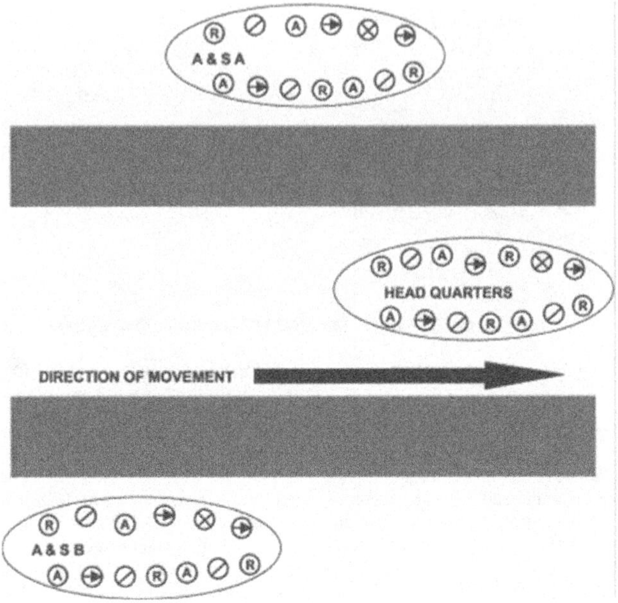

Figure 13-2. Platoon-sized Dismounted Urban Patrol Formations.

Exit points, routes from the base, departure techniques, and locations of IRPs should be varied constantly to avoid setting a pattern. This pattern avoidance may also include using vehicles to insert patrols away from the urban patrol base and employing empty vehicles as part of a deception plan.

Danger Areas

Urban patrols may encounter hundreds of danger areas during a single patrol. The three-dimensional threat requires keen situation awareness by every patrol member. Many danger areas can be dealt with simply by avoidance, while others require an adjustment of patrol formation, movement rate, etc. In the urban environment, places to be treated as danger areas are points that pose a major threat to the patrol, such as local political and religious headquarters, weapons containment areas, roads and routes that canalize movement and direct fire, and any area with a history of repeated contact.

Near and far side rally points are designated and briefed during the issuance of the patrol order. Squad-sized formations may use the A&S teams to provide flank security for the headquarters element and for each other. The headquarters element identifies the danger area and takes up a position on the near side of intersection. Individuals provide all-around security (see fig. 13-3 on page 13-14).

Two Marines (one from each A&S team) are designated to move through the headquarters element and establish respective firing positions on the near side of the danger area covering the patrol's near side flanks. They are followed by a second pair (again, one Marine from each A&S team) that moves across to the far side of the danger area and establishes respective firing positions covering the patrol's far side flanks (see fig. 13-4 on page 13-14).

Once near and far side flank security is established, the headquarters element moves across to the far side of danger area (see fig. 13-5 on page 13-15).

The remaining A&S team members then cross the danger area and join the headquarters element on the far side (see fig. 13-6 on page 13-15).

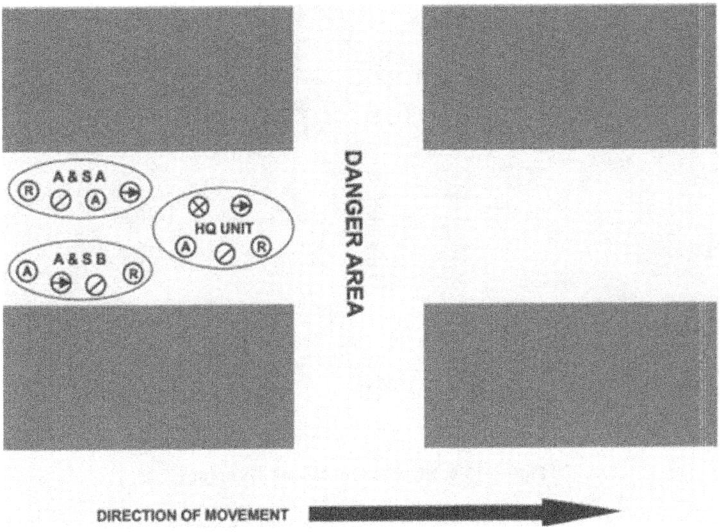

Figure 13-3. Approaching Danger Area.

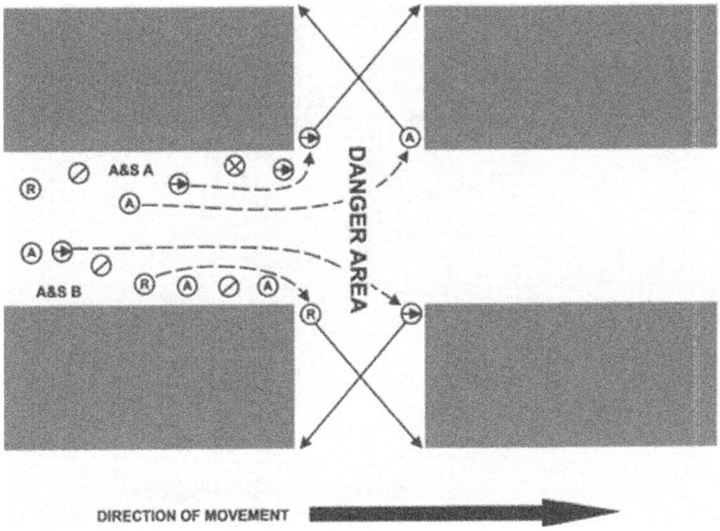

Figure 13-4. Securing Flanks.

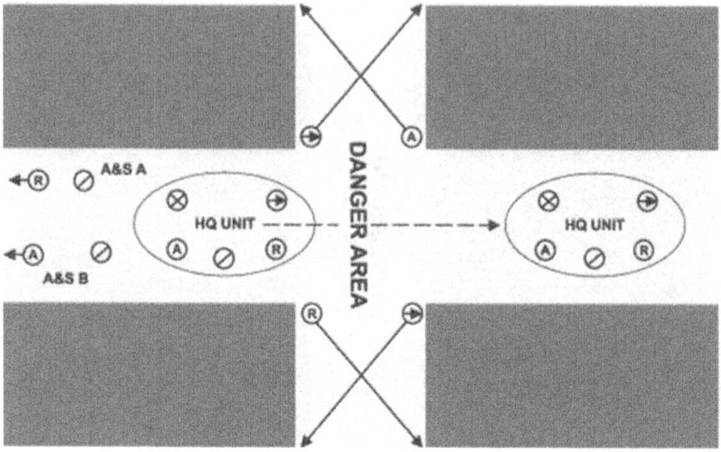

Figure 13-5. HQ Element Moves Through.

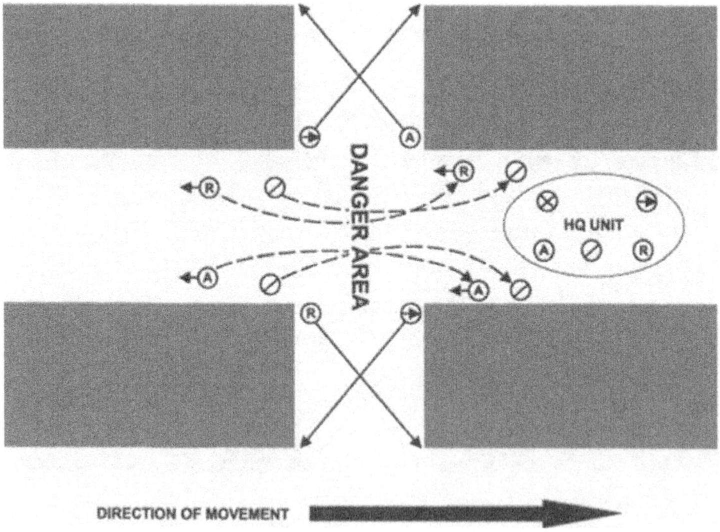

Figure 13-6. A&S Elements Move Through.

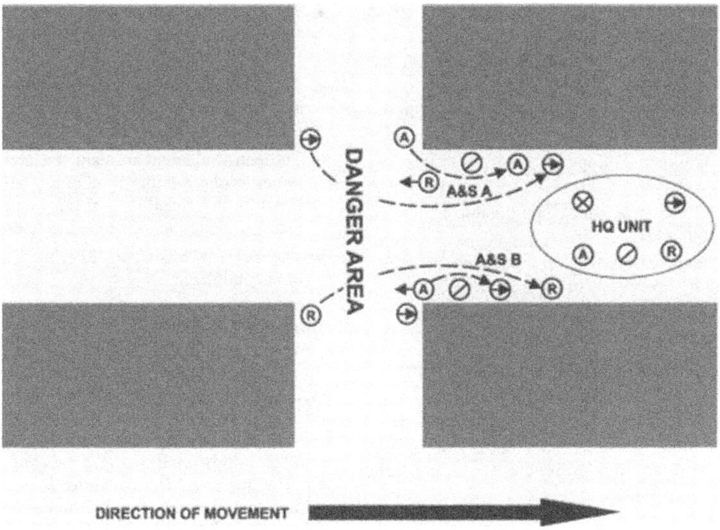

Figure 13-7. Patrol Resumes Movement Along Intended Route.

Once the trail A&S team members cross the danger area, they take up rear security and cover the movements of the near and far side flank security teams as they return to their positions in the patrol formation. The near side security team should collapse back first, followed by their far side counterparts. The patrol then resumes its advance away from the danger area (see fig. 13-7).

Interaction with Local Populace

Urban patrols must interact with the local populace. Patrols are at first a novelty to the civilians but can quickly become an unwelcome intrusion. The movement of the patrol must be fast enough to prevent the enemy from massing their fires upon it, but deliberate enough to ensure adequate security and mutual support. Patrol members must realize that they are usually the only Marines the local populace will encounter and that an inappropriate gesture, comment or act could lead to the deterioration of rapport between U.S. forces and the general population. Marines must remember that the vast majority of the individuals with whom they come in contact will be noncombatants attempting to survive in trying political, economic, and social situations.

Hostile incidents often seem to occur spontaneously, but there are usually indications that can alert Marines to imminent danger. The most obvious are the sudden alteration of normal routines, patterns, and attitudes of the local populace or other unusual activity. Some examples include—

- Observers on rooftops, in windows, etc., who are obviously tracking the patrol.
- The unusual absence of pedestrian traffic and people on porches.
- Stores, markets or street vendors closed suddenly or without explanation.
- Changes in civilian attitude toward patrol members.
- Unknown individuals or vehicles in the patrol area.
- Unfamiliar vehicles parked in the patrol area (possible car bomb).
- Roadblocks.
- Children throwing rocks at patrols to possibly draw the patrol's attention away from a more serious danger, such as a deliberate ambush.
- Vehicles riding unusually low due to overloading (possibly ferrying people, weapons, explosives).
- Agitators trying to provoke an incident with patrol members.

- Absence of the usual stray dogs (dogs are adept at sensing danger and avoiding it).
- Anti-American graffiti suddenly appearing in the patrol area.
- Pictures of enemy leaders and martyrs posted in the patrol area.
- Civilian workers failing to appear at U.S. or friendly bases.
- Normal deliveries and pick-ups conducted late or early without reason.
- Sudden change of civilian sentiment in newspaper articles, radio broadcasts or other media.
- Women and children leaving to live elsewhere.

Immediate Actions Upon Enemy Contact

Reaction to Sniping

Snipings are often executed from a single firing point, but coordinated snipings delievered from multiple points are not uncommon.

Patrol element or team leaders should constantly try to identify likely firing points and anticipate their own reactions to a possible shooting. Normally, snipers in an urban environment have a detailed withdrawal plan. Once a patrol comes into contact with a sniper, the patrol leader must immediately assess the situation and maneuver his patrol accordingly. The patrol's mission, location, size, ROE, and location of the threat often determine whether the patrol will attempt to neutralize the targeted sniper. If the patrol leader decides to kill or capture the sniper, he uses planned and rehearsed immediate actions to maneuver and counter the sniper's assault. The goal is to kill the sniper or cut off his escape and capture him. There are three immediate reactions to neutralizing a sniper: initial contact, immediate follow-up, and subsequent follow-up.

Initial Contact. The initial contact is made when the sniper fires the first shot. The patrol must react immediately and positively to get behind the firing position in order to kill or capture the gunman. The period of contact ends when the gunman is killed or captured, or the patrol element or team leader on the scene ends it. The following technique is the same for both squad- and platoon-sized patrols:

- The element or team in contact attempts to identify the firing position and maneuvers designated marksman into position to return well-aimed and controlled fire. Other members of the patrol take up positions to cover the marksman's engagement.
- The patrol element or team leader in contact sends initial contact report to the patrol leader, who notifies higher headquarters.
- The element or team leader in contact determines appropriate cut-off positions and relays them to flanking elements or teams.
- The patrol element or team leader in contact continues to observe the firing point, but does not enter it due to the possibility of booby traps. Flank elements or teams set up along likely escape routes.
- The incident ends when either the sniper ceases fire or is neutralized.

Immediate Follow-Up. Regardless of the fate of the gunman, isolation of the firing point is necessary to prevent reinforcement and preserve forensic evidence (scent, spent casings, etc.). If not under fire, members of the patrol element or team cordon off the area surrounding the firing point. Flanking elements or teams maintain their positions and prevent civilians from entering the area. The patrol leader moves to link-up with the element or team in contact (if not his own), and makes an estimate of the situation. The patrol leader sends a SPOTREP to higher headquarters.

Subsequent Follow-Up. The aim of the subsequent follow-up is to use follow-on forces to clear the building of remaining resistance or to obtain evidence that can be used to capture the gunman. The patrol leader establishes a position where he can brief arriving units (reaction force commander, S-2 representative, EOD personnel, etc.). Once the arriving units have been briefed, recommendations are made to higher headquarters via radio. No one is allowed into the cordon without the patrol leader's approval.

Reaction to Becoming Decisively Engaged

If a patrol becomes decisively engaged from numerous firing positions, the following immediate action should be taken:

- All patrol members move to available cover and return accurate fire on identified firing points.
- The patrol leader assesses the situation and makes a decision to either request the reaction force or break contact.
- If the reaction force is requested, the patrol will maintain its position until the reaction force arrives. The patrol should use fire and maneuver to gain better tactical positioning and support the arrival of the reaction force. When the reaction force arrives,

its commander may decide to either clear occupied buildings or cover the patrol during its extract.

Reaction to Bomb Threat or Discovery

The use of command-detonated explosive devices is a common ambush tactic employed by a terrorist or insurgent in an urban environment. The appropriate response to a reported threat or an actual discovery is generally involves four steps (known as the four Cs)—

1. The patrol leader **CONFIRMS** the presence of the suspicious item.

2. Without touching or moving anything suspicious, patrol elements or teams **CLEAR** the immediate danger area to a minimum of 100 meters. The area is cleared from the suspected device outward, inform civilians as to the reason for evacuation.

3. A&S teams establish a **CORDON** to secure the cleared area. Avenues of approach are cordoned off to keep people out and to protect EOD or engineer personnel clearing the device. The assistant patrol leader acts as the cordon commander and informs the patrol leader when the cordon is secure. An effective cordon technique is to tape off the area with engineer tape, creating both a physical and psychological boundary.

4. **CONTROL** of the area is maintained throughout the bomb clearing operation by the patrol leader. The patrol leader sends a report to a superior concerning details of the device (if known) and the area affected. The patrol leader coordinates with arriving personnel (EOD, engineers, etc.). The patrol leader maintains communication with the assistant patrol leader and keeps the Marines informed of the progress of the clearing operation.

Reaction to a Bomb Detonation

Bombs may be used by an insurgent as a means of initiating an ambush on mounted or dismounted patrols, in which case the actions for decisive engagement apply. Immediate action in response to an isolated explosion is similar to that used in reaction to a sniping and breaks down into the same three phases:

Initial Contact. The patrol leader attempts to identify the likely initiation point and sends an initial contact report to higher headquarters. If the bomb was command-detonated, the patrol leader sends his A&S teams deep to cut off the bombers' escape routes. Any casualties are moved a minimum of 100 meters from explosion and out of the line of sight to it.

Immediate Follow-Up. The A&S teams may need to maneuver to positions behind the likely initiation point to cut off escape. Once in position, personnel checks are conducted and any suspects are detained. The patrol leader coordinates requests for required support (MEDEVAC, reaction force, etc.)

Subsequent Follow-Up. Due to the possibility of secondary detonations, the four Cs (confirm, clear, cordon, and control) can be conducted as in reaction to a bomb discovery or bomb threat.

Civil Disturbances

Urban patrols must prepare to react to spontaneous aggression by the local populace. In many cases, civil disturbances are organized by the enemy to draw dismounted patrols into a targeted area, or to distract them from enemy activity occurring elsewhere. Civil disturbances are generally divided into two categories: minor aggressive actions, and full-scale rioting.

Minor aggressive actions are activities characterized by rock-throwing or use of devices such as Molotov cocktails and may either be directed at the patrol or take place between different ethnic factions of the population. Minor aggressive actions are normally spontaneous in nature and may have minimal or limited objectives for the insurgents.

Full-scale rioting events are usually in response to another major event or incident that may enflame the populace. Full-scale riots are well-planned and orchestrated, with clear objectives or targets in mind. At times, patrols will need to attempt to maintain control of a civil disturbance situation; however, dismounted and mobile small unit patrols should generally avoid potential flashpoints. Procedures to handle civil disturbances are as follows:

- The patrol leader reports the incident to headquarters and attempts to diffuse the crowd by talking to crowd leaders.
- If the patrol leader determines the size of the disturbance is too large for the force to handle, the patrol should move away from the disturbance to a safer, more remote covered area and occupy positions to observe and report the situation to higher headquarters. To prevent the patrol from being pursued by the crowd, the patrol should move quickly and change direction, often at road junctions, to gain distance from it.

Patrol members should maintain dispersion to create a more difficult target. They should face the crowd at all times to see and avoid any projectiles thrown. Individual self-discipline must be maintained throughout the disturbance. Marines charging into the crowd or throwing objects back at the crowd will only worsen the situation. If pursued or trapped, the patrol leader may consider using riot control agents (combat support, pepper gas, etc.) to disperse or slow a crowd's movement. The use of riot control agents must be authorized under the established ROE. If the patrol leader believes the situation is deteriorating beyond the patrol's ability to control it, the patrol leader should request the reaction force, which may be better equipped to handle a large riot or mob.

Break Contact

As with patrols in rural areas, the patrol leader may be forced to break contact as a result of decisive engagement with the enemy. On the basis of his estimate of the situation, the patrol leader will normally break contact in one of the following ways:

- As a patrol, with elements providing cover for movement as defined by clock direction and distance.
- As individual units/teams taking separate routes out of the area, then linking up at a designated rally point a safe distance away from the engagement.

As in any contact with enemy forces, smoke may be employed to screen movement. Fire support agencies can be utilized to suppress targets; riot control agents can be employed to disrupt enemy movement.

Reentry of Friendly Lines

The reentry of a dismounted patrol into an urban patrol base is no different from that of a patrol conducted in a rural area. The same planning considerations and control methods apply.

Missions Related to Urban Patrolling

House Calls

House calls missions are usually part of a coordinated effort to collect information within the area of operations. They involve obtaining up-to-date information on particular houses and occupants. When possible, local police should accompany patrols to do the actual talking to the occupants. If this is not possible, a technique that may be used by an urban patrol is—

- A&S elements move to provide cover around the target house.
- The headquarters element provides security just outside the house.
- The patrol leader and one Marine for security enter the house, if invited. If not invited, they talk to the occupants on the doorstep. Patience and tact are required in requesting information. An interpreter should be present when language differences exist.

Vehicle Checkpoints

Urban terrorists or insurgents commonly use vehicles to transport personnel, weapons, explosives, and equipment. Civilian vehicles are often used for these purposes, creating the requirement to check as many vehicles as possible. While permanent, fortified checkpoints may be conducted along approaches into an urbanized area, dismounted patrols can be employed to establish hasty vehicle checkpoints to stop vehicles and to keep the enemy off-guard. The two common types of vehicle checkpoints patrols establish are hasty and deliberate.

Hasty checkpoints are deployed anywhere based upon the decision of the patrol leader. Patrols must not set patterns through the frequent use of the same sites.

Deliberate checkpoints are tasked by higher headquarters to achieve a specific purpose. Time and locations are carefully considered to avoid setting patterns.

The general layout for a squad-sized, two-way dismounted checkpoint is depicted in figure 13-8 on page 13-20. The technique is as follows:

- The A&S teams are positioned stealthily in blocking positions on both sides of the road.
- Both the patrol leader and assistant patrol leader act as "talkers" for each direction of traffic (with local police or interrogator or translator Marines acting as interpreters) while a two-man team from the headquarters element physically searches the vehicles.
- Obstacles or parked vehicles may be employed to create a staggered roadblock in center of the checkpoint to slow approaching vehicles.
- The checkpoint location should be sited so that approaching vehicles cannot see it until they have passed a security team, and they have no escape route then available.
- Signs announcing the checkpoint should be displayed a safe distance from the search area for safety to both drivers and Marines.
- Normally higher headquarters will issue criteria that determines which vehicles are searched, but random checks of cars should normally be made as

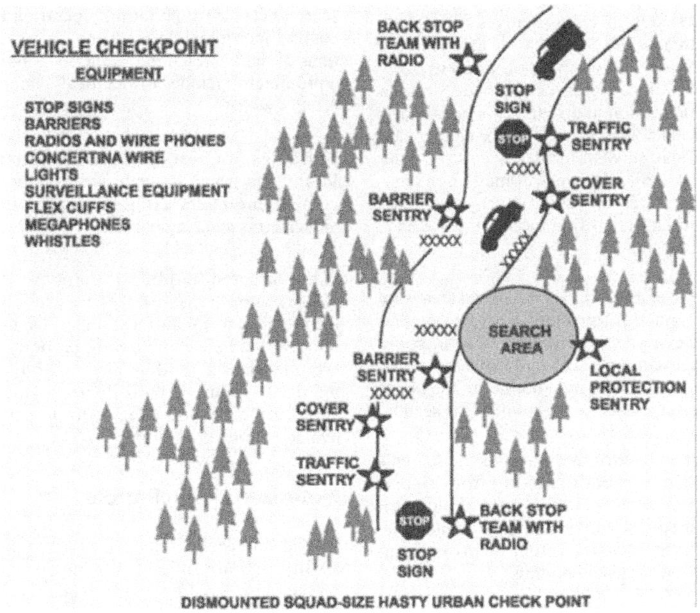

Figure 13-8. Dismounted Squad-sized Hasty Urban Vehicle Checkpoint.

well. While the vehicle is being searched the driver should accompany the searcher around the vehicle.

- Vehicle occupants should be made to exit the vehicle and then searched. Whenever possible, women should be searched by female police or female Marines.
- All patrol members should conduct themselves with courtesy and politeness. If nothing is found, an apology for the inconvenience is recommended.
- A hasty vehicle checkpoint should not be conducted any longer than 30 minutes for security reasons.
- The ROE should dictate whether or not action should be taken against vehicles that fail to stop at the checkpoint. Failure of a vehicle to stop does not automatically give authorization to fire.

Observation Posts

Urban observation posts are established to provide extended security, not only for patrol bases but also for patrols operating within the observation post's sector of observation. Observation posts can be established in conjunction with sniper operations and for providing overwatch for patrols operating within their sector of observation. Observation posts are nor-mally positioned on dominating terrain or in buildings outside the patrol base itself.

Insertion to the observation posts and conduct of observation may be either overt or clandestine in nature. Overt observation posts usually will be hardened positions to increase security. A patrol provides cover while the observation post is being inserted. A clandestine observation post relies on stealth of insertion and occupation for protection. It is normally positioned in abandoned buildings to cover sectors of observation that overt observation posts cannot. Because of their nature, clandestine observation posts are difficult to successfully establish and should not be manned for an extended period of time. Orders establishing observation posts (and patrols) must address the method of extraction as well as actions upon compromise/attack.

Cordon and Search

The cordon and search mission involves isolating a predesignated area by cordoning it off and systematically searching for enemy personnel, weapons, supplies, explosives or communications equipment. While large-scale cordon and search operations are planned and rehearsed in advance and normally entail extensive coordination with local law enforcement agencies, a squad-sized urban patrol may often conduct a cordon and search of a point target—searching one house or building identified by intelligence as a possible weapons cache.

The basic principle of a search of a populated area is to conduct it with limited inconvenience to the population. The populace may be inconvenienced to the point where they will discourage urban guerillas or insurgents from remaining in the area, but not to the point that they will assist the enemy as a result of the search.

Upon receiving intelligence that warrants the searching of a building or a specific tasking from higher headquarters, A&S elements of the patrol move to establish an inner cordon around the target building to seal it off, with the primary intent of preventing movement out of the targeted building.

On order, the designated reaction force deploys to establish an outer cordon, oriented outward some distance from the inner cordon and covering routes leading into the area in order to prevent outside interference/reinforcement. The reaction force maintains a reserve to reinforce either cordon or react to unfolding events (civil disturbance in response to the operation).

Once the cordons have been established, the patrol leader, with the assistance of local police or interpreters, informs the local populace that a building is about to be searched, that a house curfew is in effect (if permitted by higher headquarters), and that all occupants should remain indoors. Occupants of the target house are instructed to gather at a central location to stay out of the way of the search party.

The headquarters element, having linked up with any required assistance (explosive ordnance disposal (EOD), ITT, etc.) now acts as the search party and accompanies local police. A female searcher should be included in the party, if necessary.

Occupants are searched and screened first for possible enemy personnel. Apprehended persons are evacuated as soon as possible.

The head of the household should accompany the search party throughout the operation to be able to counter incriminating evidence and possible accusations of theft and looting against Marines. If possible, a prominent member of the local community should act as a witness.

Buildings are best searched from top to bottom. Ideally, the search is conducted with the assistance of combat engineers using mine detectors to locate hidden arms and ammunition.

If the targeted building is empty or the occupant refuses entry, it may be necessary to forcefully enter the premises to conduct a search of the dwelling. If an unoccupied house containing property is searched, arrangements should be made with the local community to secure it until its occupants return. Unnecessary force and damage to property should be avoided during the search.

Motorized Urban Patrols

The advantages of a motorized urban patrols is their ability to capitalize on the speed, mobility, and protection offered by various vehicles. They may be motorized, mechanized or armored vehicles or a combination. Generally, motorized urban patrols possess greater combat power than dismounted patrols and can cover larger areas faster than dismounted patrols.

The disadvantages of motorized urban patrols is that they are restricted to roads and are vulnerable to ambush by the enemy. They are also restricted in their ability to interact with the local populace.

Motorized patrols are generally organized in the same manner as dismounted patrols (see fig. 13-9). Unit integrity is maintained when assigning personnel to specific vehicles.

The urban patrolling principles apply to motorized patrols in much the same manner as dismounted patrols. Mutual support and depth are achieved by maintaining constant observation between vehicles and coordinating support with any dismounted patrols in the area. All-around security is achieved through the use of constant observation as well as the vehicle's mobility and firepower. Positive communications between units or teams are maintained through vehicle radios. Patrol routes and speeds are varied to promote deception or pattern avoidance.

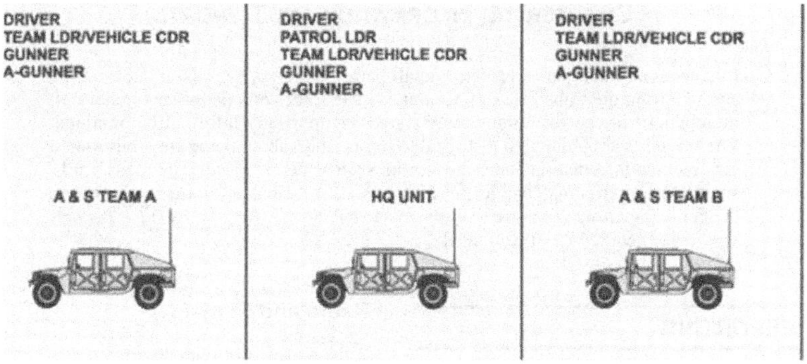

Figure 13-9. Organization of a Squad-sized Motorized Patrol.

The canalizing nature of streets and alleys force vehicular patrols to use a traveling overwatch movement technique to reduce vulnerability to ambushes. All vehicles travel at a moderate rate of speed with the lead vehicle stopping only to investigate potential danger areas. If vehicles must stop in danger areas, designated crew members' will dismount to provide security. The gunner will remain at the ready and in the turret while the driver remains in the driver's seat with the vehicle running.

Vehicles should move at a high rate of speed only when responding to an incident. At all other times, vehicle speed should be between 15 to 20 mph to allow for quick reaction and good observation. Distances between vehicles should be approximately 50 meters (one half to one city block) or such that visual contact and mutual support are ensured. Particular care is taken at major road junctions and other danger areas to ensure individual vehicles do not become isolated.

Vehicles with doors removed generally enhance observation and overall security, yet expose Marines to thrown objects, theft and concealment.

Chapter 14. Information and Reports

It is necessary that patrol leaders and all patrol members be trained in observing and reporting their observations accurately. The leader of a patrol should have all members of the patrol immediately signal or report any information obtained. These reports should not be restricted to information about the enemy, but should also include information about the terrain, such as newly discovered roads, trails, swamps, and streams. The leader includes all information in the report to the officer dispatching the patrol. (Refer to MCRP 2-15.3, *Reconnaissance Reports Guide,* for detailed report formats.)

14001. REPORTING

The officer dispatching the patrol instructs the patrol leader on whether and when messages are to be sent back during the patrol and what communication means to use. Messages may be oral or written. They must be accurate, clear, and complete. Every message should answer the question what, where, and when. For a detailed discussion on reporting, see paragraph 6002.

Verbal Messages

A patrol leader sending a verbal message should make it simple, brief, and avoid using numbers and names. The messenger should accurately repeat the message back to the patrol leader before leaving.

Written Messages

In preparing written messages, the patrol leader must distinguish between fact and opinion. Information about the enemy should include: strength; armament and equipment; actions; location and direction of movement; unit destination, if known; time enemy was observed; and the patrol's location when the observation was made. Use of an overlay or sketch may often simplify the message.

Messengers

A messenger team is given exact instructions as to where to deliver the message and the route to take. Any information obtained along the route should be reported at the time the message is delivered. Messengers must be given all practical assistance. If in danger of capture, the messenger immediately destroys the message.

Radio and Signals

If the patrol is provided with a radio, a definite radio schedule for checking in must be arranged before departure of the patrol. The patrol leader takes every precaution to ensure that codes and copies of messages are not captured by the enemy. If a close reconnaissance of enemy lines is required, the radio should be left in a concealed location at a safe distance from the enemy. Once a report is sent by radio, the patrol should immediately leave the area to avoid the possibility of detection by enemy locating devices. Pyrotechnics (flares, colored smoke, grenades) and air-ground panels may also be used by patrols for reporting information by a prearranged signal.

SALUTE Report

Information must be reported as quickly, accurately, and as completely as possible. An established method to remember how and what to report about the enemy is to use the acronym SALUTE:

Size
Activity
Location
Unit
Time
Equipment

An example of such a report is: "Seven enemy soldiers, unit unknown, traveling SW, crossed road junction on BLACK RIDGE at 211300 August carrying one machine gun and one rocket launcher."

14002. CAPTURED ITEMS

Every patrol should make a practice of searching enemy casualties, prisoners, and installations first for booby traps, then for equipment, papers, maps, messages, orders, diaries, and codes. Search techniques should be practiced often in order to reduce the time exposed to potentially dangerous situations. Items found are collected by the patrol leader and turned in with the patrol report. The items found are marked as to time and place of capture. When possible, captured items should be linked to a specific prisoner who possessed the items or to the place where the items were found. When this is done, the enemy prisoner of war tag and item tag are marked accordingly. The patrol leader must impress upon the members of the patrol the importance of turning in all documents and equipment. Furthermore, the patrol leader must ensure all information gathered by the patrol is quickly disseminated.

14003. PRISONERS

A patrol normally does not capture prisoners unless required by the mission. If prisoners are taken, the "5S and T" rule applies. This memory aid stands for: search, segregate, silence, speed, safeguard, and tag.

Search. Prisoners are body-searched thoroughly for weapons and documents as soon as they have been captured. This search must include the helmet, body armor, and gas mask. These items are left with the prisoner for protection until the patrol is completed. Weapons, equipment, and documents are tagged and immediately sent to the patrol leader.

Segregate. Prisoners are segregated into isolated groups: officers, noncommissioned officers (NCOs), privates, deserters, and civilians. By segregating prisoners, it makes it more difficult for leaders to organize escapes and issue orders to subordinates.

Silence. Silence is essential. Do not allow prisoners to talk to each other.

Speed. Speed is required in getting prisoners to the commander who dispatched the patrol. Timely information secured from prisoners is essential.

Safeguard. Prisoners are safeguarded as they are moved. They are restrained, but not abused. If the patrol will soon reach friendly positions, prisoners are not given cigarettes, food or water until they have been questioned by interrogators. If the patrol will not return to a friendly position for a lengthy time, the prisoners are given food, water and medical aid.

Tag. Equipment and personal effects recovered from prisoners are tagged so that they may later be matched back to specific individuals.

14004. PATROL REPORT

Every patrol leader makes a report when the patrol returns. Unless otherwise directed, the report is made to the person ordering the patrol. If the situation permits, the report is written and supported by overlays and/or sketches. The patrol leader's report should be a complete account of everything of military importance observed or encountered by the patrol while on the assigned mission. It should include the following information:

- Size and composition of patrol.
- Tasks and purpose (mission).
- Time of departure.
- Time of return.
- Routes, out and back (show by sketch, azimuth, trace on map).
- Terrain (general description to include any man-made or natural obstacles and critical terrain features which, if occupied by either enemy or friendly forces, would allow them to control the surrounding area).
- Enemy (size, activity, location, unit, time, equipment).
- Any map corrections (show on map).
- Miscellaneous information not covered elsewhere in report.
- Results of enemy encounters.
- Condition of patrol, including disposition of any dead or wounded.
- Conclusion and recommendations.

14005. PATROL CRITIQUE

After the patrol has rested and eaten, the patrol leader should hold a critique. Constructive criticism is made. It is an excellent time to prepare for future patrols by going over lessons learned as a result of the patrol.

APPENDIX A. PATROL WARNING ORDER

The warning order is issued as soon as practical with all available information included to assist patrol members in preparation.

1. Situation. Friendly and enemy situation information necessary for initial preparation.

2. Mission. Statement of what the patrol is to accomplish, and the purpose for accomplishing it. When, how, and where will be discussed in the patrol order.

3. Execution

a. Task Organization. General patrol organization and assignment of responsibilities if known; otherwise, promulgate in the patrol order.

b. Tasks. Alert subordinate leaders to patrol tasks (e.g., stream crossing, helicopter rappelling, demolitions) requiring preparation prior to departure.

c. Coordinating Instructions

(1) Time schedule and location for individual preparation, rest, briefings, in-spection, rehearsal, and departure.

(2) Time, place, uniform, and equipment for receiving the patrol order.

(3) Tasks for subordinate leaders to direct and supervise the initial preparation which may include drawing ammunition, rations, and special equipment; conducting immediate action drills or other necessary individual or unit training; meeting and briefing attachment personnel; reconnoitering the area for passage of lines; and coordinating with the necessary unit leaders.

(4) Preliminary guidance to specialists and key individuals regarding their roles and organization within the patrol.

4. Administration and Logistics

a. Individual uniform, equipment, weapons, and prescribed load of rations, water, and ammunition.

b. Crew-served weapons (if required) and guidance regarding distribution of weapons and ammunition during movement.

c. Special equipment requirements (wire cutters, demolitions, radios, flashlights, infrared equipment, mines, binoculars) and their distribution during movement.

d. Restricted or prohibited items.

5. Command and Signal

a. Designation of assistant patrol leader and his role in preparation.

b. Designation of navigators and radio operators (if required).

c. Brief outline of patrol leader's schedule for preparation, and where he can be reached.

APPENDIX B. PATROL ORDER

A patrol order follows a warning order. Any of the following subjects that have been addressed in the warning order may be omitted with the exception of the mission statement. A patrol order is more detailed than a 5-paragraph squad order in that a great deal of attention is given to individual duties.

The patrol order also provides orientation information, both enemy and friendly, that impacts the patrol and enemy forces. This information can include, but is not limited to, weather, terrain, visibility, NBC considerations, local population situation, terrain model and/or map orientation, and behavior.

1. Situation

   a. Enemy Forces

      (1) Composition, disposition, and strength are based on size, activity, location, unit, time, and equipment (SALUTE).

      (2) Capabilities and limitations to defend, reinforce, attack, withdraw, and delay (DRAW-D).

      (3) Enemy's most probable course of action.

   b. Friendly Forces

      (1) Mission of next higher unit (task and commander's intent).

      (2) Adjacent unit missions (task and intent). Also identify left, front, right, and rear.

   c. Attachments and Detachments. (date and time effective).

2. Mission. The mission is the task to be accomplished, and its purpose (who, what, where, when, and why). For patrols, specify if the mission or time has priority.

3. Execution

   a. Commander's Intent and Concept of Operations

      (1) Commander's intent.

      (2) The concept of operations tells the where, how, and who and lays out the patrol leader's general scheme of maneuver and fire support plan. It outlines the following:

(a) Task organization of the patrol.

(b) Movement to the objective area, to include navigation method.

(c) Actions in the objective area.

(d) The return movement, to include navigation method.

(e) Use of supporting forces (including illumination, if required).

b. Subordinate Element Missions. Subordinate element missions (task and purpose) are assigned to elements, teams, and individuals, as required.

c. Coordinating Instructions. This paragraph contains instructions common to two or more elements, coordinating details, and control measures applicable to the patrol as a whole. At a minimum, it includes—

(1) Time of assembly in the assembly area.

(2) Time of inspections and rehearsals (if not already conducted).

(3) Time of departure and estimated time of return.

(4) Location of departure and reentry of friendly lines and the actions associated with departure and reentry.

(5) Details on the primary and alternate routes to and from the objective area.

(6) Details on formations and order of movement.

(7) Rally points and actions at rally points.

(8) Final preparation position and actions at this position.

(9) Objective rally point and actions at this point.

(10) Actions at danger areas.

(11) Actions in the event of enemy contact.

(12) Details on actions in the objective area not covered elsewhere.

(13) Estimated time of patrol debriefing upon return.

4. Administration and Logistics

a. Changes/additions to uniform, equipment, and prescribed loads from that given in the warning order.

b. Instructions for handling wounded and prisoners.

5. Command and Signal

a. Command Relationships. Identify key leaders and chain of command.

b. Signal. Challenge and password, arm and hand signals, special signals, and radio frequencies and call signs.

Appendix C. Patrol Evaluation Checklist

This appendix provides a comprehensive checklist of critical patrolling steps, techniques, and procedures to aid unit leaders to critique patrol performance during training. Unit leaders may use the list as they observe the performance of a patrol to provide the patrol leader with a detailed analysis of the performance.

Warning Order	Yes	No
Did the PL ensure all patrol members were present before issuing the warning order?		
Did the PL issue a brief statement on the enemy situation?		
Did the PL issue a brief statement on the friendly situation?		
Did the PL state the mission in a clear, positive manner?		
Did the PL list all members of the patrol including attachments?		
Was the chain of command for the patrol covered fully?		
Were all members of the patrol asigned positions and duties in particular squads and teams?		
Were all the necessary individual duties assigned?		
Did the PL follow established principles in organizing the patrol into squads and teams?		
Was each patrol member assigned a particular weapon to carry on patrol?		
Did the PL list all special equipment needed to accomplish the mission?		
Was required special equipment assigned to the proper element to carry?		
Did the PL select uniform and equiment common to all based on METT-T?		
Coordination with Adjacent Units		
Did the PL coordinate with other patrols operating to the right and left?		
Was the route out and back coordinated?		
Was the time of departure and return cocordinated?		
Were call signs and frequencies coordinated?		
Was a signal for the FPF coordinated so as not to approach friendly lines during this time?		
Coordination With Front Line Units		
Did the coordinator pass the size of the patrol?		
Was the time of departure and return coordinated?		
Did the coordinator give a general area of oeprations for the patrol?		
Did the coordinator ask for information on known or suspected enemy positions and/or obstacles?		
Did the coordinator ask about information on the latest enemy activity?		
Did the coordinator ask for detailed information on friendly fire support available and the unit's barrier plan?		
Was the location of the IRP established and coordinated?		
Did the coordinator ask the forward unit to monitor their patrol frequency?		
Was the current challenge and password confirmed?		
Did the coordinator request that all information coordinated be passed on to any relieving unti?		
Patrol Order		
General		
Did the PL check to ensure all patrol members were present before issuing the patrol order?		

	Yes	No
Did the PL issue the patrol order in a forceful, confident manner?		
Did the PL make maximum use of available visual aids in issuing the patrol order (i.e., terrain model, sand table, map board, chalkboard)?		
Did the PL issue the patrol order in correct sequence?		
Did the PL issue the entire patrol order without allowing interruptions by patrol members?		
Did the PL adequately answer all questions asked by patrol members		
Did paragraph 1a (Enemy Situation) include—		
A weather forecast for the period of operation?		
A description of the terrain over which the patrol was to operate?		
Identification or description of enemy units known to bein the area of operations?		
Known locations of enemy units?		
Recent activity of enemy units?		
Stength of enemy units in the operating area?		
Did paragraph 1b (Friendly Situation) include—		
Mission of the next higher unit?		
Locations, missions, and planned actions of units on right and left?		
Fire support available to support the patrol (as per the coordination)?		
Missions and routes of other patrols operating in the immediate area?		
Did paragraph 1c (Commander's Intent) include—		
All views and ides of what the commander wants to be accomplished?		
The final result desired by the commander?		
Did paragraph 1c (Attachments and Detachments) include—		
All attachments to patrol and effective time of attachment?		
All detachments from patrol and effective time of detachment?		
Did paragraph 2 (Mission) include, at a minimum—		
Who was to conduct the patrol?		
What the patrol was tasked to do (i.e., conduct a point recon patrol)?		
Where the action was to take place?		
Why the action was to take place (i.e., the purpose)?		
Did paragraph 3 a (Commander's Intent and Concept of Operations) include—		
A complete concept of operations?		
A detailed description of the mission of elements?		
A detailed description of the mission of teams?		
A detailed description of the duties of specific individuals (i.e., navigator, compass man, pace m an, corpsman, APL)?		
Time of departure and time of return?		

	Yes	No
Type of formations and order of movement to be used?		
Description of the route and alternate routes(s) to include azimuths and distances between checkpoints as a minimum?		
Techniques to be used in the departure from friendly areas?		
Techniques to be used for the reentry into friendly areas?		
Location of the IRP and a tentative ORP, to include grid coordinates and recognizable terrain features?		
Method for designating and passing rally points?		
Actions to be taken at rally points if their use became necessary?		
Actions to be taken in the event of enemy contact?		
Actions to be taken at danger areas?		
Complete, detailed description of actions to be taken at the objective?		
Times and locations for rehearsals and the order of priority for rehearsals?		
Times and lcoations for inspections and the methods of conduct (to include uniform and equipment to be worn and/or carried)?		
Where, when, and by whom the debriefing is to be conducted?		
Did paragraph 4 (Administration and Logistics) include—		
Rations to be carried or reference to the warning order if there is no change?		
Arms and ammunition to be carried or reference to the warning order if there is no change?		
Uniform and equipment to be worn and carried or reference to the warning order if there is no change?		
Method for handling wounded or dead?		
Method for handling prisoners?		
Did paragraph 5a (Signal) include—		
Complete description of all signals, code words, frequencies, and call signs to be used within the patrol?		
Were signals adequate?		
Radio call signs to be used?		
Primary and alternate frequencies to be used?		
Call signs and frequencies to be used with other units in the area of oeprations (if applicable)?		
Required reports to be transmitted to higher headquarters, along with occasions for reporting?		
Code words and brevity codes to be used between the patrol and higher headquarters?		
Current challenge and password to be used in friendly held areas?		
Current challenge and password to be used forward of FEBA?		
Did paragraph 5b (Command) include—		
Chain of command or reference to the warning order if there is not change?		
Location of the PL during all stages of operation?		

	Yes	No
Location of the APL during all stages of operation?		
Inspection Rehearsal		
General		
Did the P L adequately use the allotted inspection time?		
Was the inspection conducted in an orderly manner?		
During the inspection, did the PL check the following:		
Completeness and correctness of uniform?		
Completeness of all equipment necessary to accomplish the mission?		
Operational condition of equipment?		
Did the PL question members to ensure they knew—		
The mission of the patrol?		
The concept of operations?		
Their individual duties and responsibilities?		
Chain of command and succession?		
Duties and responsibilities of key personnel?		
Did the PL utilize the rehearsal area to conform as much as possible to area of operations?		
Were the following major actions rehearsed:		
Actions at the objective?		
Actions at danger areas?		
Actions on enemy contacts?		
Departure and reentry of friendly lines?		
Did the PL critique each action after it had been rehearsed?		
Were interpatrol communications and control measures checked at rehearsal?		
Did the PL maintain control of the patrol during the rehearsal?		
Movement		
Did the PL use proper formations for movement?		
Did the PL adequately control the point team?		
Did the PL check the compass men?		
Was the PL aware of his position at all times?		
Was light and noise discipline enforced?		
Was the rate of movement appropriate for the mission?		
Did the PL make full use of pace?		
Did the PL make full use of count?		
Was proper security maintained during movement?		
Did the PL select and pass proper rallying points?		
Did the PL ensure that all members knew location of rallying points?		

	Yes	No
Did the PL properly use arm and hand signals?		
Did the PL recognize and halt the patrol a safe distance from a danger area?		
If the PL conducted a map check, did he conduct it properly (light discipline, security, etc.)		
Did designated individuals reconnoiter the far side of a danger area properly?		
Were support teams emplaced properly?		
Did the PL use a proper formation in crossing the danger area?		
Were reports made on enemy contact and at checkpoints?		
Actions on Enemy Contact		
Did the PL halt patrol a safe distance from a tentative ORP?		
Did the PL issue an adequate frag order before departing to look for a tentative ORP?		
Did the PL take appropriate personnel to look for the ORP?		
Did the PL select a suitable ORP?		
Did the PL secure the site adequately?		
Did the patrol occupy the ORP as stated in the patrol order or subsequent frag order?		
Did the patrol occupy the ORP in an orderly manner?		
Did the PL maintain control of the patrol during occupation of the ORP?		
Did the PL issue a satisfactory contingency plan to the APL before departing on a leader's recon?		
Did the PL take appropriate personnel on a leader's recon?		
Did the PL maintain or ensure that the patrol leader had communications with the patrol?		
Did the patrol avoid being detected by the enemy during the leader's recon?		
If contact was made, did the PL take appropriate action?		
Was security maintained during the leader's recon?		
Did the PL ensure the objective was kept under surveillance?		
Were appropriate orders given to surveillance teams?		
If a leader recon proved a tentative ORP unsuitable, did the PL move the ORP?		
Was a satisfactory leader's recon conducted?		
Was the ORP move conducted in a satisfactory and orderly manner?		
On return to the ORP, did the PL issue appropriate frag order and allow enough time for dissemination?		
Actions at Objective		
Did the PL issue a frag order for action at the objective?		
Was security emplaced prior to the teams departing from the ORP?		
Did the PL employ the terrain at the objective to the best advantage?		
Was stealth maintained while moving into the objective?		

	Yes	No
Did the Pl exercise positive control of squads, teams, and individuals at the objective?		
Were actions at the objective in accordance with the details outlined in the patrol order?		
Did the PL make use of supporting arms at the objective?		
Was the action at the objective successful?		
Was withdrawal from the objective accomplished quickly and orderly?		
Did units withdraw according to the details reflected in the patrol order?		
Was reorganization at the ORP completed in an expeditious and orderly manner?		
Did the patrol withdraw from the ORP quickly and quietly?		

Appendix D. Ambush Formations

Formations discussed here are identified by names which correspond to the general pattern formed on the ground by deployment of the assault element.

Line Formation

The assault element is deployed generally parallel to the route of movement of the enemy. This positions the assault element parallel to the long axis of the killing zone and subjects the target to heavy flanking fire. The size of the force which can be trapped in the killing zone is limited by the area the assault element can effectively cover. The enemy is trapped in the killing zone by natural obstacles, mines, demolitions, and direct fires. (See fig. D-l.)

A disadvantage of the line formation is the chance that lateral dispersion of the target may be too great for

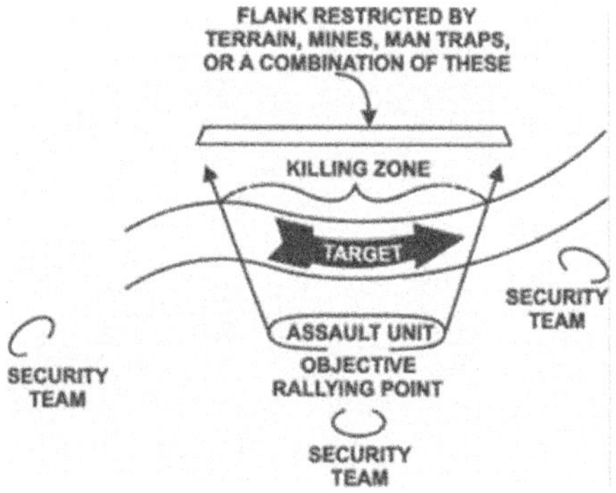

Figure D-1. Line Formation (Harassing or Destruction Ambush).

effective coverage. The line formation is appropriate in close terrain that restricts enemy maneuver, and in open terrain where one flank is protected by natural obstacles or can be protected by mines and demolitions. Similar obstacles can be placed between the attack force and the killing zone to provide protection from possible enemy counter attack. When an ambush is deployed in this manner, access lanes are left so that the enemy can be assaulted. (See fig. D-2.)

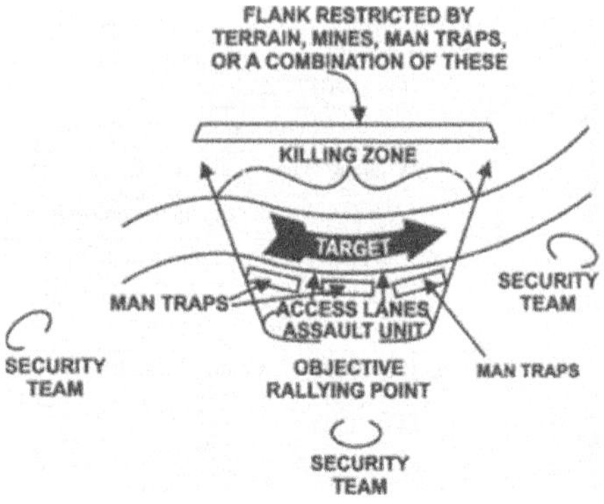

Figure D-2. Line Formation (Access Lanes for Assault of Target).

The main advantage of the line formation is its relative ease of control under all conditions of visibility.

L Formation

The "L" shaped formation is a variation of the line formation. This formation is very flexible because it can be established on a straight stretch of a trail or road (see fig. D-3) or at a sharp bend in a trail or a road (See fig. D-4 on page D-2).The long side of the assault element is parallel to the killing zone and delivers flanking fire. The short side of the attack force is at the

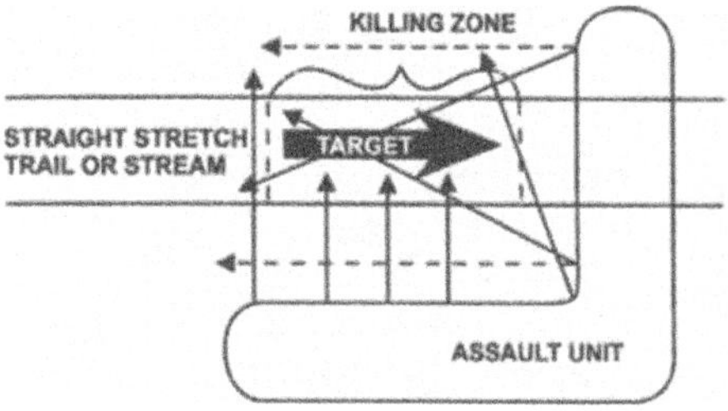

Figure D-3. "L" Formation (Destruction Ambush).

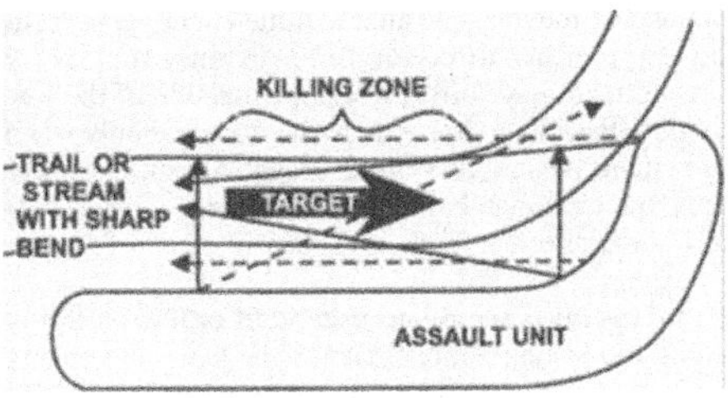

Figure D-4. "L" Formation (Bend of Trail or Stream).

end of, and at right angles to, the killing zone and delivers enfilading fire that interlocks with fire from the long side. When appropriate, fire from the short side can be shifted to parallel the long side if the enemy attempts to assault or escape in the opposite direction. In addition, the short side prevents escape and reinforcement. (See fig. D-5.)

"Z" Formation

The "Z" shaped formation is another variation of the line formation.

The assault element is deployed as in the "L" formation, but with an additional side so that the formation resembles a "Z". (See fig. D-6).

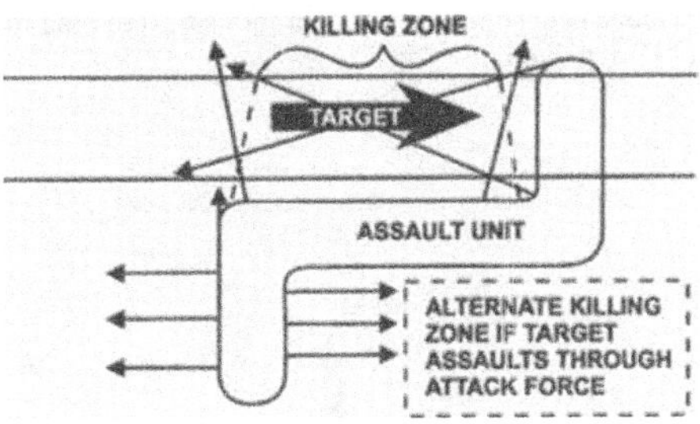

Figure D-6. "Z" Formation.

The additional wing may serve any of the following purposes:

- To engage an enemy force attempting to relieve or reinforce the enemy unit engaged in the kill zone.
- To seal the end of the killing zone.
- To restrict a flank.
- To prevent an envelopment.

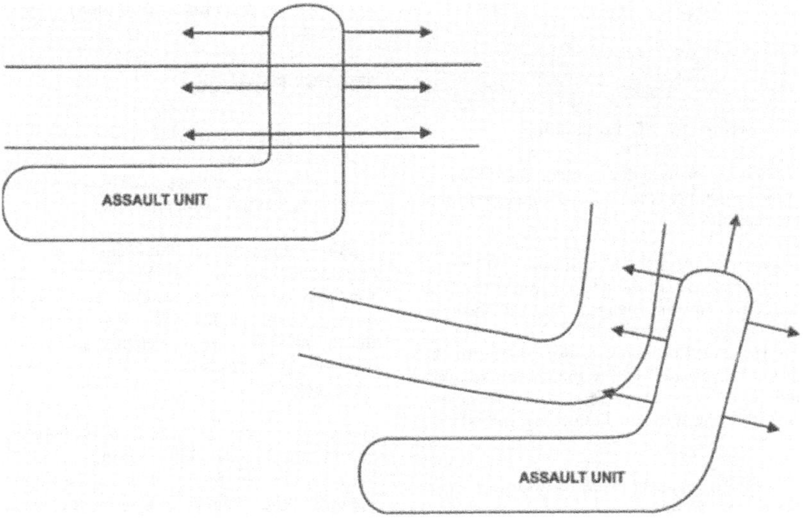

Figure D-5. "L" Formation.

"T" Formation

In the "T" shaped formation, the assault element is deployed across (perpendicular to) the enemy's route of movement so that its position forms the crossing of a "T" at the top. (See fig. D-7.)

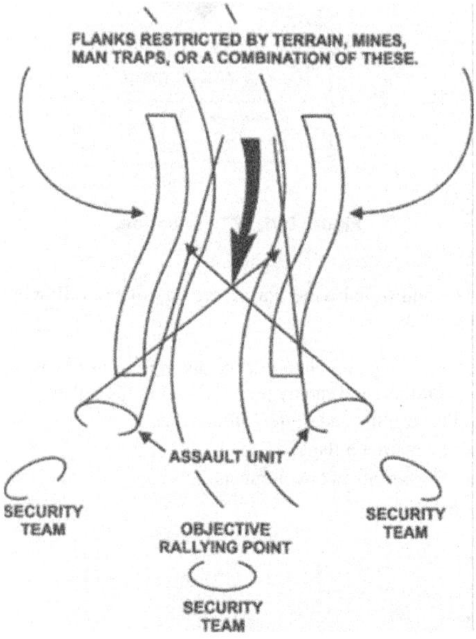

Figure D-7. "T" Formation.

This formation can be used day or night to establish an ambush to interdict movement through open areas that are hard to seal off.

A small force can use the "T" formation to harass, slow, and disorganize a larger force. When the lead elements of the enemy are engaged, they will normally attempt to maneuver right or left to close with the ambush. Mines and other obstacles placed to the flanks of the killing zones slow the enemy's movements and permit the ambush force to deliver heavy fire and withdraw without becoming decisively engaged.

The "T" formation can be used to interdict small groups attempting night movement across open areas. For example, the assault element is deployed along an avenue of approach with every second man facing the opposite direction. The attack of the enemy approaching from either direction requires only that every second man may shift to the opposite side of the formation. Each man fires only to his front and only when the enemy is at a very close range. Attack is by fire only and each man keeps the enemy under fire as long as it remains to his front.

If the enemy attempts to escape in either direction along the killing zone, each Marine takes the enemy under fire as the enemy comes into the Marine's sector of fire. The "T" formation is very effective at halting infiltration. But it has one chief disadvantage; there is a possibility that the ambush will engage a superior force at night while spread out. (See fig. D-8.)

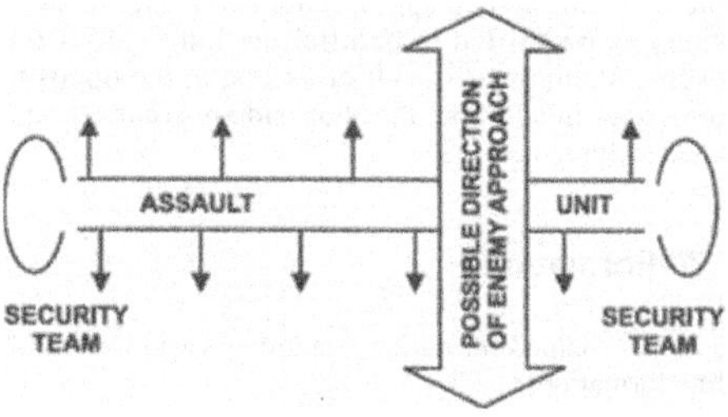

Figure D-8. "T" Formation (Target Approaching from Either Direction).

The "V" Formation

The "V" shaped formation is deployed along both sides of the enemy's route of movement so that it forms a "V"; care is taken to ensure that neither group (within the "V") fires into the other.

This formation subjects the enemy to both enfilading and interlocking fire. The "V" formation is best suited for fairly open terrain but can also be used in close terrain. When established in close terrain, the legs of the "V" close in as the lead element of the enemy force approaches the apex of the "V", and opens fire at a close range.

Here, even more than in open terrain, all movement and fire must be carefully coordinated and controlled to ensure that the fire of one wing does not endanger the other wing. The wider separation of forces makes this formation difficult to control, and there are few sites that favor its use. Its main advantage is that it is difficult for the enemy to detect the ambush until it is well into the killing zone. (See figs. D-9 and D-10.)

Triangle Formation

The triangle is a variation of the "V" formation and can be varied in three ways:

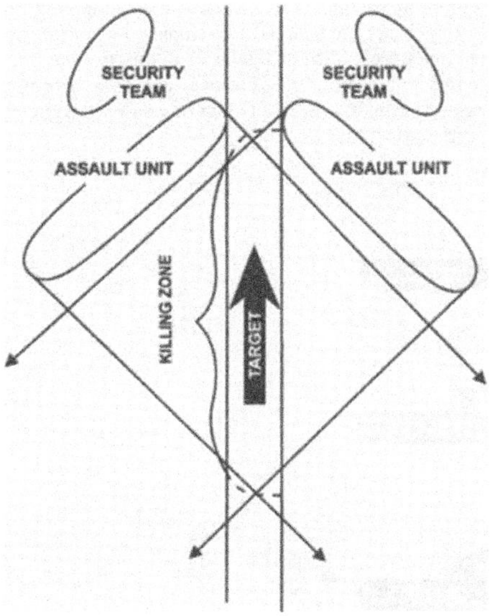

Figure D-9. "V" Formation (Open Terrain).

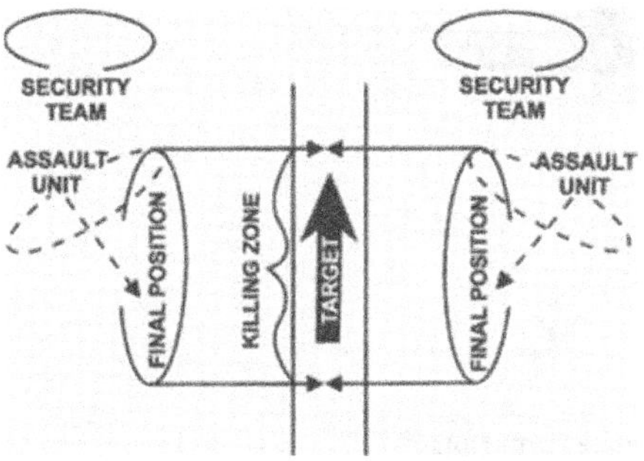

Figure D-10. "V" Formation (Close Terrain).

Closed Triangle Formation

(See fig. D-11.) The assault element is deployed in three teams, positioned so that they form a triangle (or closed "V"). An automatic weapon is placed at each point of the triangle and positioned so that it can be shifted quickly to interlock with either of the others. Men are positioned so that their sectors of fire overlap. Mortars may be positioned inside the triangle.

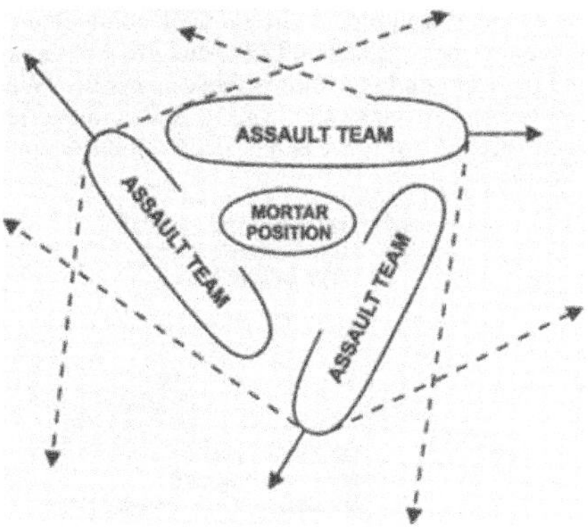

Figure D-11. Closed Triangle Formation (Night Ambush).

When deployed in this manner, the triangle ambush is used to interdict night movement through open areas. When enemy approach is likely to be from any direction, this formation provides all-around security, and security forces are deployed only when they can be positioned so that if detected by an approaching enemy, they will not compromise the ambush. Attack is by fire only, and the enemy is allowed to approach within close range before fire is initiated.

The advantages of the closed triangle formation are ease of control, all-around security, and the enemy can be brought under the fire of at least two automatic weapons, regardless of the direction they approach.

Disadvantages are that it requires a force of platoon size or larger to reduce the danger of being overrun by an unexpectedly large force and that one or more sides of the triangle may come under enfilade fire. The lack of dispersion, particularly at the points, increases the danger from enemy mortar fire.

Open Triangle Harassing Formation

This variation of the triangle ambush enables a small force to harass, slow, and inflict heavy casualties upon a large force without itself being decisively engaged. The assault element is deployed in three teams, positioned so that each team becomes a corner of a triangle containing the killing zone. (See fig. D-12.)

When the enemy enters the killing zone, the team to the enemy's front opens fire on the leading element. When the enemy counterattacks, the group withdraws and the team on the enemy's flank opens fire. When this team is attacked, the team to the opposite flank opens fire. This process is repeated until the enemy is pulled apart. Each team reoccupies its position, if possible, and continues to inflict the maximum damage possible without becoming decisively engaged.

Open Triangle Destruction Formation

The assault element is again deployed in three teams, positioned so that each team is a point of the triangle, 200 to 300 meters apart. The killing zone is the area within the triangle. The enemy is allowed to enter the killing zone; the nearest team attacks by fire. As the enemy attempts to maneuver or withdraw, the other teams open fire. One or more teams, as directed, assault or maneuver to envelop or destroy the enemy. (See fig. D-13 on page D-6.)

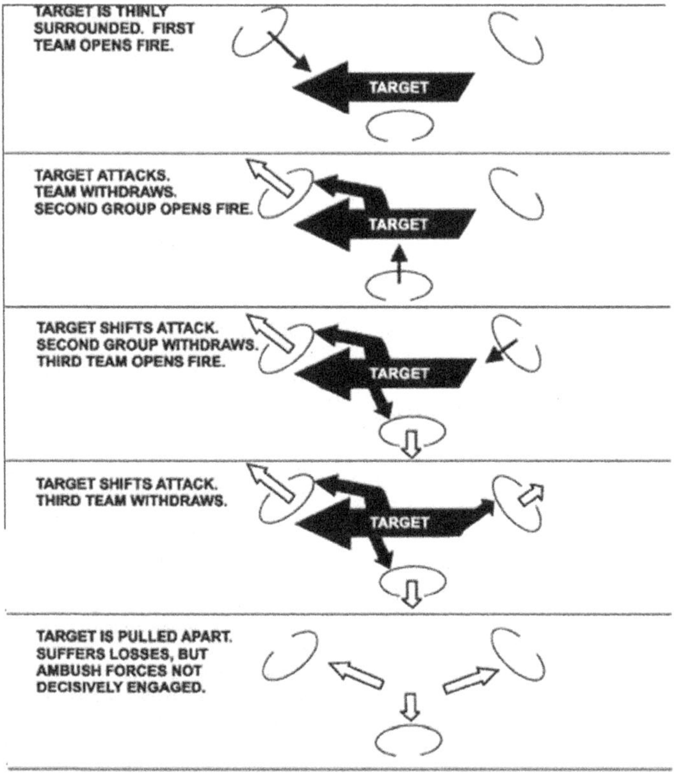

Figure D-12. Open Triangle Formation.

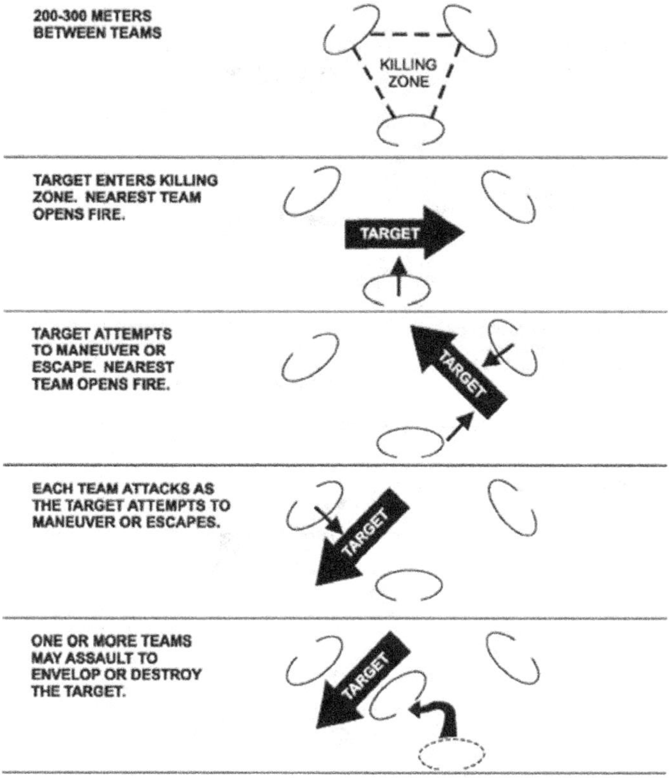

Figure D-13. Open Triangle Formation (Destruction Ambush).

This formation is suitable for platoon-size or larger ambush forces. A smaller force would be in too great a danger of being overrun. Another disadvantage is that control, in assaulting or maneuvering, is very difficult. Very close coordination and control are necessary to ensure that assaulting or maneuvering teams are not fired on by another team. The ambush site must be a fairly level open area that provides concealment around its border for the ambush force.

The Box Formation

The "box" formation is similar in purpose to the open triangle ambush. The assault element is deployed in four teams, positioned so that each team becomes a corner of a square or rectangle containing the killing zone. It can be used for a harassing or destruction ambush in the same manner as the variations of the open triangle formation. (See figs D-14 on page D-7 and D-15 on page D-8.)

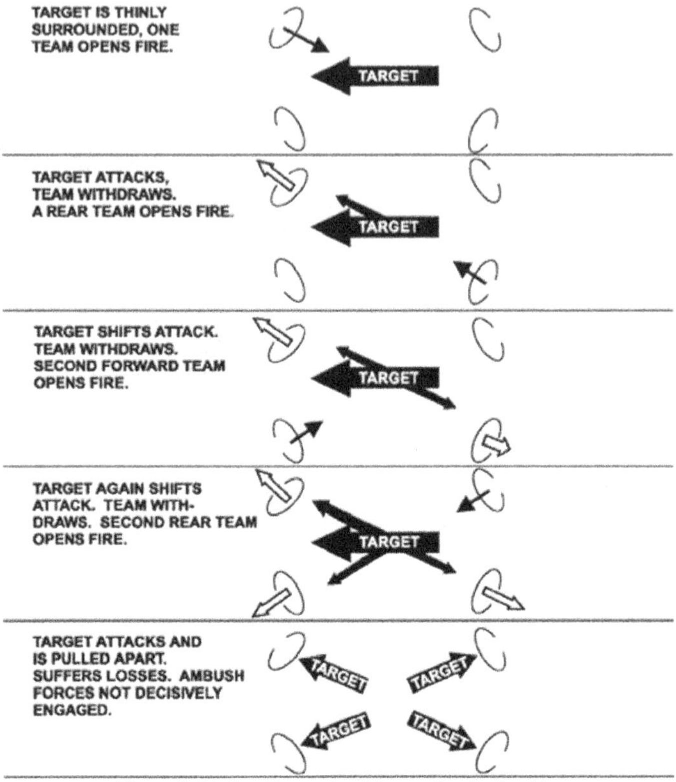

Figure D-14. Box Formation (Harrassing Ambush).

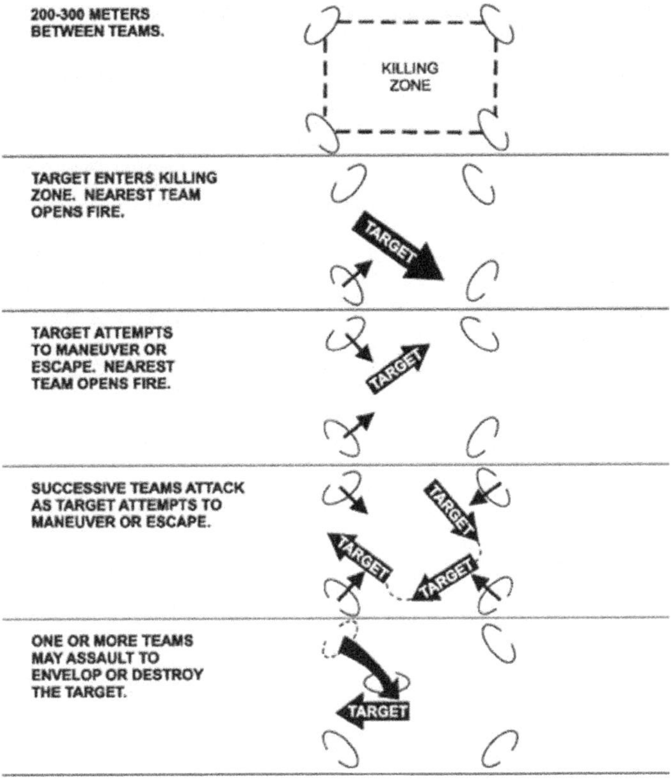

Figure D-15. Box Formation (Destruction Ambush).

APPENDIX E. ACRONYMS

A&S . . . . assault and security

CIT . . . . counterintelligence team

EOD . . . . explosive ordnance disposal

ITT . . . . interrogator-translator team

MAGTF . . . . Marine air-ground task force
MEDEVAC . . . . medical evacuation
METT-T . . . . mission, enemy, terrain and weather, troops and support available

NBC . . . . nuclear, biological, chemical
NCO . . . . noncommissioned officer

OTL . . . . overserver-target line

RCA . . . . riot control agents
ROE . . . . rules of engagement

SMAW . . . . shoulder-launched multipurpose assault weapon
SMEAC . . . . situation, mission, execution, administration and logistics, and command and signal
SOP . . . . standing operating procedures
SPOTREP . . . . spot report

TTP . . . . tactics, techniques, and procedures

Appendix F. References

Joint Publication (JP)

1-02 DOD Dictionary of Military and Associated Terms

Fleet Marine Force Manual (FMFM)

6-5 Marine Rifle Squad (proposed MCWP 3-11.2)

Marine Corps Warfighting Publications (MCWP)

3-15.3 Scout Sniping (under development)

3-35.3 Military Operations on Urbanized Terrain (MOUT)

3-41.2 Raids (under development)

Marine Corps Reference Publications (MCRP)

2-15.3A Reconnaissance Patrol Leader's Planning Handbook (under development)

2-15.3B Reconnaissance Reports Guide

3-02C Water Survival Handbook (under development, currently exists as Fleet Marine Force Manual 0-13)

5-12A Operational Terms and Graphics

5-12C Marine Corps Supplement to the DOD Dictionary of Military and Associated Terms

Made in the USA
Monee, IL
05 November 2026

67808312R00184